A Positron Named Priscilla

SCIENTIFIC DISCOVERY AT THE FRONTIER

Marcia Bartusiak, Barbara Burke,
Andrew Chaikin, Addison Greenwood,
T. A. Heppenheimer, Michelle Hoffman,
David Holzman, Elizabeth J. Maggio,
and Anne Simon Moffat

NATIONAL ACADEMY OF SCIENCES

NATIONAL ACADEMY PRESS
Washington, D.C. 1994

NATIONAL ACADEMY PRESS • 2101 Constitution Ave., N.W. • Washington, D.C. 20418

The National Academy of Sciences is a private, nonprofit, self-perpetuating society of distinguished scholars engaged in scientific and engineering research, dedicated to the furtherance of science and technology and to their use for the general welfare. Upon the authority of the charter granted to it by the Congress in 1863, the Academy has a mandate that requires it to advise the federal government on scientific and technical matters. Dr. Bruce M. Alberts is president of the National Academy of Sciences.

The National Academy of Engineering was established in 1964, under the charter of the National Academy of Sciences, as a parallel organization of outstanding engineers. It is autonomous in its administration and in the selection of its members, sharing with the National Academy of Sciences the responsibility for advising the federal government. The National Academy of Engineering also sponsors engineering programs aimed at meeting national needs, encourages education and research, and recognizes the superior achievements of engineers. Dr. Robert M. White is president of the National Academy of Engineering.

The Institute of Medicine was established in 1970 by the National Academy of Sciences to secure the services of eminent members of appropriate professions in the examination of policy matters pertaining to the health of the public. The Institute acts under the responsibility given to the National Academy of Sciences by its congressional charter to be an adviser to the federal government and, upon its own initiative, to identify issues of medical care, research, and education. Dr. Kenneth I. Shine is president of the Institute of Medicine.

The National Research Council was organized by the National Academy of Sciences in 1916 to associate the broad community of science and technology with the Academy's purposes of furthering knowledge and advising the federal government. Functioning in accordance with general policies determined by the Academy, the Council has become the principal operating agency of both the National Academy of Sciences and the National Academy of Engineering in providing services to the government, the public, and the scientific and engineering communities. The Council is administered jointly by both Academies and the Institute of Medicine. Dr. Bruce M. Alberts and Dr. Robert M. White are chairman and vice-chairman, respectively, of the National Research Council.

Support for this project was provided by the National Research Council's Basic Science Fund.

Library of Congress Cataloging-in-Publication Data

A Positron named Priscilla / by Marcia Bartusiak ... [et al.].
p. cm.
Includes bibliographical references and index.
ISBN 0-309-04893-1
1. Discoveries in science. 2. Creative ability in science.
3. Research. I. Bartusiak, Marcia, 1950- .
Q180.55.D57P67 1994
500—dc20 93-49495
CIP

Printed in the United States of America

Foreword

A Positron Named Priscilla is about exciting new vistas in science. Not only does it chronicle scientific progress, it affords a snapshot of the true frontiers of science across a range of fast-paced research fields.

The topics presented in these pages matter. Gene replication, earthquake prediction, the medical battle against AIDS, nanotechnology—everyone is affected by advances in these fields. Other breakthroughs—our ability to glimpse the craggy face of Venus or manipulate individual atoms on a surface — also appeal to our sense of wonder.

A Positron Named Priscilla is about a fascination with science. It both poses and answers the questions: "How did we get to this point?" and "Where must we go from here?" A fitting place to embark on the story of scientific endeavor at the leading edge.

Each year, beginning in 1989, the National Academy of Sciences has hosted an extraordinary symposium called the Frontiers of Science. The meeting is about communication among and between scientists. For three days, some of the nation's and the world's top young researchers—Packard and Sloan fellows, Waterman and MacArthur award winners, Fields medalists, younger members of the National Academy of Sciences—meet at the Academy's Beckman Center in Irvine, California, to report on their current work to peers *outside* their discipline. For many here, that can be an unfamiliar task. All at once, these scientists cannot assume their audience will understand their specialized terminology or appreciate the rationale for their experimental approaches.

As one might expect, they remain undaunted. Over the course of the symposium, these extraordinary scientists explain what they do and why they do it. They surprise themselves by discovering how an idea can bridge disciplines. They compare notes on the challenges they face in the laboratory, or behind the telescope, or in the field. Finally they revel in their enterprise. These are, after all, adventurers and explorers.

A Positron Named Priscilla allows a larger audience to join in this adventure. Like the Frontiers of Science symposium itself, the creation of this book has been an effort at communication. Top science writers have worked diligently to fill in the background and lend cohesiveness to the stories of research at the leading edge. Now everyone interested in the course of science can better appreciate the direction of scientific discovery while also learning something about the human dimension of science—that mixture of skill, dedication, and serendipity that sometimes leads to breakthroughs. In this volume, scientists no longer speak only among themselves, but to everyone. It is a discussion well worth having.

Bruce Alberts, *President*
National Academy of Sciences

Acknowledgments

The authors would like to thank the following individuals for their generous assistance in the preparation of this volume:

Duncan Agnew, Tom Alber, Walter Alvarez, Laszlo Babai,
Donald Bethune, Gregory Beylkin, Robert Cava, Sylvia Ceyer,
Shirley Chiang, Steven Chu, Ralph Cicerone, Rick Dahlquist,
Ingrid Daubechies, Mark Davis, David Donoho, Adrien Douady,
Dennis Dougherty, Donald Eigler, Kevin Einsweiler, William Ellsworth,
David Fahey, Marie Farge, Michael Frazier, Michael Freedman,
Douglas Gough, Robert Grimm, Jerome Groopman, Alex Grossmann,
Ashley Haase, Scott Hammer, Dennis Healy, Jr., Thomas Heaton,
Michael Hopkins, John Hubbard, Eric Hunter, Piet Hut, John Huth,
Harold Jaffee, Thomas James, Vaughan Jones, Maria Klawe, Eric Lander,
Thorne Lay, F. Thomson Leighton, Nathan Lewis, Ken Libbrecht,
Stéphane Mallat, Seth Marder, Marcia McNutt, Yves Meyer,
Mario Molina, Alessandro Montanari, Richard Muller, Noriyuki Namiki,
Erin O'Shea, Roger Phillips, William Press, David Rudman,
Gerald Schaber, Jim Siegrist, Kerry Sieh, Jan Smit, Suzanne Smrekar,
Sean Solomon, Michael Steigerwald, Bruce Stillman, Joann Stock,
Robert Strichartz, Kevin Struhl, Robert Tarjan, Robert Tjian,
William Thurston, Donald Turcotte, Randy Updike, Kerry Vahala,
Robert Whetten, Victor Wickerhauser, John Wilkerson, and
Richard Young

Table of Contents

Appendixes

A POSITRON NAMED Priscilla

SHAKE, RATTLE, AND SHINE

New Methods of Probing the Sun's Interior

by Marcia Bartusiak

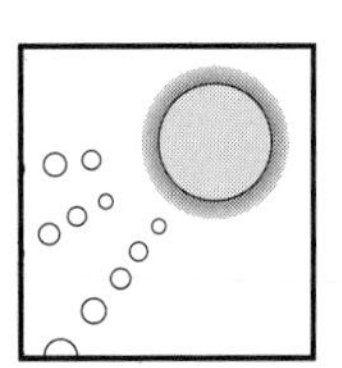

Some 5 billion years ago a dense clump of interstellar gas, possibly nudged by a shock wave coursing through space, began to condense and slowly spin. Over time a star was born, a relatively ordinary yellow dwarf situated in the outer fringes of the Milky Way galaxy. Ancient astronomers, who looked up at this star several aeons later from one of its attendant planets, called it the sun, and they were both awed and fascinated by its mysterious brilliance. To these early celestial observers the sun was perfect, an unblemished orb of fire. And despite reports that dark splotches occasionally appeared on the face of the sun, this ancient conception of a flawless solar globe held firm even into the Middle Ages.

This illusion, of course, was shattered in the early seventeenth century when Galileo in Italy, as well as observers in Holland, Germany, and England, pointed a newfangled instrument, called a telescope, at the sun and confirmed that the solar surface was indeed spotted. "For the most part they are of irregular shape, and their shapes continually change, some quickly and violently, others more slowly and moderately," wrote Galileo of his sighting.

Since then, the science of solar astronomy has largely evolved as an extension of Galileo's first effort. What is known about the sun essentially comes from examination of its outer features, although modern-day instruments, both on the ground and in space, have revealed a solar surface more turbulent and varied than seventeenth-century astronomers could ever have imagined: High-speed streams of solar particles emanate from dark coronal "holes"; solar prominences, immense arches of glowing gas, soar for hundreds of thousands of kilometers above the solar surface; and solar flares, lightning-like cataclysmic explosions, can flash across a region of the sun in a matter of minutes.

Nearly all these effects reflect complicated and tumultous activities inside the sun itself. But an exact description of what lies beneath the sun's fiery surface has been based more on conjecture than explicit measurement. The sun may be the star closest to Earth, yet at the same time, it is very remote. Its center lies some 700,000 kilometers from its surface, and all that fiery hot material in between acts as an effective shield, keeping solar astronomers from directly viewing the sun's interior. The laws of physics, however, do enable scientists to make some educated guesses on what they would see. Theoretical modeling and computer simulations have established that the sun is powered at its core, the inner 20 percent, by the thermonuclear conversion of hydrogen into helium. The resulting energy slowly makes its way out of the core, first by radiative diffusion, and then, starting about seven-tenths of the way out, by convection as the heated gases physically flow upward within the sun's outer layers. The gases subsequently release their energy at the surface, bursting through like bubbles in a pot of boiling fudge, only to recirculate downward to be heated once again. In this way a regular pattern of convection cells—columns of hot gas rising, cooling off, and then descending—is created within the sun (see Figure 1.1).

Many elements in this description, however, are far from secure. Certainty can arrive only if the sun is probed directly. Given the very nature of the sun—its 1.5-million-kilometer width and scorching temperatures—such an endeavor always seemed like an impossible dream—but no longer. In recent decades, solar astronomers have noticed that the sun quivers and shakes. It continually rings, in fact, like a well-hit gong. These reverberations, which carry information about the sun's deep

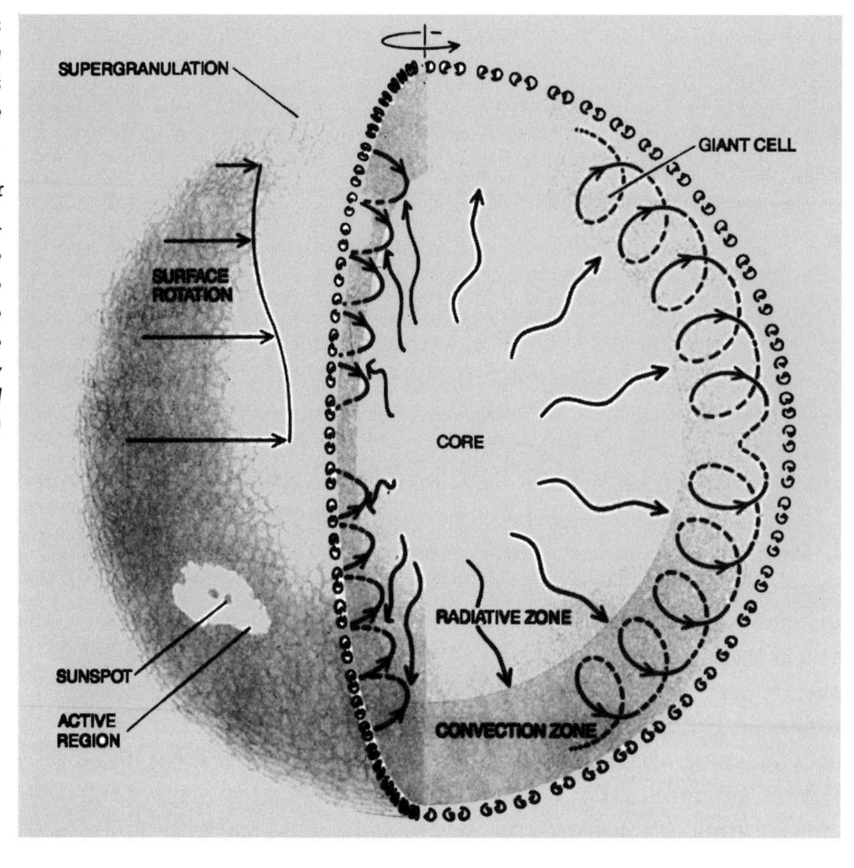

FIGURE 1.1 *Current models of the sun propose that a chain of fusion reactions within the sun's core release energy that moves outward, first by radiative diffusion, then via a hierarchy of convective cells of progressively smaller size. The many granules that mottle the sun's surface are the topmost parts of these convection cells. (Courtesy of National Optical Astronomy Observatories.)*

interior, are allowing solar observers to begin to examine the sun's hidden layers, much the way seismic tremors rumbling through our planet permit geophysicists to scan the earth's interior. Appropriately enough, the name of this new field is helioseismology.

"Two decades ago few people would have believed that it would have become possible to make measurements of conditions inside stars," says Cambridge University's Douglas Gough, a pioneering theorist in this up-and-coming field. "Yet the advent and the rapid development of helioseismology [have] provided accurate probes that have made seeing inside at least one star a reality."

Although still in its infancy, helioseismology has already challenged and revised several long-held conceptions of the solar interior, such as the depth of the convection zone and the way in which the inner sun rotates. Astronomers expect additional revisions as an international helioseismological network, presently in the process of being established, attempts to measure the solar quivers more accurately than ever before.

The information that is gleaned may affect far more than solar models. Knowledge of the sun's inner composition affects calculations of the age of the universe, as well as the amount of helium forged in the Big Bang. Moreover, knowing exactly how the sun spins internally is important in testing Einstein's theory of general relativity, which is the anchor for most of modern cosmology. The domain of helioseismology is broad and far reaching. Yet like so many other significant findings in astronomy, discovery of the quivering sun was totally unexpected.

THE MUSIC OF THE SUN

In 1960, using the 60-foot tower solar telescope atop Mount Wilson in southern California, Robert Leighton of the California Institute of Technology, together with Robert Noyes (now at the Harvard-Smithsonian Center for Astrophysics) and George Simon (now at the National Solar Observatory in Tucson), set out to measure changes in certain absorption lines in the solar spectrum. The lines were observed to Doppler shift, to move to higher or lower frequencies, as gases at the surface of the sun moved either toward or away from the observers. By measuring this shift the observing team hoped to discern the bobbing motions of individual solar granules, the cells of upwelling and sinking gases that cover the solar surface. A spectral-line feature would move toward the blue end of the spectrum whenever a cell heaved upward; the line would shift toward the red end as the cell dove back down. To the surprise of the Caltech astronomers, these velocity patterns were not chaotic, which is what they had expected, but instead were fairly oscillatory. Like a churning sea, the entire surface of the sun was found to be awash with periodic waves, not discernable to the naked eye, each rising and falling with a period of about 5 minutes. Moving at a speed of nearly 2000 kilometers per hour, any one patch can rise and then fall more than 70 kilometers over the 5-minute cycle.

For a while these pulsations, which continually grow and die away at any given site, were thought to be merely a local phenomenon, possibly eruptions from the roiling convection zone just beneath. However, that assumption began to change in 1970 when Roger Ulrich at the University of California in Los Angeles, and, independently, John Leibacher and Robert Stein, who is now with Michigan State University, provided a more global interpretation. Leibacher, now the director of the National Solar Observatory, says that his insight was due to a bit of theoretical serendipity, a stroke of good fortune that occurred as he was trying to simulate the 5-minute solar oscillation on a computer. "Hard as

I tried, I couldn't get my model to yield the answer that I wanted," he recalls. "Another mode kept overwhelming it. I tried and tried to get rid of what I thought was an error, but nature, or in this case the computer's simulation of nature, would not yield. It gave us the right answer."

Ulrich, Leibacher, and Stein came to realize that the 5-minute oscillation was not a local effect but rather the superposition of millions of acoustic or sound vibrations ringing throughout the sun. Since the sun is a spherical cavity with set dimensions, only particular wavelengths can be trapped inside and resonate, much the way an organ pipe resonates at specific frequencies. At any given spot on the sun, the 5-minute oscillation thus grows and decays as these myriad modes, each with its own period, velocity, and strength, move in and out of phase. It is as if the sun were a symphony orchestra, with all the instruments being raucously played at the same time. It sounds like a cacaphony, but when the noise is properly analyzed the separate instrumental tones emerge (although the sounds are far below audible frequencies). All these vibrations combine at times to produce a net oscillation on the solar surface that is thousands of times stronger than any one vibration.

The physics of these solar acoustic waves was already well understood from studies of the earth's atmosphere. Such waves propagate at the speed of sound by means of alternating compression and rarefaction of the solar gas, with pressure as the restoring force. Hence, these waves are also known as *p* modes. If you squeeze the sun, it rebounds under pressure. The individual *p* modes have periods ranging from a few minutes to nearly an hour.

Helioseismologists believe that the sun should also exhibit gravity waves, or *g* modes, just as seen in the earth's atmosphere. In this mode the solar material would be oscillated by the pull of gases of different density upon one another. Here, buoyancy is the principal restoring force. For example, if a mass of gas is displaced downward, entering a denser medium, it is buoyed upward. Once this rising mass becomes heavier than its surrounding medium, though, it falls downward once again, ready to repeat the cycle. Primarily originating in the sun's central regions, these longer-period waves (40 minutes or longer) do not propagate very well through the convection zone and are therefore expected to have extremely small amplitudes at the surface. The unstable convection zone does not support buoyancy oscillations very well. So far, reported sightings of *g* modes in the sun have not been confirmed. There are also modes, called *f* modes, that exist at the surface of the sun and travel about much like waves on the ocean. Such waves are essentially surface gravity waves and are virtually compressionless.

Since the sun is three dimensional, each solar acoustic wave is a bit more complicated than a simple wave resonating in an organ pipe. What is known as the "degree" of the wave, a parameter conventionally labeled *l* by helioseismologists, can loosely be thought of as the total number of horizontal wavelengths that encircle the sun's surface. These wavelengths range from the width of an individual solar granule—a few thousand kilometers, resulting in a high degree number—to the entire solar circumference, a number approaching unity. Each degree, in turn, can have varied frequencies and overtones, which reflect the variety of resonances possible in the other directions as well. Being three dimensional, the nodes of these standing waves—the regions where nothing moves—are not points but either concentric spheres or planes that slice through the sun parallel and perpendicular to each other. The quantity *l*, in fact, is actually the total number of nodal planes, both parallel and perpendicular to the solar equator, belonging to each oscillation.

Why should solar acoustic waves exist at all? That is not known with certainty, but helioseismologists have their suspicions. "Something is driving these modes," says solar astronomer Ken Libbrecht of Caltech. "We think it is due to the sun's turbulent convection. The turbulence in the surface layers generates acoustic noise, and if you generate noise within a cavity, then you excite the normal modes of that cavity." What is better understood is the method by which these waves travel around the sun.

Imagine a sound wave diving into the depths of the sun. Both temperature and density increase as the wave, traveling at around 200 kilometers per second, penetrates deeper and deeper, and this causes the wave to refract, or bend, as it travels inward. "The surface of the sun is cold, some 5800 degrees. The center is hot, around 15 million degrees. So, the sound speed actually increases as you go down into the sun, because sound speed increases with temperature," explains Gough. "A wave propagating downward into the sun thus experiences a faster speed deeper in than it does near the surface. As a result, it gets refracted." Eventually, the wave turns completely around and heads back up to the surface, where, because of the sharp drop in density at that boundary, it is reflected downward once again. In this way, the acoustic wave can travel around the sun many times, establishing a standing-wave pattern that lasts for days or weeks.

Acoustic modes above a certain frequency (roughly 5.5 millihertz) cannot be reflected by the sun's photosphere. These higher-frequency waves simply move into the sun's chromosphere and dissipate their energy. Thus, there exists a finite number of *p* modes that can be trapped inside the sun, about 10 million. Not all are detectable, but a good fraction of these modes are excited to observable amplitudes.

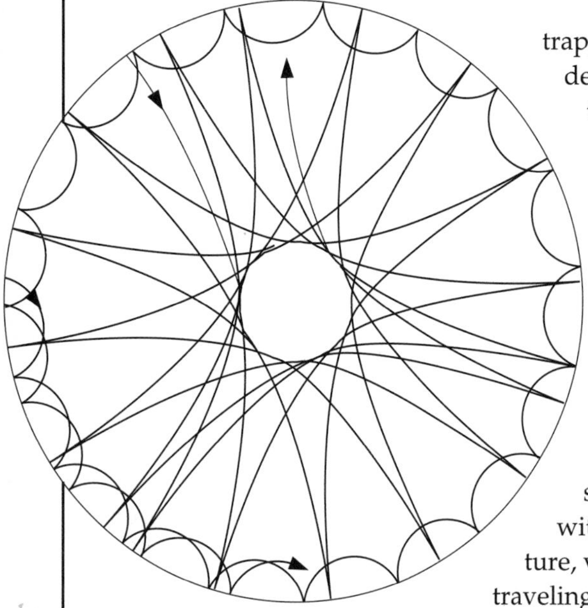

In general, the longer the horizontal length of the wave (in other words, the lower its degree), the deeper its plunge into the solar interior. This is because acoustic waves with longer wavelengths are refracted more gradually and so propagate more steeply into the sun. Shorter wavelengths, on the other hand, stay near the surface. Thus, by studying a wide range of modes, from high to low degree, solar physicists can effectively "enter" the sun in a steplike fashion, peeling away each of the star's layers as if it were an onion (see Figure 1.2). And with a wave's propagation dependent on the temperature, velocity, and density of the medium through which it is traveling, each mode offers valuable clues on the makeup and structure of the solar interior, much the way the resonant tone of a musical instrument provides hints as to the instrument's design—for instance, whether it's shaped like a flute or clarinet. Such effects afford solar astronomers with enormous diagnostic capabilities. For example, waves traveling in the same direction as the sun's fluid material will move a bit faster relative to a fixed observer, shifting their apparent frequency upward. Conversely, waves traveling against the flow will decrease in apparent frequency. Analysis of these splits in frequencies therefore offers a means of mapping the sun's large-scale internal motions.

FIGURE 1.2 *An illustration of two of the millions of possible acoustic waves refracting within the sun. Note that the low-degree or long-wavelength oscillation penetrates to the sun's deep interior. Short-wavelength or high-degree waves stay near the surface. (Courtesy of Douglas Gough, Cambridge University.)*

In 1970, when Ulrich, Leibacher, and Stein first formulated their idea of acoustic modes, it was only one of many possible explanations of the sun's 5-minute oscillation. The notion didn't take firm hold until 5 years later. At that time, in 1975, Franz-Ludwig Deubner of West Germany was able to separate his observations of the 5-minute oscillation into neat differentiated modes. A power spectrum

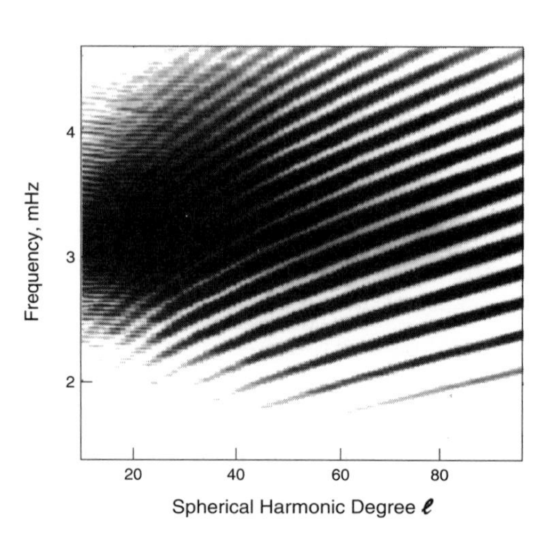

FIGURE 1.3 *A plot of the power measured in solar oscillations as a function of the degree of the wave, l, and its frequency. Each ridge in the diagram corresponds to those oscillations with a fixed number of radial nodes, n; the lowest-frequency ridge has n = 1. (Courtesy of Ken Libbrecht, Caltech.)*

of his Doppler velocity data, recorded for many hours along a strip of the solar equator, took the form of narrow and strikingly regular bands, a representation of the allowed frequency values for the resonant modes. The sum of these values was the 5-minute oscillation. A theory had at last been transformed into a valuable new tool for solar astronomy (see Figure 1.3).

Over the years, solar observers have refined their techniques for detecting both high- and low-degree modes. For the lowest modes, waves with values from zero to three, whose lengths are comparable to the size of the sun, observers look at the collective Doppler shift of a spectral line averaged over all or much of the solar disk. Since day/night gaps introduce spurious signals that make analysis difficult, investigators at the University of Birmingham in Great Britain and the Observatory of Nice in France established field stations around the globe to obtain an uninterrupted record of the sun's activity. Researchers have also made (and continue to make) long-term observations at the South Pole, where the sun never sets during austral summers.

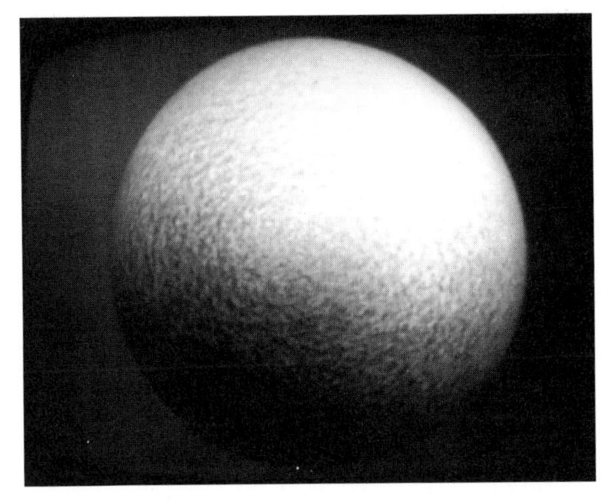

FIGURE 1.4 *Very localized Doppler motions across the face of the sun. The light regions indicate material rising from the sun's surface; darker shades indicate sinking. The side-to-side brightness variation is due to the sun's rotation. (Courtesy of Ken Libbrecht, Caltech.)*

Modes with degrees in the tens to hundreds, however, do not show up in such globe-spanning Doppler shift data. The wavelengths of these modes are relatively small compared to the size of the sun and so are averaged out. These higher-degree modes are effectively discerned in spectrograms that register very localized Doppler motions across the face of the sun, much like the first Caltech measurements. Ground-based instruments can observe parcels as small as 1000 kilometers across, about the width of the state of Texas. When processed, pictures of these parcels, known as velocity images, look like salt and pepper strewn over the solar disk (see Figure 1.4). The dark areas depict the regions on the solar surface that are sinking; the bright spots, conversely, are rising or moving toward the earth. It is these oscillations that are differentiated into the high-degree components. This method has enabled observers to differentiate modes with degrees up to a few thousand, although atmospheric distortion does play havoc with degrees that measure above 400. "The surface of the sun is seething with movement. There's a background noise of about a kilometer per second," points out Libbrecht. "But, on top of that, there are oscillations as little as a millimeter per second, which we can see. It's quite remarkable."

REWRITING THE TEXTBOOKS

Analysis of these data can be handled in one of two ways. Traditionally, researchers have constructed a set of solar models and then adjusted certain parameters, such as the temperature and density of various solar elements, until they best fit the *p* modes observed ringing through the sun. More recently, however, theorists have been developing mathematical techniques known collectively as inversion, which extract the solar parameters directly from the modes themselves. Inversion, in essence, converts the solar "tones" into a map of its essential features. This second approach is far more challenging than the first. Supercomputers are often used to deal with tens of thousands of modes at one time.

Interpretation of the *p* modes began soon after they were discovered. Deubner, for example, reported that the modal frequencies he had uncovered were actually lower than theoretical predictions. These observations led Gough in 1975 to deduce that the sun's convection zone must be deeper than previously estimated. Additional observations prompted Edward Rhodes (now at the University of Southern California), Roger Ulrich, and George Simon to draw the same conclusion. A deeper convection zone could account for the unexpected signal. It had long been assumed that the convection zone's depth was 20 to 25 percent of the solar radius. Calculations by Rhodes and his group determined that it was more like 30 percent. It was the first major solar parameter to be adjusted based on helioseismological data. The convection zone's greater depth means that convection can transport heat from the bowels of the sun more efficiently than once thought.

By profiling the speed of sound throughout the sun, helioseismologists have also been able to profile its composition. As pointed out earlier, the speed of sound generally increases at greater and greater depths as temperatures increase, but that trend changes when the wave approaches the very center of the sun. At that point the speed of the wave actually dips back down a bit, because thermonuclear reactions in

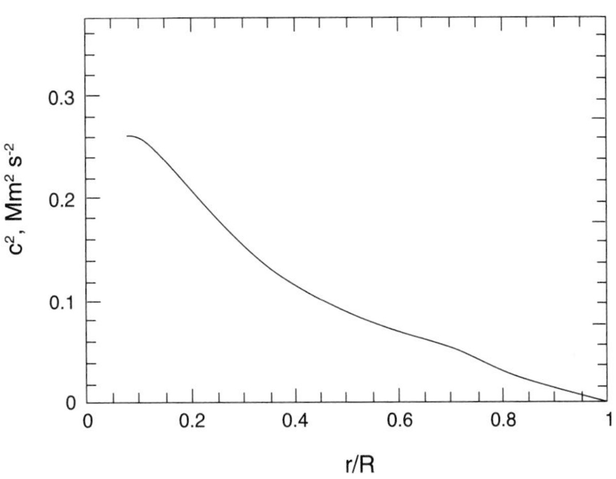

FIGURE 1.5 *This diagram, based on helioseismological data, shows how the speed of sound steadily increases (right to left: from one solar radius to fractions thereof) as temperatures increase toward the center of the sun. But the rise diminishes near the core, where hydrogen is being converted into helium, because the speed of sound decreases in denser material. (Courtesy of Douglas Gough, Cambridge University.)*

the core have converted about half of the original hydrogen into heavier helium. The speed of sound decreases, gets more sluggish, in denser materials (see Figure 1.5).

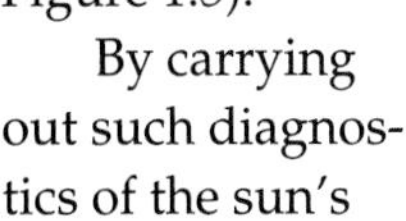

By carrying out such diagnostics of the sun's interior, helioseismologists turned the sun into a physics laboratory. One of their experiments, in fact, caught an error in some previous calculations of the sun's properties. In completing their profile of the speed of sound through the sun, helioseismologists discovered a discrepancy. The acoustic observations suggested that the solar sound speed was greater in the sun's mid-regions (roughly halfway in) than theoretical models had expected. Since this increase (nearly 1 percent) was difficult to explain, the helioseismologists began to wonder if solar physicists were underestimating the sun's opacity in that region (opacity being a crucial parameter in the helioseismological calculations). Gough and several colleagues predicted the opacity might differ by as much as 10 to 15 percent. "This suggestion," says Gough, "motivated a group at the Lawrence Livermore National Laboratory to look into the matter with care. And indeed, our prediction was confirmed." The opacity was recomputed, and, with the new parameter plugged in, the speed-of-sound discrepancy disappeared. Theory and observation came into line.

A view of the sun that has changed most dramatically since the birth of helioseismology is the overall profile of the sun's internal rotation. It turned out to differ considerably from the picture previously generated by theorists in their computer simulations. It has long been known, from observations of sunspot movements, that the sun's rate of rotation steadily declines from the solar equator to the poles. The poles complete a circuit in about 36 days, the equator in just 25. (Being a ball of gas, the sun is not constrained to rotate like a rigid body.) "This is poorly understood," notes Libbrecht, "but possibly linked to both convection processes and coriolis forces." Numerical simulations of this process had led to a model of the sun's differential rotation that was commonly referred to as "constancy on cylinders." The sun in this picture, at least through the convection zone, was supposedly composed of a set of nested cylinders that extended from pole to pole, aligned with the sun's axis of rotation. The inner cylinders, which surfaced at the higher latitudes, rotated more slowly than the outer ones, which met the surface at the more rapidly rotating lower latitudes. This also meant that the angular velocity at a particular latitude should have gradually decreased with depth.

But this picture failed to fit the observations of a number of helioseismologists, including Timothy Brown at the National Center for Atmospheric Research in Colorado and Cherilynn Morrow, then a student at the University of Colorado. Morrow and Brown began to show that the sun's rotation rate at a given latitude actually remains fairly constant down through the convection zone. Past that zone, angular velocities at the poles and equator shift toward the same rate. Halfway into the sun, beyond the convection zone and into the radiative interior, the sun rotates somewhat like a rigid body. These observations confirm the suspicion that the sun's differential rotation at the surface, long a mystery, is somehow generated by convection rather than processes deeper in the interior.

Brown and Morrow's model was sustained and extended by a wealth of new data gathered by Ken Libbrecht. For 6 months in 1986 at Caltech's Big Bear Solar Observatory, located in the center of Southern California's Big Bear Lake, Libbrecht and his students took a Doppler image of the sun each minute, gathering a total of around 70,000 pictures. The team then extracted vibrational modes from these images after some 40 hours of supercomputer time. "We were interested in measuring as many modes as we could," says Libbrecht, "because each mode has its own story to tell about the medium in which it was trapped." Lastly, inversions of these modes, performed by Jorgen Christensen-Dalsgaard of Denmark's Aarhus University and others, mapped the sun's rotation down to a depth of about 450,000 kilometers, 60 percent of the way to the sun's center. Similar sets of images were taken again in 1988, 1989, and 1990 and added to the data base.

"We find that the rotation persists almost independent of radius, down to the base of the convection zone," says Libbrecht. "Then, there is a fairly sharp transition to solid body rotation. This is one of the biggest outstanding questions concerning the sun—why does it rotate in this manner? We couldn't go down to the very core—we went down about six-tenths of the way—but, dynamically, it seems to make sense that the whole interior is rotating uniformly with about a 27-day period."

These and other measurements, particularly data taken earlier at the South Pole by Thomas Duvall of the NASA Goddard Space Flight Center, John Harvey of the National Solar Observatory, and Martin Pomerantz of the Bartol Research Institute, have served as a valuable check on the theory of general relativity, Einstein's revolutionary view of gravity. A competing theory of general relativity, introduced in the 1960s, had suggested that Einstein's calculation of a general relativistic effect that perturbs the orbit of the planet Mercury might be wrong. Supporters of this alternate theory of general relativity argued that a large portion of the inner sun was spinning much faster than the solar surface, causing the sun's core to flatten. If the solar interior is rotating very fast, Einstein's explanation for Mercury's peculiar orbital behavior would be in jeopardy, since his calculations assumed a fairly spherical sun. "The sun was rotating faster in the past than it is now. The sun is slowing down," points out Gough. "But has the inside of the sun slowed down as well? The discussion hinges on how strongly the inside is coupled to the outside." While helioseismologists are not yet able to make an exact measurement ("We can't totally rule out a fast-spinning nugget at the solar center," notes Libbrecht), the data do strongly suggest that the innermost core, a few percent of the sun's total volume, is not rotating fast enough to squish the sun and disrupt Einstein's theory.

Also affected by the changing profile of the inner sun has been astronomers' understanding of the solar dynamo, the "engine" that drives the ebb and flow of activity over a solar cycle by inducing immense electrical currents and magnetic fields. A decade or so ago, astronomers thought that the dynamo resided in and was driven by the turbulent convection zone as a whole. However, now that angular velocities are seen to remain fairly constant through the convective regions, that idea is now ruled out. In its place, theorists are suggesting that the dynamo occupies a more narrow zone between the bottom of the convection layer and the top of the deep interior, the region of transition where rotation rates change most sharply.

THE CHICKEN OR THE EGG?

Interpretation of helioseismological data can, in some ways, be likened to that old conundrum, "Which came first, the chicken or the egg?" Theorists turn to current models of the sun to differentiate and analyze the various acoustic modes; the modes, in turn, help refine the standard model of the sun. It's a tricky business, as a number of uncertainties are incorporated in the standard model of the sun. As noted earlier, estimating the sun's opacity, a feat that requires massive computing, can generate uncertainties of several percent or more, and this affects how modal frequencies are calculated from the Doppler information. Convection near the surface, the structure of the sun's atmosphere, nuclear reaction rates, and the resultant element abundances all affect helioseismologists' interpretations of the oscillations. Theorists and observers thus work hand in hand, each responding to the findings of the other in hope of converging on the correct model of the sun.

Helioseismologists are setting an ambitious agenda for themselves. A top priority is understanding the solar cycle, that 11-year period over which sunspot counts and solar flares wax and wane and solar magnetic field strengths build up and decline. Solar astronomers already suspect that the sun's magnetic fields interact with convection, indeed at times might suppress it, but helioseismology is needed to see such an effect below the solar surface. Helioseismologists have already noticed that solar oscillations behave differently in the presence of strong magnetic fields. The stronger the field, the stronger the effect, which can be either an increase or decrease in the frequency of a mode. Moreover, investigators have seen oscillations around single sunspots exhibit an intriguing phenomenon: up to 100 percent more vibrational power appears to move into a spot than moves out of it, as if the sunspot is absorbing *p* modes.

Such effects should provide observers with valuable tools in studying the sun's mysterious magnetic interior. For instance, how far down into the solar interior do flux tubes—threads of hot, highly magnetized gases first seen just 20 years ago—extend into the solar interior? Is the sun tunneled with these tiny sunspotlike features? And what is the vertical structure of a sunspot? Does its magnetic field branch like the roots of a bush, or does the field remain bound as one massive trunk that extends more deeply?

Since the propagation of acoustic modes is dependent on temperature and density, helioseismologists are also using the modes to follow the sun's temperature changes. After dissecting modes for nearly a decade, observers have now seen temperature gradients shift over the length of a solar cycle. Having analyzed this effect with helioseismic observations extending back to 1980, Jeffrey Kuhn of Michigan State University suspects it to be a reflection of changes in large-scale magnetic fields as the sun's activity waxes and wanes every 11 years.

Modal frequencies, too, appear to change synchronously with the solar cycle, a modulation that may also be slowly driven by varying magnetic fields. In analyzing his four large data sets, taken over the last half of the 1980s, Libbrecht saw the modal frequencies shift by one part in 10,000, due to changing magnetic activity on the sun's surface as the solar cycle progressed. Moreover, the shape of the sun's acoustic cavity itself appears to alter subtly over a full cycle. All of these observations suggest that the solar cycle may have deeper roots than previously suspected. Some even speculate, although it's very controversial, that the very core of the sun, where thermonuclear reactions take place, may somehow participate in the solar cycle.

The potential uses of helioseismology are legion. Some observers hope to study solar surface features, such as supergranules, and follow their movements over days and weeks. Others are hunting for large-scale convective flows. And helioseismologists are not stopping at the sun. Already a few adventurous observers have looked for seismic quivers in other stars, which offers astronomy the chance to plumb other stellar interiors. Analogs to the 5-minute oscillations of our sun have been reported in such stars as Alpha Centauri A, Procyon, and Epsilon Eridani. For the moment, however, the sun remains the top priority for specialists in this pioneering discipline.

GONG

Solar astronomers have long assumed that the sun's convection zone is lined with giant cells that act as monstrous conveyor belts, transporting solar material to and from the sun's fiery interior. But no hint of these massive structures has yet been found on the surface. To discern such a detailed solar topography, helioseismologists need long uninterrupted views of the sun, especially with instruments that resolve the high-degree modes. (The worldwide networks set up near Birmingham and Nice do not image the sun and so distinguish only low-degree modes.) With that need for high-degree data in mind, helioseismologists established the Global Oscillation Network Group, or GONG for short, a reminder of the acoustic qualities of the solar tremors. (As an additional reminder, a small gong is struck at the start of each annual GONG meeting.) Initiated by the U.S. National Solar Observatory in 1984, this international project now involves more than 100 observers and theorists from 61 institutions in 16 countries.

Principally funded by the National Science Foundation, the project is setting up six helioseismological field stations—identical and highly sensitive Doppler imaging instruments—around the globe at roughly equal spacing. "We will be able to increase our resolution 10 times over by observing, not at one site for 6 months, but at several sites around the world," points out Libbrecht. "Going beyond a basic understanding of the structure of the sun, these new helioseismological measurements are expected to turn the sun into a precision laboratory for learning about the physics of high-temperature plasmas and magnetohydrodynamics, neutrino oscillations, radiative transfer, and the dynamics of large-scale stratified convection and rotation."

The GONG instruments are being placed in California, Chile, the Canary Islands, India, Australia, and Hawaii. Contrary to many astronomical sites, a lack of air turbulence, or good "seeing," is not a prime consideration, considering the resolutions at which they will be working. Each station will be housed in a refurbished commercial cargo container and automated to the fullest extent possible, as if it were a spacecraft on the ground. "Mission Control," in this case, will be the National Solar Observatory in Tucson, where the first station was erected for testing. Much like the system already established for global radio-telescope arrays, data taken at each station will be recorded on videocassette tapes, which will be mailed periodically to the central data analysis center in Tucson. With each station recording some 2 gigabytes a week, 1.5 trillion bytes of data will be acquired after 3 years of continuous sun watching, the project's planned lifetime (although there are hopes to extend that to 5 years). Such prolific data gathering will make it one of the largest data sets in astronomy, after the Hubble Space Telescope. With special computer software, users will be able to browse, view, and select the data sets they wish to interpret within the archives.

Helioseismologists utilize a number of techniques to take high-resolution Doppler measurements. GONG scientists have chosen Fourier tachometry, because of its ability to measure Doppler velocities quickly and accurately. A given mode, one of many that make up a solar oscillation, moves some 10 centimeters per second or less. To obtain a good signal above the noise, a GONG instrument should then detect movements over the sun with a precision as small as a centimeter per second.

Timothy Brown of NCAR pioneered Fourier tachometry, and the GONG instrument is a product of the evolution of that technology. Sunlight first enters through an 8-centimeter telescope, or light-feed, that automatically tracks the sun over the course of a day. This light passes through a filter that isolates a specific spectral line, in this case the un-ionized nickel absorption line at 6768 angstroms. These red rays then enter the heart of the Fourier tachometer, a compact Michelson interferometer in the form of a cube 2.5 centimeters on a side. As with many interferometers, this cube splits the incoming light into two parts, routes each beam along a different pathway, and in the end recombines the two beams. If the light waves are in phase, or in step (peak matching peak), the two beams will add up to a bright signal; if out of phase (peak matching trough), the two will combine to produce a dark image.

The Fourier tachometer's keen ability to detect small Doppler shifts on the solar surface results from the makeup of the cube: one pathway or arm is solid glass; the other is air. With such different pathways (the glass arm is 30,000 wavelengths longer than the air arm), a tiny change in the wavelength of light entering the interferometer results in a measurable change of phase when the two beams are recombined. These phase shifts thus indicate the way in which motions on the sun are increasing or decreasing the original 6768-angstrom wavelength.

The output image of the interferometer, which encompasses the entire solar disk, will ultimately be focused on an electronic detector with an array of more than 65,000 pixels. Each pixel will record the signal intensity and phase at that point on the sun. At the field stations, a complete image will be recorded once a minute. In this way GONG scientists should detect any solar oscillations with durations of 3 minutes or more.

Ultimately, atmospheric fluctuations prevent observers from studying the highest-degree modes, which are tinier and require very good resolution to resolve. Therefore, the field of helioseismology will soon take to space, free from turbulent air, as well as disruptive day/night gaps. Helioseismic instruments will be included on SOHO, the Solar and Heliospheric Observatory, a joint project of the European Space Agency and the National Aeronautics and Space Administration. Scheduled for launch later this decade, SOHO will be placed at a Lagrangian point 1.5 million kilometers sunward from Earth. There a variety of detectors will be trained on the sun for at least 2 years, although the observatory could operate for 6 years, a major slice of the solar cycle. SOHO scientists are developing an instrument called the Michelson Doppler Imager, or MDI, which will perform both long-term Doppler scans, up to 2 months of continuous coverage at one image per minute, and daily readings. In this way SOHO investigators hope to measure solar vibrational modes with degrees from 1 to 3000. With such long-term high resolution, MDI's data could enable helioseismologists to focus on the topography of active regions, zoom in on granulation, and track the movement of sunspots.

Other detectors aboard SOHO will concentrate on low-degree modes and possibly *g* modes, whose extremely tiny amplitudes on the sun's surface make them the most elusive of the sun's oscillations. Years of data collection, either in space or by GONG, may be needed to discern the faint *g*-mode signal above the noise. But the payoff will be big if *g* modes are firmly discovered, for the detection will enable solar astronomers to peer directly into the sun's core.

"The very core of the sun where nuclear reactions take place, that's the biggest prize in helioseismology," declares Gough. At the sun's core, helioseismologists will find a laboratory that is irreproducible on this planet, where matter is pulled apart, ionized, and fused at unearthly temperatures. Indeed, one of astronomy's most nagging mysteries lies at the heart of the sun, where a flood of ghostly particles called neutrinos are continually and copiously generated and released into space. Sensitive detectors on Earth, located deep underground, are capturing only one-third to one-half of the neutrinos predicted by physicists' standard model of the sun. New findings in particle physics may eventually explain this discrepancy. Perhaps the neutrino, now assumed to be massless, has a bit of mass after all, affecting its detection on Earth. Or maybe other undiscovered particles huddling in the sun's core somehow temper the solar nuclear fire. If not, solar physicists will be forced to amend their ideas on stellar structure. If answers are not forthcoming from helioseismology, they will surely arrive as investigators advance the art of neutrino astronomy, another very powerful means of probing the sun's interior.

THE ELUSIVE NEUTRINO

The neutrino was one of the first elementary particles to be postulated before it was observed. For Wolfgang Pauli, who originated the idea, it was a desperate remedy. During the 1920s, particle physicists had begun to notice a troubling anomaly. Whenever a radioactive nucleus decayed by ejecting an electron, a process called beta decay, something went awry. In carefully controlled experiments conducted at Great Britain's famous Cavendish Laboratory, investigators saw that the energy of the nucleus before it radioactively decayed was more than the energy of the system afterward (i.e., the combined energy of the depleted nucleus and the fleeing electron). It looked as if energy were actually disappearing during beta decay, and this violated one of the bulwarks of physics—the law of conservation of energy, which states that energy is neither created nor destroyed. The laws of physics governing the conservation of linear and angular momentum were being violated as well. Atoms, it seemed, weren't playing by those rules. Energy and momentum were somehow getting lost along the way. So distressing was this finding that the distinguished physicist Niels Bohr openly discussed the possibility that atoms might at times violate some of the standard laws of physics.

Pauli wasn't as pessimistic as his Danish colleague. The Viennese physicist had an abiding faith that atoms were obeying the physical laws of the land. But to maintain that allegiance, Pauli took the rather radical step in 1930 of suggesting that an entirely new particle, invisible to ordinary instruments, had to exist to explain the discrepancies seen in the Cavendish experiments. Every time a nucleus undergoes beta decay, he proposed to some colleagues by letter, this neutral phantomlike particle is emitted and vanishes off into the night carrying off that extra bit of energy and momentum. Usually full of chutzpah, the young Pauli was actually intimidated by the outrageousness of this idea. "Dear radioactive ladies and gentlemen . . . ," he teasingly wrote his friends, who were then attending a meeting in Tübingen, Germany. "For the time being I dare not publish anything about this idea and address myself confidentially first to you, dear radioactive ones, with the question of how it would be with the experimental proof of such a particle." It was only after the chargeless neutron was discovered in 1932 that Pauli finally got the courage to publish his unusual suggestion. The noted physicist Enrico Fermi soon dubbed Pauli's hypothetical mote the *neutrino,* Italian for "little neutral one." The name was apt. The neutrino seemingly had no mass, and it had no charge. Indeed, it was nothing more than a spot of energy that flew from a radioactive atom at the speed of light.

Fermi perceptively recognized that Pauli's idea would also require a whole new force, what came to be known as the weak nuclear force. It is this force that enables a neutron to convert into a proton, releasing the electron and the neutrino (actually an antineutrino) seen in beta decay. (Interestingly enough, the prestigious journal *Nature* rejected Fermi's idea on the weak force as too speculative and too remote to be of any interest to practicing scientists; a small Italian review, however, did publish the hypothesis. Thus, the weak force entered the world of physics with little fanfare.)

It took so long to prove that the neutrino was more than a figment of Pauli's imagination that some physicists began to call his neutrino "the little one who was not there." In fact, Pauli began to wonder whether he had committed the theorist's ultimate sin: postulating a particle that could not possibly be detected. He had good reason to be fearful.

Capturing a neutrino is an extremely difficult task. The neutrino, which only interacts via the weak force, is terribly oblivious to matter. "A neutrino could actually go through 10^{20} centimeters of water, on average, before it scatters," explains John Wilkerson of the Los Alamos National Laboratory, a nuclear physicist who came to specialize in neutrino physics. "It's pretty hard to put such a number in terms we can comprehend. It says that the neutrino could travel through a pool of water the diameter of 100,000 of our solar systems put end to end, about a hundred light-years, before it interacted with another particle." A neutrino could bolt right through the entire earth and feel nary a thing, as if the earth were no more substantial than a cloudy mist.

The odds of catching a neutrino are considerably increased, however, if there is a *flood* of neutrinos coming at you. In that way, a few out of the hordes have a chance of bumping into an atom closeby and triggering a weak interaction. As a result, the atomic nuclei are altered, sending out flashes of radiation that can be spotted with photomultiplier tubes. The construction of nuclear reactors in the 1950s at last provided the necessary neutrino spigot for carrying out this tricky endeavor.[1] By 1956 Clyde Cowan and Frederick Reines, in a difficult but elegant experiment conducted at a South Carolina nuclear power plant, were finally able to corner the ghostly neutrino, or at least see its telltale footprints of light. The experiment was appropriately named Project Poltergeist.

Over each second that the giant Savannah River reactor operated, hundreds of trillions of neutrinos were generated in the reactor's core, whereupon they immediately escaped and raced through Cowan and Reines's 10-ton detector set nearby. Only a few of the slippery particles got stopped each hour. With such a sparse harvest, it took more than 3 years for the two young researchers to gather enough evidence to declare the neutrino a certified member of the particle zoo—to Pauli's great relief. Receiving news of the verification while attending a conference in Geneva, Pauli, by then in his 50s, held the telegram high and gleefully announced to his fellow physicists, "The neutrino exists!" He had lived to see his "sin" forgiven.

The particle that Cowan and Reines detected is specifically known as the electron neutrino because of its appearance in nuclear reactions involving ordinary electrons. Since then, physicists have found a second type of neutrino, the muon neutrino, which is associated with interactions involving the muon, a more massive relative of the electron. The muon is about 207 times heavier than the electron. A third neutrino, the tau neutrino, is confidently assumed to exist as well. It would be linked to a still heavier cousin of the electron called the tau that was first seen in particle accelerator experiments in the mid-1970s. The tau is nearly 3500 times more massive than the electron. To use the jargon of physics, neutrinos come in three "flavors": electron, muon, and tau.

In the standard model of particle physics, the rest mass of each neutrino type is arbitrarily set to zero. But theories that go beyond that standard model suggest that might not be true; neutrinos may have mass. And that could be a reason why solar physicists see fewer neutrinos flying out of the sun than they expect. Why this might be so requires a short review of physicists' understanding of how the sun shines.

THE SOLAR FURNACE

In the 1910s and 1920s, before a full theory of quantum mechanics was developed, physicists could only speculate on the sun's power source. Sir Arthur Eddington, among others, suggested that atoms must somehow be fusing and releasing energy. When Eddington's critics responded that the sun's interior was simply not hot enough to bring about these transmutations, Eddington readily responded that they should "go and find a hotter place."

By 1938 Hans Bethe and Charles Critchfield in the United States and Carl von Weizsäcker in Germany at last independently showed, in a series of steps that elegantly moved from one nuclear reaction to another, exactly how the sun and stars can be powered by the welding of 4 nuclei of hydrogen into helium. Normally, hydrogen nuclei, or protons, strongly repel each other, since they are positively charged. But deep in the core of a star, where temperatures reach up to 15 million degrees and densities are 12 times that of lead, protons can gain enough energy to collide with a force that sometimes overcomes that electromagnetic repulsion. This enables some of the protons, following an involved chain of reactions as outlined by Bethe and von Weizsäcker, to stick together with a powerful nuclear glue called the strong force. In the sun about one in every 10,000 billion billion proton collisions gives rise to a nuclear reaction. Long odds, but the firestorm under way in the sun's heart still enables it to convert approximately half a billion tons of hydrogen into helium with each tick of the clock. In the process, about 4 million tons of mass is transformed each second into pure energy, which eventually bathes our entire solar system in heat and light. The sun has been doing this for nearly 5 billion years and has enough hydrogen fuel in its core to continue for about 5 billion more.

The proton-proton reaction is the first step in a series of nuclear reactions that lead to helium production in the sun. First, two protons collide; one releases a positron and electron neutrino to become a neutron, which immediately combines with the remaining proton to form deuteron, a nucleus of heavy hydrogen.

$$p + p \rightarrow {}^{2}\mathrm{H} + e^{+} + v_{e}$$

This proton-proton reaction is the primary fusion reaction thought to occur in the sun. The deuteron proceeds to latch on to another proton, forming a light isotope of helium, helium-3, and releasing a gamma ray.

$${}^{2}\mathrm{H} + p \rightarrow {}^{3}\mathrm{He} + \text{gamma ray}$$

Two of these helium nuclei then collide, ejecting two protons and leaving a nucleus of helium-4.

$${}^{3}\mathrm{He} + {}^{3}\mathrm{He} \rightarrow {}^{4}\mathrm{He} + 2p$$

According to calculations by John Bahcall at the Institute for Advanced Study in Princeton, New Jersey, and Roger Ulrich, over 90 percent of the neutrinos emanating from the sun are produced by the proton-proton reaction, the first step in the chain above. As much as 2 percent of the sun's energy is emitted as these electron neutrinos, whose energies span some 400,000 electron volts or less. "Given its weak interaction with matter, a neutrino can escape directly from the sun's core, whereas a photon created at the center can take 10,000 years to escape," notes Wilkerson. Here was the perfect tool for looking into the sun, as physicists estimate that every second about 66 billion of these solar neutrinos rain down on each square centimeter of the earth's surface.

There are additional nuclear reactions going on inside the sun, besides the proton-to-helium chain, and each of these side reactions gives off neutrinos as well—neutrinos with characteristic energies. For instance, particularly energetic neutrinos (some of them with energies of more than 10 million electron volts), are produced in a minor reaction involving nuclei of boron-8 (a rare isotope composed of five protons and three neutrons). Specifically, it is a reaction in which a nucleus of boron-8 decays into beryllium-7, releasing a positron and a neutrino in the process.

$${}^{8}\mathrm{B} \rightarrow {}^{7}\mathrm{Be} + e^{+} + v_{e}$$

Though far fewer in number than the neutrinos generated by the sun's proton-proton reaction (they represent only a small fraction, a mere 0.01 percent, of the total flux of neutrinos coming out of the sun), these boron-reaction neutrinos are easier to detect. In 1946 Bruno Pontecorvo, a student of Fermi's, first suggested how such high-energy neutrinos might be captured. "His idea," says Wilkerson, "was to take an atom of chlorine-37, which is a stable nucleus with 17 protons and 20 neutrons, and add a neutrino to it. That turns the chlorine into argon-37, which is radioactive and can be watched for its decay." More than 20 years went by, though, before physicists could actually embark on such an ambitious endeavor.

UNDERGROUND TELESCOPES

The world's first neutrino observatory was finally established in America's heartland in 1967, and it has been gathering vital clues on the quirky nature of the neutrino ever since. The observatory's "telescope" is a huge tank of chlorine-rich cleaning fluid, 100,000 gallons of perchloroethylene, set in the Homestake gold mine situated nearly a mile beneath the Black Hills of South Dakota. Such a depth is required to keep the measurements free from disruptive cosmic rays. Additional shielding was added to eliminate interference from natural sources of radioactivity within the deep chamber. It took 20 railroad tanker cars to fill the tank, roughly the amount of stain remover American consumers use in a day (see Figure 1.6).

With this gargantuan apparatus, University of Pennsylvania radiochemist Raymond Davis, the founding father of neutrino astronomy, has been catching a few electron neutrinos out of the legions that are continually spewed into the solar system as the sun burns its nuclear fuel. For more than two decades now, the chlorine atoms in his cleaning fluid have been occasionally stopping some of the cagy particles. Following Pontecorvo's original scheme, the electron neutrino gives itself away by turning an atom of chlorine into traceable radioactive argon. The argon atoms, suspended within the perchloroethylene, are extracted after 2 months of exposure by bubbling helium gas through the cleaning fluid. Once the gas is collected, it is sent through a cold trap, where the argon freezes out. The extracted argon is then placed in low-background proportional counters, which record any radioactive decays of the argon-37 atoms.

But the results over the years suggest that the chlorine is capturing neutrinos (and subsequently transforming into argon) at an unexpect-

FIGURE 1.6 *For more than two decades, this huge tank, filled with chlorine-rich cleaning fluid and situated 1 mile underground in the Homestake Gold Mine in South Dakota, has been capturing fewer solar neutrinos than theory predicts. (Courtesy of Brookhaven National Laboratory.)*

edly low rate—only about one neutrino is captured every couple of days.[2] This is just about one-third the number of solar electron neutrinos that theorists, such as Bahcall, estimate should be snared, given the rate of nuclear reactions thought to be going on within the sun. Was it a real effect or a bug in the experimental setup?

Not until 1986 did another experiment at last come on line to provide additional data on the solar neutrino flux. It served as a valuable check on whether the Homestake measurements were truly recording a neutrino signal three times lower than the signal predicted by solar models. Japan's Kamiokande detector, a huge vat of water originally built to look for the proton decays anticipated by physicists' grand unified theories, was reconfigured to search for solar neutrinos. The Kamiokande detector is fundamentally different than the radiochemical type used by Davis. Unlike the Homestake detector, the Kamiokande detector is surrounded by more than 900 photomultiplier tubes, which record the distinctive light given off when a neutrino interacts with an electron. The electron, upon being scattered by the solar neutrino, emits a characteristic cone of light, known as Cerenkov radiation, as it moves through the water. This light is then detected by the photomultiplier tubes, yielding data on an event-by-event basis. Within a few years, this neutrino-detection scheme recorded a deficit of solar electron neutrinos, similar to the shortfall seen for two decades in South Dakota. Since it's been running, Kamiokande observes, on average, about one neutrino scattering event every 8 days, or some 45 signals over a year. "The Kamiokande detector has the nice advantage in that it can actually point to the sun," says Wilkerson. "It became the first experiment to convincingly demonstrate that neutrinos were being emitted from the sun."

But both the Homestake and Kamiokande detectors have limitations: they can only register fairly high energy neutrinos, most of them produced by that minor boron reaction in the sun, a nuclear reaction that is very sensitive to the sun's internal temperature. It's quite possible that a slight change in the solar model, turning down the sun's core temperature a bit, for example, can explain the undersupply of high-energy neutrinos. There are several ways such a temperature change could come about; perhaps, the proportions of hydrogen and helium in the sun, the two elements that make up 98 percent of the sun's mass, are different than solar physicists think. "One of the many suggestions that have been made to explain the neutrino deficit is that the sun's helium abundance is very low, leading to a comparatively low central temperature," says Gough. Solar models show that if the abundance of helium in the sun were as low as 19 percent by mass the sun's internal temperature should register about 13 million degrees Kelvin, causing the production of high-energy neutrinos to be curtailed—enough to explain the current deficit. According to Gough, though, a look at how the speed of sound varies through the sun, a helioseismological calculation based on an examination of the sun's acoustic modes, convincingly rules out such a helium deficit. A helium abundance of around 23 to 25 percent is needed to explain the oscillations detected in the sun's outer layers.

But there are other possibilities for explaining the neutrino deficit, conditions that are open to examination by helioseismological techniques. Gough notes that there is some evidence that the sun's rotation, just beneath the convection zone, varies on the time scale of the solar cycle. Whether the energy-generating core also varies is not yet known. There are theoretical reasons why it might, but observations have not yet been able to detect it. If future measurements do observe such an effect, it could greatly alter theorists' calculations of neutrino production rates. Such are the future questions for helioseismology.

Other ideas for reducing the sun's temperature are being discussed in the astrophysics community. Some theorists have wondered if there is a new class of weakly interacting massive particles (WIMPs) sitting in the sun's core. Such particles, as yet hypothetical, would have the effect of lowering the core temperature and thus cutting back the number of high-energy neutrinos reaching earth's detectors. Experiments conducted so far seem to rule out a WIMP solution, but physicists won't know for sure until they can measure the lower-energy neutrinos that are generated by the sun's predominant nuclear process, the proton-proton reaction, a process not so sensitive to the sun's inner temperature. If a shortfall is seen in those neutrinos as well, though, physicists suspect that the solution to the neutrino-deficit mystery will lie in particle physics, the nature of the slippery neutrino, rather than in WIMPs or solar physics.

To help decide, more sensitive neutrino observatories are being erected around the world. "We have now gone from a field that went for 20 years with one lone experiment to many experiments," says Wilkerson. "The decade of the 1990s should prove to be a landmark period for the study of solar neutrino physics. New experiments are poised to discover whether this deficit is a problem in our understanding of how the sun works, is a hint of new neutrino properties beyond those predicted by the standard model of particle physics, or perhaps a combination of both."

Collaborators from the United States and the former Soviet Union have initiated a search, called the Soviet-American Gallium Experiment (SAGE), deep underground in the heart of the Caucasus Mountains, near the Russian town of Baksan. There several dozen tons of liquid gallium (gallium melts at 30°C) await the shower of solar neutrinos passing through the earth to turn some of the gallium atoms into radioactive germanium. "You take gallium metal, add a neutrino to it, and you then produce germanium-71 atoms, which are radioactive," explains Wilkerson, who is a participating scientist in the SAGE experiment. More important, half of the neutrinos expected to interact with the gallium will be of lower energy, particles produced by the proton-proton reaction, the sun's primary energy source. "It's very difficult to make these lower-energy neutrinos go away based on models of the sun," notes Wilkerson. Simply changing the sun's core temperature a bit, for example, would have little effect on their production rate. In solar physics the strength of the proton-proton reaction is fundamentally linked to the sun's observed luminosity; uncertainties in the flux of neutrinos expected from the reaction are minimal, around 2 percent. So any deficit in those lower-energy neutrinos would be better explained through particle physics (see Figure 1.7).

SAGE started out with 30 tons of gallium and has since increased its size to nearly 60 tons. With 30 tons of gallium, SAGE investigators expected to produce one atom of germanium-71 per day. Since this germanium has a short half-life, less than 12 days, it starts decaying. "After a month of running, you have about 16 atoms remaining," states Wilkerson. Through an intricate chemical extraction process, this very sparse harvest of germanium atoms is eventually isolated from the gallium and counted. After running half a year in 1990, SAGE scientists, like their counterparts in South Dakota and Japan, found considerably fewer neutrinos than theory predicted. By 1992, using data from both the 30- and the 57-ton setups, they calculated that they were finding only 44 percent of the amount expected (with the statistical uncertainty ranging from –21 to +17 percent). "We aren't claiming any new physics," cautions Wilkerson, "but the experiment seems to be hinting that something is happening. It is nearly impossible for solar physics models to account for this result and, when taken in conjunction with the Davis and Kamiokande data, offers the hint that the low solar neutrino flux may be due to new neutrino properties."

A similar gallium experiment, called GALLEX, is under way in a laboratory set within Italy's Gran Sasso tunnel in the Apennines, a mountain range northeast of Rome. Operating since 1991, the GALLEX

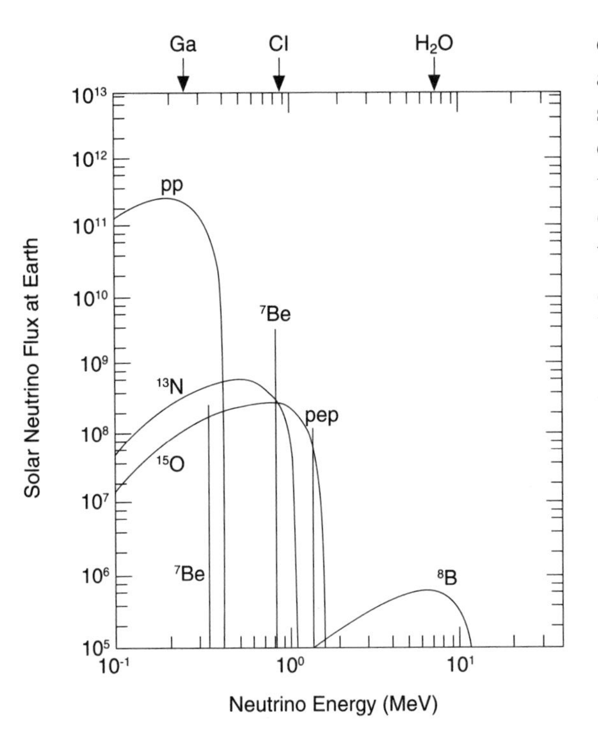

FIGURE 1.7 *Energy spectra for neutrinos released in various solar nuclear reactions. The chlorine and heavy water detectors on earth preferentially capture the high-energy neutrinos, while the gallium detectors are able to detect the lower-energy neutrinos produced during the vital proton-proton reaction taking place in the sun's core. (Courtesy of John Wilkerson, Los Alamos National Laboratory.)*

detector consists of a 100-ton aqueous solution of gallium chloride—a setup that GALLEX designers believed would simplify the germanium extraction process (gallium atoms make up 30 tons of that solution). Its initial findings, although very tentative (both SAGE and GALLEX must still undergo extensive calibrations), are intriguing. After running nearly a year, GALLEX captured 63 ± 16 percent of the neutrinos expected to be seen based on the standard solar model—more than the other detectors and just within shooting distance (given the statistical uncertainties) of the standard solar model, yet still a shortfall. The only way to reconcile the various GALLEX, SAGE, Kamiokande, and Homestake results with conventional solar physics is to lower the sun's core temperature by some 5 to 15 percent. But that's a conclusion that astrophysicists believe would play havoc with other well-measured solar parameters. What may be easier to tweak are the properties of the neutrinos themselves (see Figure 1.8).

The neutrino deficiency, so long seen in South Dakota and later in Japan, Russia, and Italy, might be explained if electron neutrinos have mass and so are able to "oscillate" on their way out of the sun; with mass, a fraction of them might be able to transform themselves into the other two neutrino types, the muon neutrino and the tau neutrino. Since the perchloroethylene and gallium detectors can only "see" electron neutrinos and not the other two types, this might easily explain the shortfall, why only a fraction of the expected neutrino signal is detected. By the time the solar neutrinos reach Earth, a large portion of them would have been transformed into either muon or tau neutrinos. *Physics*

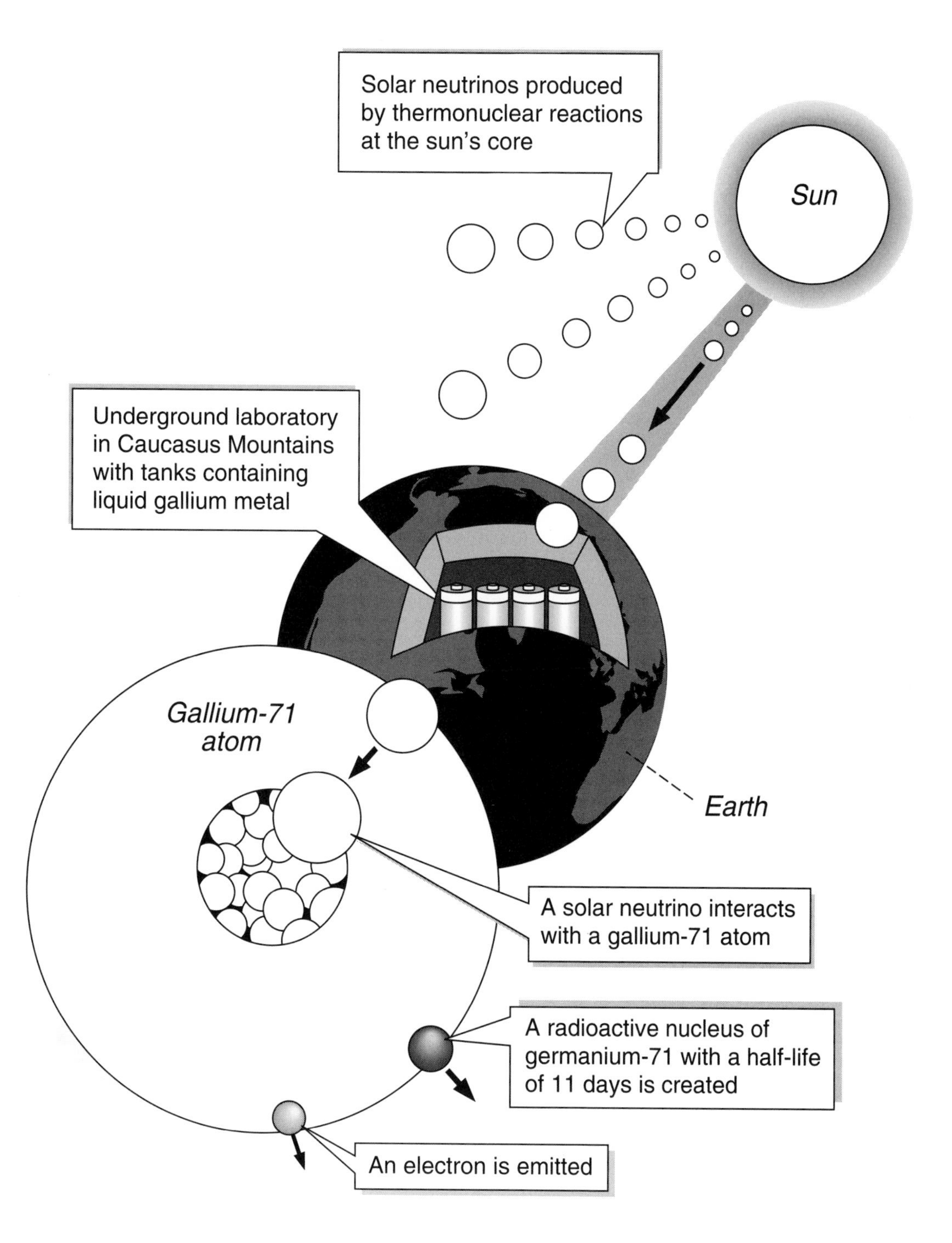

FIGURE 1.8 *Tracking a solar neutrino. (Courtesy of Los Alamos National Laboratory)*

Today editor Barbara Goss Levi, intrigued by this possible metamorphosis, once poetically described it this way:

> In caverns deep under the ground
> They hunt SNUs like hungry bloodhounds.
> But maybe the prey
> Can change 'long the way
> And sneak by without being found.[3]

The idea that neutrinos might oscillate was first discussed by Pontecorvo in 1957. More recently, the idea has been extended by physicists Stanislav Mikheyev and Alexi Smirnov of the former Soviet Union, building on earlier work by Carnegie-Mellon physicist Lincoln Wolfenstein. According to this theory, known by its originators' last initials as the MSW model, neutrino oscillations would rarely occur in the vacuum of space, but the effect would be enhanced as neutrinos speed through more dense collections of celestial matter, such as the sun. As they propagate through this matter, the various neutrino types could "mix," making it easier for electron neutrinos to switch their identity and turn into either muon or tau neutrinos. "One of the reasons that the MSW solution is so popular," adds Wilkerson, "is that in addition to being able to accommodate the data from all four neutrino experiments, Homestake, Kamiokande, SAGE, and GALLEX, it also yields values of neutrino masses that are reasonable in most grand unified theories that go beyond today's particle physics model." The masses would not be very sizable; the electron neutrino, for instance, would still be hundreds of thousands of times less weighty than an electron. Some theorists are suggesting that electron neutrinos must have energies greater than 500,000 electron volts to experience the conversion, which may explain why the Homestake and Kamiokande detectors are noticing the biggest shortfall; each is particularly sensitive to neutrinos in that energy range.

"The MSW model is definitely the leading contender, and it's a beautiful theory," notes Wilkerson, "but it only takes a few ugly facts to kill a beautiful theory." Neutrino astronomers will have a better chance at testing the MSW model later this decade with the opening of the Sudbury Neutrino Observatory, a collaborative venture sponsored by Canada, Great Britain, and the United States. This ambitious detector is now under construction in a deep nickel mine more than 200 miles north of Toronto and will consist of 1000 tons of deuterium oxide, or "heavy" water (where the proton in each hydrogen nucleus is accompanied by a neutron), encased in a bubble of clear plastic. This casing, in turn, will be surrounded by 7000 tons of ordinary water and a vast array of sensitive photon detectors. Numbering 9500, these phototubes will spot the bursts of Cerenkov light released whenever a neutrino interacts in the detector. More important, this heavy water detector will be able to distinguish all three flavors of neutrinos—the electron, muon, and tau. Unlike previous neutrino "telescopes," the Sudbury Observatory will be able to detect the distinctive signals released whenever an electron, muon, or tau neutrino hits a deuterium nucleus. No longer would the electron neutrino alone be recognized. Thus, the Sudbury detector should be able to prove whether the MSW model is the likely solution to the solar neutrino problem. "The Sudbury Neutrino Observatory can essentially find the smoking gun," says Wilkerson.

In the meantime, other neutrino detectors are scheduled to come on line in the near future. The Japanese are starting to build a facility similar to their current Kamiokande detector, except with 10 times more volume. This Super Kamiokande detector may be able to register up to 18,000 events per year, an efficiency hundreds of times better than the current instrument. Similar counts may be achievable in another planned experiment known as BOREXINO, a collaborative effort of Italian, U.S., and Russian scientists. Its design calls for a 100-ton liquid scintillator doped with boron, which would also be very sensitive to neutrino oscillations.

With the pace of neutrino astronomy increasing yearly, Wilkerson is optimistic that all these new detectors may at last provide the long-anticipated breakthrough. "It seems likely that the longstanding mystery of missing solar neutrino flux will be solved during the decade of the 90s," he predicts. "It will be interesting to see if the answers will come from solar physics or particle physics. If the current hint of new neutrino physics endures, it will be a curious occurrence that observing neutrinos from the sun has shed new light on physics beyond the current model of elementary particle physics."

NOTES

1. Atom bombs were briefly considered a source of neutrinos. But, as Fermi wryly noted, nuclear reactors were certainly better because they didn't require setting off an A bomb everytime someone wanted to check the experiment's results.

2. It is important to note that this capture rate is extrapolated from the raw data, which have lower values because of detection efficiencies and the fact that argon-37 has a relatively short half-life (35 days). The actual number of observed decays for each 2-month extraction is about four, or 23 to 24 counts per year. Thus, the Homestake experiment directly measured fewer than 500 counts over its first 20 years of operation.

3. Neutrino astronomers often report their results in terms of solar neutrino units, or SNUs. One SNU equals one neutrino capture per second for every 10^{36} atoms of the target material.

BIBLIOGRAPHY

Bahcall, J.N. 1989. Neutrino Astrophysics. Cambridge University Press, Cambridge.

Bahcall, J.N., and R. Ulrich. 1988. Solar models, neutrino experiments, and helioseismology. Reviews of Modern Physics 60:297.

Bartusiak, Marcia. 1993. Through a Universe Darkly. Harper Collins, New York.

Bethe, H.A. 1939. Energy production in stars. Physical Review 55:434.

Bethe, H.A., and C.L. Critchfield. 1938. The formation of deuterons by proton combination. Physical Review 54:248.

Brown, T.M., J. Christensen-Dalsgaard, W.A. Dziembowski, P.R. Goode, D.O. Gough, and C.A. Morrow. 1989. Inferring the sun's internal angular velocity from observed *p*-mode frequency splittings. The Astrophysical Journal 343:526–546.

Christensen-Dalsgaard, J., T.L. Duvall, Jr., D.O. Gough, J.W. Harvey, and E.J. Rhodes, Jr. 1985. The speed of sound in the solar interior. Nature 315:378–382.

Christensen-Dalsgaard, J., D. Gough, and J. Toomre. 1985. Seismology of the sun. Science 229(Sept. 6):923–931.

Deubner, F.L. 1975. Observations of low wavenumber nonradial eigenmodes of the sun. Astronomy and Astrophysics 44:371.

Gough, D.O. (ed.). 1986. Seismology of the Sun and Distant Stars. D. Reidel Publishing Co., Dordrecht, Holland.

Gough, D.O., and J. Toomre. 1991. Seismic observations of the solar interior. Annual Review of Astronomy and Astrophysics 29:627–684.

Harvey, J., and the GONG Instrument Development Team. 1988. The GONG instrument. Pp. 203–208 in Proceedings of the Symposium on the Seismology of the Sun and Sun-like Stars, Tenerife, Spain, 26–30 September, 1988. ESA SP–286 December.

Harvey, J. W., J. R. Kennedy, and J. W. Leibacher. 1987. GONG: to see inside our sun. Sky & Telescope (Nov.):470–476.

Leighton, R. B., R. W. Noyes, and G. W. Simon. 1962. Velocity fields in the solar atmosphere: I. Preliminary report. Astrophysical Journal 135:474–499.

Libbrecht, K. G. 1988. Solar and stellar seismology. Space Sciences Review 47:275–301.

Libbrecht, K.G., and M.F. Woodard. 1991. Advances in helioseismology. Science 253(July 12):152–157.

Schwarzschild, B. 1992. GALLEX data can't quite lay the solar neutrino problem to rest. Physics Today XX(Aug.):17–20.

Waldrop, M. M. 1989. Gazing into the interior of the sun. Science 244(April 7): 31.

RECOMMENDED READING

Bahcall, J.N. 1989. Neutrino Astrophysics. Cambridge University Press, Cambridge.

Gough, D.O., and J. Toomre. 1991. Seismic observations of the solar interior. Annual Review of Astronomy and Astrophysics 29:627–684.

Harvey, J. W., J. R. Kennedy, and J. W. Leibacher. 1987. GONG: to see inside our sun. Sky & Telescope (Nov.):470–476.

Leibacher, J., R. Noyes, J. Toomre, and R. Ulrich. 1985. Helioseismology. Scientific American (Sept.):48–57.

Libbrecht, K.G., and M.F. Woodard. 1991. Advances in helioseismology. Science 253(July 12):152–157.

Freedman, D. H. 1991. The ghost particle mystery. Discover (May):66–72.

Freedman, H. 1986. Sun and Earth. Scientific American Books, New York.

A POSITRON NAMED PRISCILLA

Trapping and Manipulating Atoms

by T. A. Heppenheimer

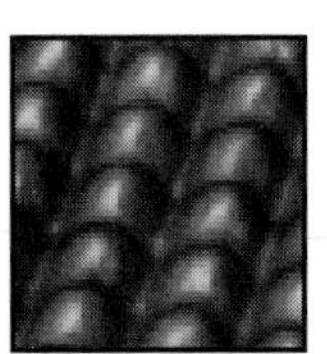

"There's plenty of room at the bottom," the physicist Richard Feynman declared in 1959. Speaking at the annual meeting of the American Physical Society, he envisioned an advanced technology whereby small machines would build yet smaller ones that would go on to build tinier ones still. Each of these generations might minister in turn to the diminishingly small fleas that have smaller ones to bite them, in the poem by Jonathan Swift, but Feynman had a different goal in mind. He raised, in his words, "the possibility of maneuvering things atom by atom. . . . It is something, in principle, that can be done; but, in practice, it has not been done because we are too big."

Today, however, we are beginning to take significant steps toward achieving Feynman's vision. A new instrument, the scanning tunneling microscope, already has picked up and dropped single atoms into specific locations. It has even dragged such atoms across an underlying surface. Physicists, working with lasers, have caused individual atoms to move slowly and to stop on demand, permitting measurements of exquisite precision.

THE SCANNING TUNNELING MICROSCOPE

How is it possible to manipulate atoms on a surface, as if they were a child's toy blocks? The beginning of an answer lies in appreciating that, when seen at close quarters, an atom is not a hard-shelled object like a billiard ball. It lacks a distinct surface; instead, its surface is fuzzy. The fuzziness, in turn, results because the properties of an atom's perimeter derive from the behavior of its outer electrons. If those electrons could move outward only to a distinct radius and no farther, the atom indeed would resemble a billiard ball. In fact, these electrons have some likelihood of being found at greater distances from the nucleus. This likelihood or probability drops off rapidly as the radius increases, but that merely means that the electron density falls off quickly, rather than dropping off abruptly, as one proceeds to greater radii. The region of this falloff in electron density contains the fuzzy periphery.

The same is true of a surface formed of atoms. It does not resemble a cobblestoned pavement, with individual atoms as the stones. It, too, is fuzzy, as if the cobblestones had an overlay of cotton. Again, the fuzziness represents a region where the electron density is falling off sharply but is not terminating abruptly. In manipulating atoms on a surface, then, physicists work with the fuzziness, surrounding both the atoms and the surface.

This fuzziness has a thickness on the order of an angstrom (symbol Å), or 10^{-8} centimeters. This is a very small thickness, 10,000 times finer than the smallest details that appear in an electronic microchip. Indeed, an angstrom is as much smaller than a postage stamp as the stamp is smaller than the state of Texas. Manipulations made with atomic precision nevertheless must achieve such accuracy, and this immediately rules out any type of mechanical system with moving parts. Even if built with the care of a watchmaker, such a mechanism would amount to carrying out delicate surgery on the eye using the clanking machinery of a Mississippi steamboat. Hence, it would appear at the outset that one must accomplish the impossible: To build a movable mechanism that has no moving parts.

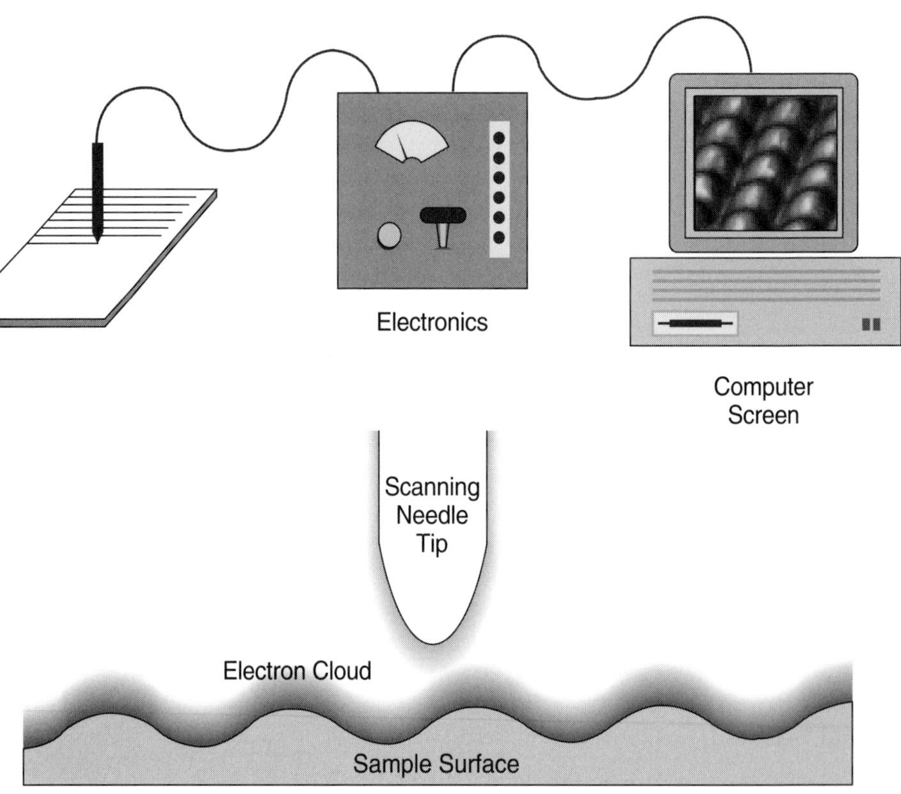

FIGURE 2.1 *Principle of the scanning tunneling microscope. Bands of electronic fuzziness surround both sample surface and needle tip, leading to a flow of current when these bands overlap. Feedback control maintains the current at constant value, raising and lowering the tip while it scans across the surface. The tip scans in a raster pattern, while its vertical motion follows the contours of surface atoms. The image on the computer screen, showing these atoms, amounts to a plot of vertical needle motion during the raster scan.*

The answer lies in piezoelectricity. There are materials with the property that, when squeezed or compressed, produce measurable voltages. The reverse is also true: When one applies even a slight voltage, they change length. The change can be no more than a fraction of an angstrom in a piezoelectric rod having a length measured in centimeters, but that is just what we want. A similar effect could be achieved by using rods that expand and contract in response to small temperature changes, but piezoelectric materials offer quick response with no need to wait while a rod warms or cools. Such rods then can push a sensor or manipulator across a surface, raising or lowering it within the fuzzy layer.

The sensor demands attention in its own right. It features a needle with a very sharp tip, able to probe individual atoms. A sewing needle is much too coarse; it would look to an atom like the Washington Monument. But tungsten, a very hard metal, offers tips of sufficient sharpness.

This needle, mounted in a holder or sensor head and guided by the piezoelectric rods, constitutes the essential feature of the scanning tunneling microscope. The needle rides above the surface to be studied at a distance of 4 to 8 Å. The fuzzy layers of the surface and needle tip then just barely overlap; they overlap a good deal more strongly if the needle approaches the surface by as little as a fraction of an angstrom.

As little as 0.1 volts, applied to the piezoelectrics, can move the needle by a full angstrom. In addition, a small voltage applied to the needle causes an electric current to flow from the surface, as a consequence of the overlapping regions of fuzziness. This is called a tunneling current. It increases rapidly as the tip comes closer to the surface; if the tip comes 1 Å closer, the current goes up by a factor of 10. Atoms are typically some 3 Å in diameter, which means that, if the tip moves downward by no more than the width of an atom, the tunneling current increases as much as a thousandfold. A feedback loop, measuring this current, keeps the needle tip at a desired height above the surface (see Figures 2.1 and 2.2).

Gerd Binnig and Heinrich Rohrer, at the IBM Zurich Research Laboratory in Switzerland, built the first scanning tunneling microscope

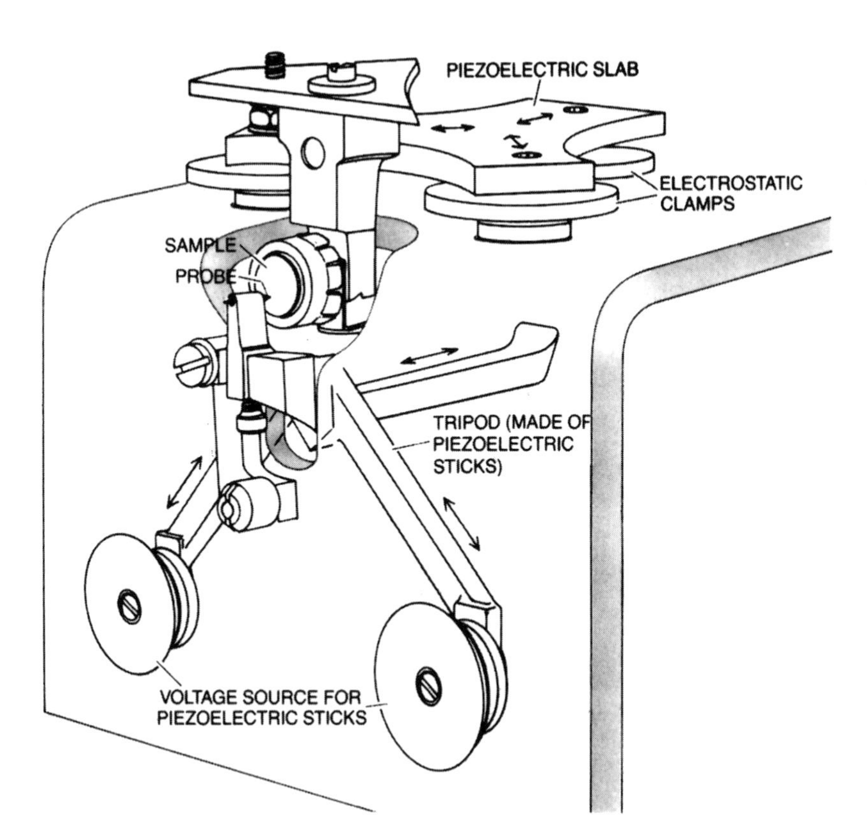

FIGURE 2.2 *Example of a scanning tunneling microscope. The probe features a fine-pointed needle. The piezoelectric tripod scans the needle over the sample surface while adjusting its height to an accuracy as great as a thousandth of an atomic diameter. The piezoelectric slab and electrostatic clamps, at the top, provide coarse adjustment by moving the sample to approximately the right location. From "The Scanning Tunneling Microscope," by G. Binnig and H. Rohrer. Copyright © 1985 by Scientific American, Inc. All rights reserved.)*

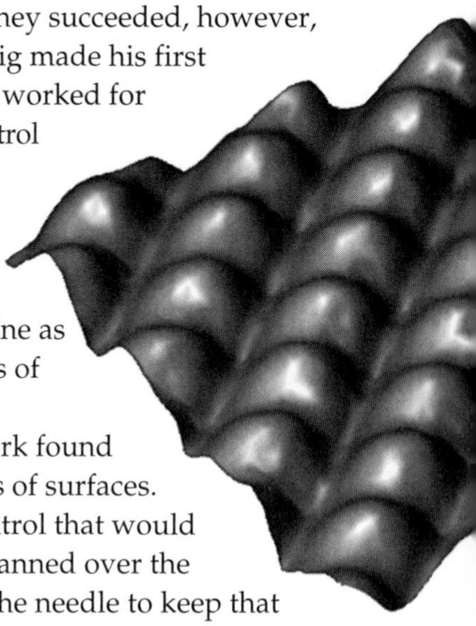

in 1981. A principal problem for them was to prevent disturbances due to vibration; footfalls of people in the lab could resemble the San Francisco earthquake at atomic scale, while the vibrations of ordinary speech could resound like Big Ben tolling the hour. They succeeded, however, and at 2 a.m. one morning in mid-March, Binnig made his first observation with the new instrument. He had worked for weeks to overcome difficulties with a balky control system, but his reward came as he observed steps of atomic scale on a surface of gold. Later, observations of silicon showed bumps corresponding to single atoms, along with vertical resolutions, normal to the surface, as fine as 0.5 Å. Similar results have been seen in studies of nickel (see Figure 2.3).

The instrument that emerged from this work found quick application in making atomic-scale maps of surfaces. A common procedure involved a feedback control that would measure the tunneling current as the needle scanned over the surface, automatically adjusting the height of the needle to keep that current constant. The varying voltage fed to the vertical piezoelectric element, to adjust the needle height, then would correspond to the surface details seen while scanning. In the words of Binnig and Rohrer, "By sweeping the tip through a pattern of parallel lines a three-dimensional image of the surface is obtained. A distance of 10 centimeters on the image represents a distance of 10 Å on the surface: A magnification of 100 million." By contrast, a conventional optical microscope achieves a magnification of 2000, which is 50,000 times coarser.

"There they were, like tennis balls lying on the floor. Atoms!" recalls Calvin Quate of Stanford University, who in 1982 became one of the first American investigators to use this device. His excitement was understandable, for just after World War II the physicist Edward Teller had written an alphabet book for the nuclear age:

> A is for atom. It is so small
> No one has ever seen it at all.

The scanning tunneling microscope was not the first instrument to give direct views of individual atoms; using specialized techniques, physicists had seen tungsten atoms during the 1950s. But with this new microscope, physicists could see atoms at will, within a wide range of materials. Binnig and Rohrer went on to share the 1986 Nobel prize in physics for their achievement.

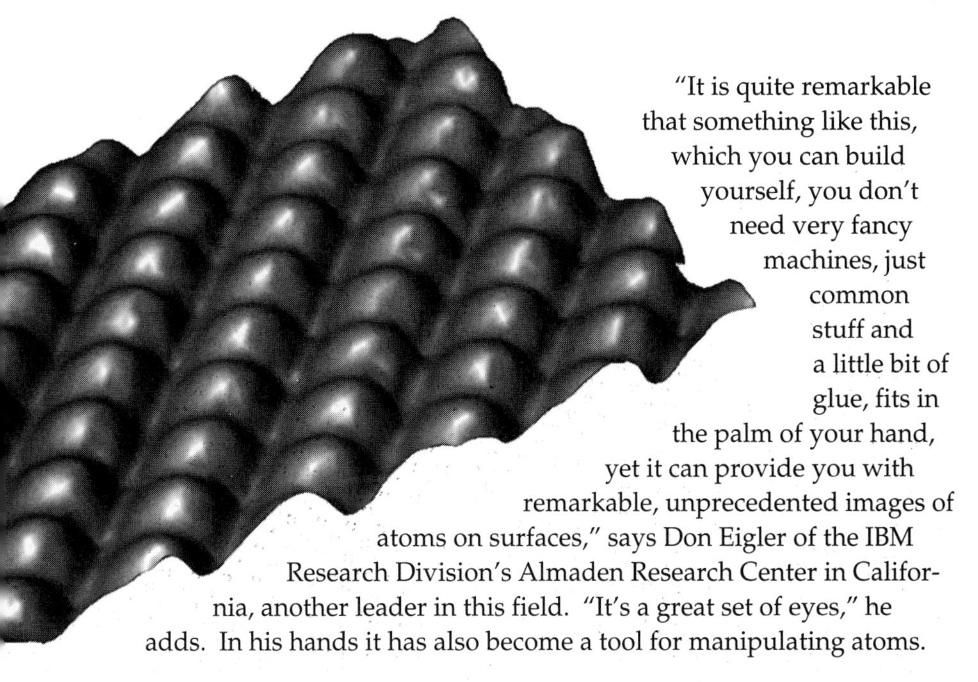

FIGURE 2.3 *Surface of nickel as seen by a scanning tunneling microscope. Bumps correspond to individual atoms. (Courtesy of Donald Eigler, IBM.)*

"It is quite remarkable that something like this, which you can build yourself, you don't need very fancy machines, just common stuff and a little bit of glue, fits in the palm of your hand, yet it can provide you with remarkable, unprecedented images of atoms on surfaces," says Don Eigler of the IBM Research Division's Almaden Research Center in California, another leader in this field. "It's a great set of eyes," he adds. In his hands it has also become a tool for manipulating atoms.

TOTE THAT ATOM, LIFT THAT MOLECULE

Eigler and his colleagues have operated their equipment under conditions of high vacuum and temperatures close to absolute zero. The vacuum protects their surfaces against contamination, allowing them to remain atomically clean. The low temperatures, in turn, encourage atoms to stay put rather than hopping around, as they would do if the equipment were warmer.

At the outset, Eigler faced the issue of identifying particular atoms. "They don't come with labels on them," he warns. "We start off with an atomically clean surface and we observe the mean number of defects and what kinds they are. Then we put down a lot of the atoms that we want to study, and we see that they all appear to be a particular kind. From that, we learn to identify what they look like."

An important set of experiments have involved xenon atoms on a surface of nickel. Xenon is an inert gas; it does not readily take part in chemical reactions. Still, that does not mean that its atoms lie loosely on the surface like ping-pong balls; they indeed experience forces, though these are weak. One of them is the Van der Waals force, which results when electrons surrounding an atom show greater density on one side than the other. Such shifts in charge density can result from interactions of atoms with their neighbors. A second force arises from overlap of the fuzziness of a particular atom with that of a neighbor. This overlap means that the two atoms share electrons to a slight degree, and the force that results resembles that of a chemical bond.

Such forces come into play both in attaching a xenon atom to the underlying surface and in attracting it to a pointed tip. Eigler notes that the tip can operate in two modes: Imaging and manipulation. In the imaging mode the tip rides above the surface at a distance of several angstroms, mapping the location of xenon atoms without disturbing them. When manipulating, the tip simply goes lower; more properly, it uses the feedback system to position itself so as to receive a stronger tunneling current, which is the measure of this closeness. "That increases the magnitude of the attractive force between the atom and the tip," Eigler adds. The consequence is that the atom remains attached to the surface but can also follow the tip as it moves about. The atom does not rise from the surface but acts somewhat like a heavy steel ball that can roll, with the tip attracting and guiding it as if the tip were a magnet.

"We find that the ability to slide a xenon atom over a nickel surface is independent of both the sign and the magnitude of the electric field, the voltage, and the current," Eigler writes. "It does, however, critically depend upon the separation between the tip and the atom." This leads to a simple procedure for repositioning such atoms, one by one: Locate them using imaging; lower the tip to attract an atom of interest; move the tip as desired; and then raise the tip. The atom will then stay behind, and the low temperatures of the surface will keep it there (see Figure 2.4).

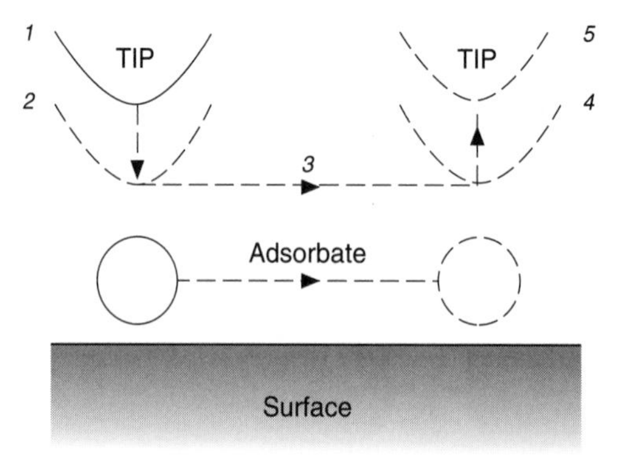

FIGURE 2.4 *Dragging an atom across a surface, in five steps: (1) locate atom and place tip over it; (2) lower tip to increase its force on the atoms; (3) move tip sideways, pulling atom with it; (4) atom is now in desired position; (5) raise tip, leaving atom in place on surface. (Courtesy of Donald Eigler, IBM.)*

In this fashion Eigler and his colleagues have placed such atoms on the surface as if they were lightbulbs on a theater marquee, spelling out the letters I B M. More in the nature of basic research, these researchers have studied the fundamental behavior of such atoms in small groups, carrying out observations that in no way could have previously been performed. In one of them they built a row of seven xenon atoms (see Figure 2.5). In Eigler's words,

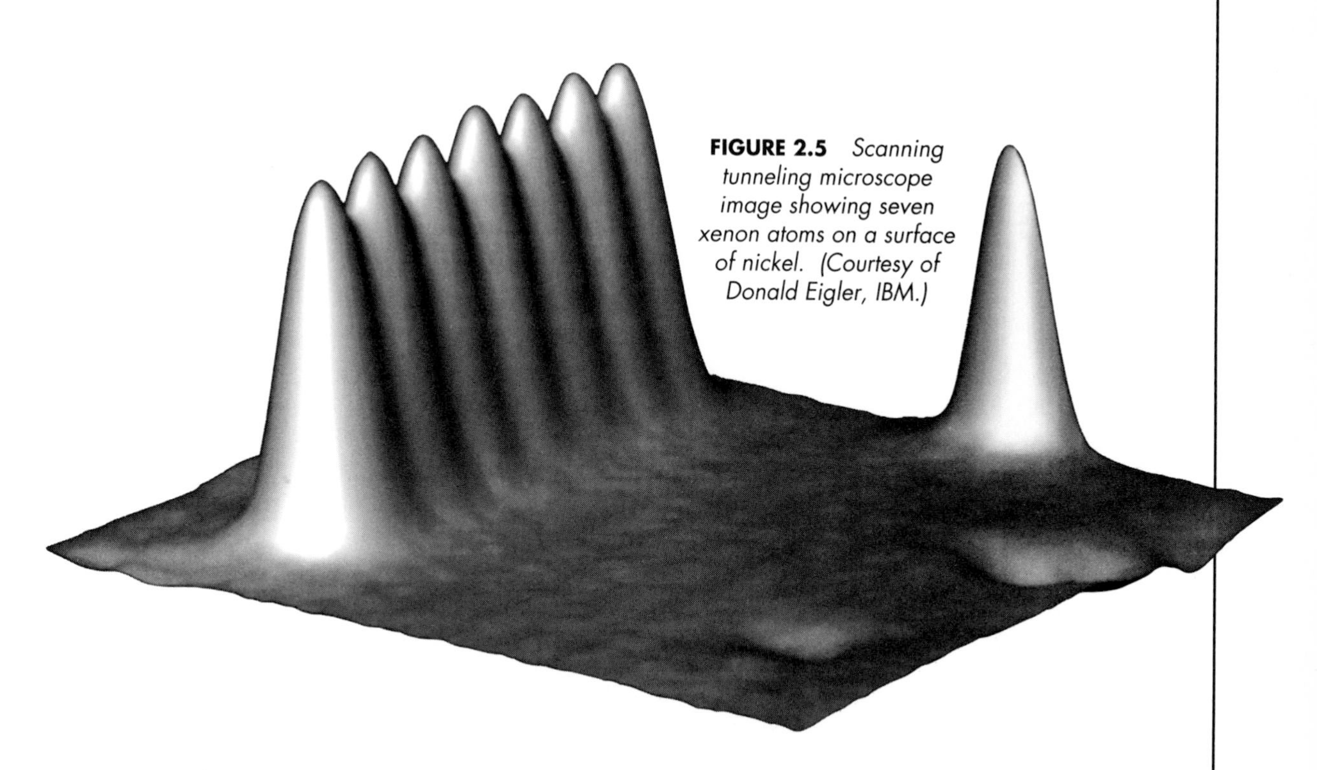

FIGURE 2.5 *Scanning tunneling microscope image showing seven xenon atoms on a surface of nickel. (Courtesy of Donald Eigler, IBM.)*

> We observe that the apparent spacing between the xenon atoms is 5.0 ± 0.2 Å. This corresponds to just twice the spacing of atoms in the underlying nickel surface. From this we deduce that such linear chains of xenon atoms order commensurately with the underlying nickel lattice. This indicates that the in-plane xenon-nickel interaction dominates over the in-plane xenon-xenon interaction. In solid xenon the spacing between atoms is just 4.4 Å. Accordingly, attempts to make a more compact linear array of xenon atoms along the rows of nickel atoms met with failure. It was found that in order to pull a xenon atom off the end of the chain the tip had to be lowered closer to the atom than was necessary to move a lone xenon atom. Thus, we learn that the xenon-xenon interaction along the chain is attractive.

This last observation, that adjacent xenon atoms attract, does not hold as a general rule for other atoms or molecules. Molecules of carbon monoxide show no such mutual attraction when adsorbed on platinum, regardless of their separation.

This technique can reposition more than single atoms, for such atoms sometimes bind weakly to form groups, resembling a rack of billiard balls. "We've been able to drag around rafts of 12 atoms," Eigler asserts. In addition, the scanning tunneling microscope can do more. It can lift an atom from the surface and drop it in a desired location.

ATOMIC SWITCHING

The pertinent discovery dates to February 1990. To pick up a xenon atom, one holds the tip close to it and applies a pulse of +1 volts. That causes the atom to jump from the surface and attach itself to the tip. The tip then moves to a new location, and one applies a pulse of –1 volts. The atom then jumps back to the surface.

Having discovered how to do this, Eigler's group has gone on to use this arrangement as a switch of ultimately small size in which the active element consists of no more than a single atom. When the atom lies on the surface, beneath the tip, the arrangement is in a low-conductance mode. When the atom jumps to the tip, the system switches into high conductance. In a representative experiment these two states, respectively, passed a current of 1.2×10^{-8} and 9×10^{-8} amperes, repeatedly turning the high current on and off as the single xenon atom jumped back and forth between tip and surface in response to the voltage pulses (see Figure 2.6).

Eigler regards this as a proof of concept for an atom switch. Such switches, in turn, offer a possible route whereby the manipulation of atoms could win a niche in the world of technology. As people at IBM generally appreciate, miniature switches are the basic elements of computers. Transistors serve this role in today's designs, but there are

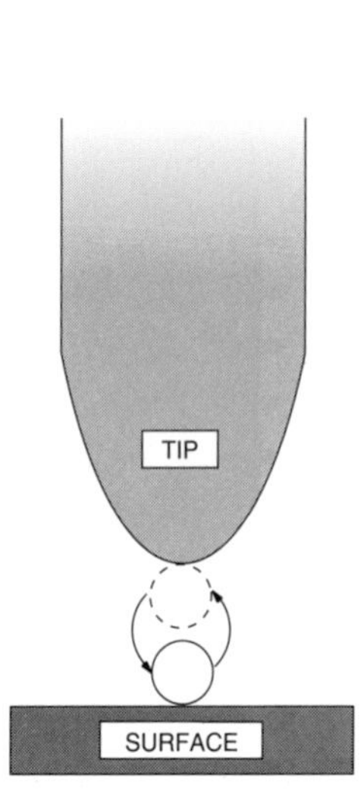

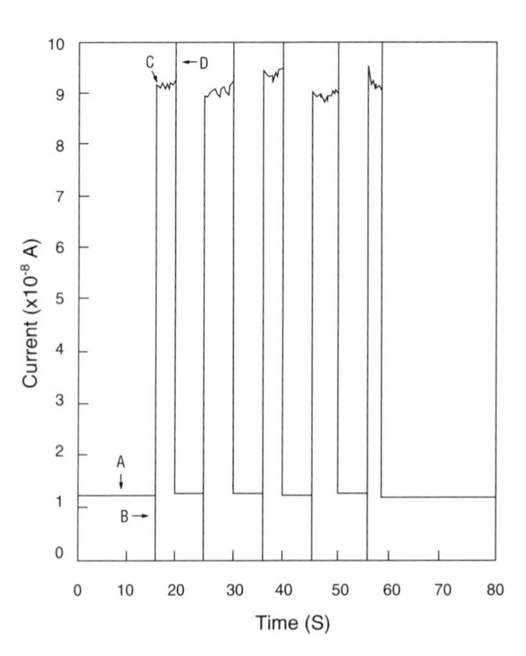

FIGURE 2.6 *An atomic switch. Left, the switch operates by transferring an atom back and forth between surface and tip, through application of voltage pulses. Right, plot of current through the switch. A, low-conductance state with xenon atom on surface. B, +0.8-volt pulse transfers atom to the tip. C, resulting high-conductance state with greatly increased current flow. D, –0.8-volt pulse returns atom to the surface and reestablishes the low-conductance state. (Courtesy of Donald Eigler, IBM.)*

limits to how small they can be, and switches of tinier size could in time become objects of great interest. Already, at least two research topics have grown out of Eigler's work.

At Stanford University, Thomas Albrecht, a graduate student working with Calvin Quate, has used standard techniques of microfabrication to create the tunneling microscope on a chip. Such a device could serve as an Eigler-type switch. Albrecht's instruments do not scan but rely on a sharp tip that rides at the end of a cantilever arm 1000 microns in length that is fabricated from layered aluminum and piezoelectric zinc oxide. Some 200 such devices can fit on a 3-inch silicon wafer, a standard in the electronics industry. Furthermore, because they are so small, they are not affected by vibrations from the outside world.

Eigler himself envisions that such switches could confine a movable atom within a molecular lattice, with zeolites and fullerenes being possible choices. Then there is the matter of the fine conducting connections that would carry flows of current to and from the switch, and these conductors or leads must dissipate the heat that will arise within them due to electrical resistance. "Will it work or will it fry?" Eigler asks. The question then becomes one of investigating the conducting and heat dissipation properties of leads built to nanometer dimensions, on a scale of perhaps tens of angstroms. Here, too, the scanning tunneling microscope offers help.

Chris Lutz, a colleague of Eigler at IBM's Almaden center, has used this instrument to move atoms of platinum on a surface of platinum. To do this, he had to bring the tip particularly close to the atoms so as to apply enough force to drag them across the surface. This represents a rather demanding experiment because platinum atoms would tend to bind somewhat strongly to a surface of the same material. It opens the way to working with atoms of different metals, assembled atop an insulator to produce microscopic patches. In Eigler's words, "We have the possibility someday of being able to study how electrons move through such small structures, and that's a very exciting direction for us."

LIMITATIONS OF THE TECHNIQUE

These techniques offer unparalleled opportunities to study the behavior of atoms adsorbed on surfaces individually or in small groups. In particular, a variant of the basic instrument, the atomic force microscope, measures the forces applied in moving such atoms and thus gives quantitative information. However, little in this area is easy. Atoms sometimes are lost while being pulled to new locations, as they become trapped in defects within the atomic structure of the surface. That is somewhat like having a steel ball drop into a knothole within a wooden table. Even when this doesn't happen, there are other problems. It has not proven possible to drag an atom over the atom-size step formed by a patch of atoms previously in place on the surface. From this perspective, physicists are like the pyramid builders of ancient Egypt, puzzled as to how to slide a stone block on top of an existing layer built of such blocks.

Sometimes the basic technique fails completely. Eigler recalls trying to move oxygen atoms on a platinum surface. In his words, the result was "no luck. In order to move the oxygen, the tip had to be so close that the forces between the tip and the surrounding platinums were enough that when we moved the tip sideways, the whole surface got destroyed. There is a case where we cannot get oxygen under control." The experiment might well have set a world record for the finest line ever scribed in a metal surface, but that was not its purpose.

Even when atoms and molecules are better behaved, there are problems in working with them. In chemistry a topic of vast interest involves reactions that take place on surfaces. Many catalysts provide such surfaces within the petrochemical industry, and chemists hope to study such reactions at the level of individual atoms and molecules by using the scanning tunneling microscope. As a prelude to such work, Eigler has tried to produce a molecule of carbon dioxide by coaxing carbon monoxide into combination with an atom of oxygen.

The two species do not react spontaneously; we see this in the polluted air of our cities, where carbon monoxide in motor vehicle exhaust resists this reaction even though there is plenty of oxygen in the surrounding air. At the molecular level, there is an energy barrier whereby one must apply energy to the oxygen atom, or to the carbon monoxide, for the reaction to proceed. "We went about trying to build carbon dioxide," Eigler reports. "We could not put enough force on the carbon monoxide to make it react with the oxygen." It was not difficult to manipulate carbon monoxide molecules; they demanded less force than platinum atoms, but that wasn't enough. Eigler sees other possibilities: Bashing the reactants into each other, catalyzing with the tip, zapping with photons or with energetic electrons from the tip. So far, however, he has yet to produce his carbon dioxide.

Further limitations come into play because there is no way to accomplish precision navigation over a surface. Typically, one deals with material samples of centimeter size. Within this vastness, one produces microscopic images of regions perhaps a hundred angstroms in length and width, or 10^{-6} centimeters. That is like examining in great detail a meter-size patch of soil in the rangeland of Texas with nothing more than a Rand McNally highway map to help you to find it again. Similarly, physicists have no reliable way to find their way back to a particular spot if they relocate the tip. As Calvin Quate put it, "We see things that seem so distinctive, but unless we leave the microscope just as it is, we'll never see them again."

Despite these caveats, the scanning tunneling microscope offers unparalleled vistas in the study of atoms on surfaces. Moreover, similar vistas exist in studying atoms in free space. Here the goal is to make highly precise measurements of fundamental properties. That is not possible when the atoms adhere to a surface for the surface disturbs these properties in complicated ways. As a result, atoms in empty space represent a topic in their own right.

A POSITRON NAMED PRISCILLA

As an example of what can be achieved, Hans Dehmelt of the University of Washington has carried out exquisitely precise studies of individual electrons. Electrons carry an electric charge, which allows them to respond readily to electric fields; they also respond easily to a magnetic field. This makes it possible to trap such particles and to hold them in place for long periods. Dehmelt has also trapped the positron, which amounts to an electron with positive charge. Indeed, because positrons do not exist naturally on earth, he has succeeded in showing that a particular positron under study has had no opportunity to change places with a different one. He has held that particle in place for as long as 3 months, and he has named it Priscilla. Writing in the journal *Science*, Dehmelt notes that "the well-defined identity of this elementary particle is something fundamentally new, which deserves to be recognized by being given a name, just as pets are given names of persons."

Working both with Priscilla and with electrons, he has made measurements of a basic quantity known as the g-factor. This is the ratio of magnetic moment to spin angular momentum. In appropriate units this quantity is a simple number; Dehmelt's value, published in 1987, is accurate to 12 decimal places:

$$g/2 = 1.001\ 159\ 652 \pm 4 \times 10^{-12}$$

A basic prediction of physics is that this value should be the same for both positron and electron. Dehmelt finds the ratio of values:

$$g(e^+)/g(e^-) = 1 + (0.5 \pm 2) \times 10^{-12}$$

Hence, within the errors of measurement, the two g-factors indeed are equal to 12 significant figures. Dehmelt describes this as "the most stringent test" of this equality and hence of the underlying theory.

This theory, known as quantum electrodynamics, is among the most accurate in all of physics. It postulates that the electron exists as a simple point, having zero diameter, and Dehmelt's work makes it possible to critique this assumption. By regarding this theory as correct in all particulars, a modified g-factor whose value is exactly 2.000 can be defined. The experimental value is

$$g = 2.000\ 000\ 000\ 110 \pm 6 \times 10^{-11}.$$

Hence, within the errors of measurement, this g-factor is *not* exactly equal to 2.000. The difference can serve to infer an experimental value for the electron radius: 10^{-20} centimeters. That is some 10 million times smaller than a proton or neutron. Yet if one believes this inference, it opens the most far-ranging prospects.

Protons and neutrons, with radii of some 10^{-13} centimeters, most definitely are not geometric points. Indeed, they have a complex internal structure, featuring three quarks that bind together through exchanges of other particles known as gluons. This binding force, in turn, is the subject of its own physical theory, known as quantum chromodynamics. And if the electron also has a finite radius, this could mean that it too might be a composite entity, having a layer of structure below what we now study and observe. Such a structure, in turn, might apply as well to the quarks themselves, for physicists regard quarks and electrons as equally elementary.

This would amount to nothing less than a new level of detail in subatomic physics. In this century we have proceeded along a succession of steps: First atoms, then constituent particles such as protons and neutrons, then quarks. The next step, if it is real, could add a domain of subquarks that combine to yield electrons and, perhaps, the quarks we study. Moreover, if the electron indeed has a dimension of 10^{-20} centimeters, one can infer that the forces between such subquarks are 10^7 times stronger than those between the quarks in a proton or neutron. This follows from a principle of quantum mechanics: Processes that take place in diminishingly small regions require correspondingly higher energies.

THE PROBLEM OF PRECISION

These measurements, and the resulting inferences, all stem from Dehmelt's ability to trap and hold individual charged particles. In dealing with atoms, however, similar accuracies in measurement demand entirely different approaches. The basic technique for the study of atoms, in use for over a century, has been spectroscopy: The detailed observation of spectral lines. The common sodium light fixture, seen along highways, has a close relation to the usual apparatus. Here an electric current vaporizes sodium metal and adds energy to the atoms, exciting them and causing them to glow brightly at characteristic wavelengths. In this case the principal spectral lines are bright yellow in color, while other elements have lines with other wavelengths and different colors. When observed through a spectroscope, excited sodium shows those lines clearly.

The lines contain fine detail. Each of them is made up of closely spaced subsidiary lines, somewhat as a newspaper photo is actually an ordered collection of small dots. However, this detail ordinarily cannot be seen, and the reason lies in the Doppler effect. This comes into play when we hear the lonesome whistle of a passing train, rising in pitch as it approaches and then falling as the train recedes. Atoms in a hot gas are in motion, some toward us and others away from us, and their emissions of light behave similarly. They shift to higher frequencies when approaching and to lower ones when receding. The result is that spectral lines broaden or smear out, destroying their fine detail. Nor can one easily get rid of these motions; they result from the fact that the gas is hot. Indeed, their velocities measure the temperature.

Physicists can ease this problem with cleverness. Rather than confine the hot gas within a vessel, it can escape through a nozzle, forming a well-collimated beam of atoms that moves at right angles to the line of sight of the spectroscope. That gets rid of the main part of the Doppler broadening but leaves two subtle effects that also broaden spectral lines. The first stems from Einstein's theory of relativity. Atoms in the beam continue to have different velocities, and relativistic effects then dictate that they experience the passage of time at slightly different rates. This leads to slightly different frequencies in the various atoms' spectral lines, and again the result is a broadening of the lines from the atomic beam. The second effect stems from the fact that within the beam individual atoms are in view only momentarily as they fly across the field of view. Quantum mechanics then dictates that there is a corresponding uncertainty in the true frequency of any particular feature in a spectral line. This uncertainty causes the features to blur and the lines, again, to broaden.

The use of this transverse atomic beam offers some help; the resulting line broadenings at least are generally less severe than those that arise from observing a simple hot gas, with its full-blown Doppler effect. But to study the lines in their full detail, one must contrive to get rid of the motion of the atoms. That means cooling them close to absolute zero. One cannot do this in a chamber, however; the atoms would condense on the cold walls and then would have different properties than those a physicist seeks to observe. Hence, one must chill the atoms as they fly through space. Lasers can do this, and the principal approaches rely on turning the Doppler effect from an adversary into an advantage.

LASER COOLING

The words "laser cooling" sound like a contradiction in terms. We ordinarily think of the laser as a source of intense heat. Those of the Pentagon's "Star Wars" program, for instance, are to have sufficient energy to destroy a missile in flight. But when directed against atoms rather than missiles, a laser indeed can act to slow them down, which is the same as cooling them.

To do this, one begins by appreciating that the frequency of a spectral line represents a condition wherein atoms absorb photons of light particularly readily. Such photons, having that frequency, come from a laser tuned to the appropriate wavelength. Then, after absorbing such a photon, the atom rapidly reemits it, which puts the atom in a condition to absorb another one.

The absorbed photons all come from one direction, that of the laser beam. The reemitted photons, however, fly off in every direction. During both absorption and reemission, the atom feels a force sufficient to change its velocity; for sodium atoms this change amounts to 3 centimeters per second. The absorbed photons produce a cumulative effect, combining to slow the atoms. The reemitted photons, by contrast, lead to no more than small changes in each atom's path through space, because they produce no combined or collective effect. Here, then, is a powerful technique for slowing and cooling a beam of atoms.

Right at the outset, though, there are problems. The usual procedure tunes the laser to a frequency just below that of a spectral line and points the laser directly into the oncoming beam. Due to the Doppler shift, atoms in flight see a laser wavelength close to that of their spectral line and indeed absorb the photons quite readily. But as the atoms slow down, two effects can break the close match between laser and spectral-line frequencies, rendering this slowing process ineffective.

The first effect applies particularly to sodium, which is a common atom used in such studies. The process of cyclic absorption and reemission of photons depends on the atom returning to a particular ground state following each photon emission. In that state an atom quickly picks up a new photon to continue the cycle. However, sodium atoms have some tendency to fall into a slightly lower ground state, associated with a different spectral line. The atoms then do not readily absorb the laser's photons and fly onward without slowing down.

The most straightforward solution is simply to use a second laser, tuned to the adjacent spectral line. Its photons then are absorbed in the usual fashion. However, at the National Institute of Standards and Technology (formerly the National Bureau of Standards), William Phillips and Harold Metcalf have introduced a different approach. They impose a magnetic field on the sodium atoms to suppress the transition to the unwanted ground state. The atoms then cycle between ground and excited states in the desired fashion, and a single laser suffices to slow them.

Still, there is a second effect that breaks the match between laser and spectral line wavelengths. This again involves the Doppler shift, for as atoms slow, this shift becomes less effective and the laser wavelength, as seen by the atoms, moves away from the desired value. The result again is that the atoms fail to absorb the oncoming photons. Again, though, both optical and magnetic methods exist to deal with this.

The optical method, introduced by V. I. Balykin at the Institute of Spectroscopy in Moscow, subjects the laser to a rapid frequency change as the atoms decelerate. This keeps the laser in tune with the spectral line as the atoms slow, with the line shifting as a result of the Doppler effect. This technique, known as "chirp cooling," produces slowed atoms in bunches. It works well with the two-laser method for avoidance of the unwanted ground state.

The magnetic method again has resulted from work by Phillips, Metcalf, and their colleagues. It relies on the Zeeman effect, whereby a magnetic field causes a spectral line to split in two. As the field increases, the two halves of the line move farther apart. This offers a means whereby a magnetic field can produce a predetermined shift in a spectral line. In practice, the field initially is strong but diminishes along the atoms' line of flight, as they slow. The atoms then stay in tune with a cooling laser of fixed frequency. This laser, in turn, yields slowed atoms in a continuous stream, rather than in bunches (see Figure 2.7).

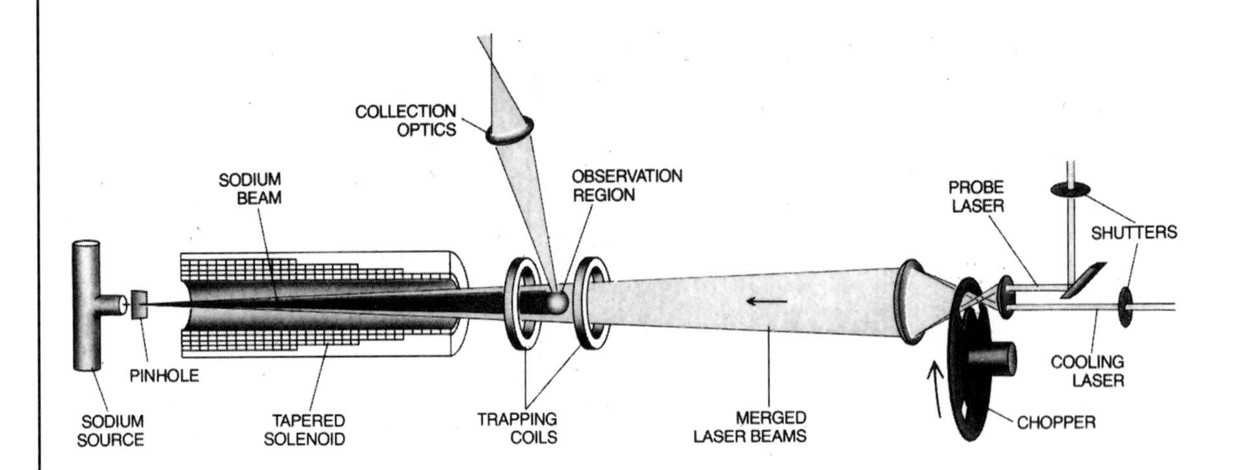

FIGURE 2.7 *Layout of apparatus for use of a magnetic field in slowing and trapping sodium atoms. The atomic beam comes from the left; the laser beam, which slows the atoms, comes from the right. A tapered solenoid applies a magnetic field that diminishes with distance, imposing a Zeeman effect that adjusts the atoms' spectral-line frequency to match the constant frequency of the laser. A pair of magnetic coils then trap the atoms, which have nearly zero velocity. From "Cooling and Trapping Atoms," by W.D. Phillips and M. J. Metcalf. Copyright © 1987 by Scientific American, Inc. All rights reserved.)*

Within a sodium beam, atoms initially fly at speeds of some 1000 meters per second. Since each encounter with a photon can slow such an atom by three centimeters per second, every atom must absorb and reemit some 30,000 photons to come close to a standstill. Fortunately, this cyclic process repeats so rapidly that the atoms decelerate at rates of some 10^6 meters per second squared, or 100,000 times greater than the acceleration due to gravity. That suffices to bring them to a stop in as little as a millisecond, along a range of only 50 centimeters.

OBSERVATIONS OF IONS

Similar techniques suffice to cool ions as well as atoms, and ions offer several advantages. Being electrically charged, it is possible to hold them for long periods by applying electric fields. Indeed, Hans Dehmelt, who has pursued his lengthy involvement with Priscilla the Positron, has similarly trapped a barium ion named Astrid. The opportunity to trap ions electrically, in turn, means that one can first trap them and then cool them, which can be easier than the cool-first, trap-later procedure with atoms. A third advantage is that ions that resist laser cooling can nevertheless reach low temperatures by storing them in a trap along with other ions that indeed respond to the laser. This technique, known as sympathetic laser cooling, treats the coolable ions as a type of ice, able to receive heat from the other ions and hence to slow their motions as well.

As an example of the measurements that then become possible, David Weinland and colleagues, at the National Institute of Standards and Technology, have carried out studies on a single positively charged ion of mercury, Hg^+. This species shows a strong spectral line with wavelength close to 2800 Å, corresponding to a frequency of some 1.07×10^{15} hertz. The natural line width is only 43 megahertz, but at room temperature Doppler effects broaden the line to 3 gigahertz, blurring out its details. Weinland, by contrast, has cooled his Hg^+ ion, which has no name, to 0.006 Kelvin, or degrees above absolute zero. More recently, he claims to have reached 0.00024 K. Working with an adjacent spectral line at 2815 Å, his group has observed line widths as narrow as 30 kilohertz, some 100,000 times finer than the room temperature line. Nor is this the limit; the 2815-Å line has a natural width of as little as 1.6 hertz, which might be approached in practice using a laser with sufficient stability in its frequency.

TRAPPING ATOMS

Atoms are electrically neutral; hence, unlike ions, one cannot trap them by using electric fields. Perhaps the simplest trap involves three sets of laser beams, oriented respectively to define x, y, and z axes and intersecting within a small region of space. These lasers are tuned to just below the frequency of a strong spectral line. An atom in motion, whatever its direction, will absorb photons more effectively and hence will experience a drag force that acts to slow it down. Indeed, provided its velocity is small, the force it feels will increase with its speed. The atom then behaves as if it were held within a viscous fluid that resists its motion. Steven Chu of Stanford University, who first demonstrated this technique while at AT&T Bell Laboratories, calls this approach "optical molasses," noting that the idea of using lasers to cool neutral atoms was first proposed by Ted Hansch and Arthur Schawlow.

A different technique takes advantage of the fact that while atoms carry no electric charge they do have electrical properties. Light, such as comes from a laser, carries a rapidly varying electric field that oscillates at its frequency. With a wavelength of 5000 Å, for instance, which is a reasonably typical value, the frequency is 6×10^{14} hertz. Electrons within an atom tend to oscillate in response to this field. The consequence is a rapidly varying redistribution of charge within the atom that indeed makes it responsive to electric forces.

This effect resembles what happens when you rub a comb with fur and use it to attract small bits of paper. Rubbing the comb gives it a charge of static electricity. The bits of paper, like atoms, carry no such charge; but when the comb approaches, it carries an electric field that redistributes the electrons within the paper. The paper then acquires an induced charge and flies toward the comb, collecting at a place on the comb where the static charge is strongest.

In 1968 the Soviet physicist Vladilen Letokhov proposed that atoms immersed in laser light, with their rapidly varying redistribution of charges, would behave like the bits of paper. This would happen if the laser were tuned to a frequency well away from that of any spectral line; the atom's electrons then indeed would oscillate in response to the light's rapidly changing electric field. By contrast, if the laser had a spectral-line frequency, the electrons would tend to absorb and reemit photons, which would quickly push them out of the trap. Then, just as the paper bits are attracted to the strongest region of charge on the comb, the atoms feel an attraction toward the strongest or most intense region of laser light. One produces such an intense region simply by bringing the light to a sharp focus.

In practice, one locates this focus within the intersection of the sets of beams that form optical molasses. The lasers then feature two different frequencies: There are molasses beams that are close to a spectral line, along with the focused beam that is well away from such a line. The resulting arrangement can then produce a large increase in the density of trapped atoms. At AT&T Bell Laboratories, for example, molasses beams alone give a density of 10^6 atoms per cubic centimeter. The focused beam then raises this density a millionfold.

These techniques, however, leave something to be desired. The volume within the focus is quite small; it can be as low as a billionth of a cubic centimeter. In addition, these traps leak atoms badly. The reason is that, although they are immersed in a vacuum, a physicist's vacuum is far from perfect. It contains residual atoms that bump into the trapped ones and knock them away. Some improvement is possible; better vacuums give longer trapping times. In addition, Chu at Stanford reports creating a "super molasses" by misaligning his laser beams in a particular way. This holds particular atoms for as long as a minute, compared with storage times with conventional molasses of closer to 1 second. But in seeking long storage times together with large volumes, researchers have turned to the use of magnetic fields or a combination of magnetic and laser fields.

Atoms, even when left alone, have a small ability to act like bar magnets and to respond to a magnetic field. This effect is weak and comes into play only at vanishingly low temperatures, but such temperatures characterize laser-cooled atoms, which means that magnetic traps can operate successfully. Such a trap features two current-carrying rings, mounted like barrel hoops, with the current in one ring flowing in a direction opposite to that of the other. This creates a magnetic field that has zero strength on the centerline midway between the two rings and that increases in all directions away from this point. An early trap of this type, built at the National Institute of Standards and Technology, held atoms within a volume the size of a golf ball. More recent traps, using stronger magnetic fields and better vacuums, have featured larger volumes. These have held atoms for a number of minutes, even hours.

THE ATOMIC FOUNTAIN

Nevertheless, whether trapped magnetically or optically, atoms are not in the pristine state that physicists would prefer to study. These trapping techniques all perturb the atoms in various ways. When caught in molasses, they are continually absorbing and reemitting photons, and their electrons are jumping up and down between energy levels. Use of the focused beam introduces rapidly varying oscillations in their electron densities. The magnetic trap, in turn, changes the spectral lines through the Zeeman effect. As a consequence, a current trend is to collect atoms in such a trap but not to study them there. Instead, a quick burst of tuned laser light can launch them out of the trap and into ballistic trajectories, within a vacuum. And just as a football has hang time when punted, these atoms have hang time within the chamber where they are available for study, away from the trap. The time for study approaches a full second if the chamber measures a meter from top to bottom. That is virtually an eternity; in principle, it suffices to permit measurement of spectral features down to line widths as narrow as 0.2 hertz. This is because longer observation times yield greater accuracy and precision.

In this fashion, Chu and his colleagues have built an atomic fountain. The apparatus collects atoms in a trap for some one-half second and then launches them upward at some 2 meters per second. Near the top of their trajectories, the instrument probes these atoms by applying two microwave pulses, separated in time by 0.25 seconds. These produce a transition between two closely spaced energy states, with the separation in time yielding superb measurement accuracy (see Figure 2.8). "In our first experiment," Chu notes, "we measured the energy difference between two states of an atom with a resolution of two parts in 100 billion." Line width was a mere 2 hertz. Furthermore, repeated observations made over 15 minutes resolved the center of the pertinent spectral line to 0.01 hertz. Current work with cesium atoms is improving the resolution still further, to as much as one part in 10^{15}.

Chu has also gone on to develop an "atomic funnel" that produces a

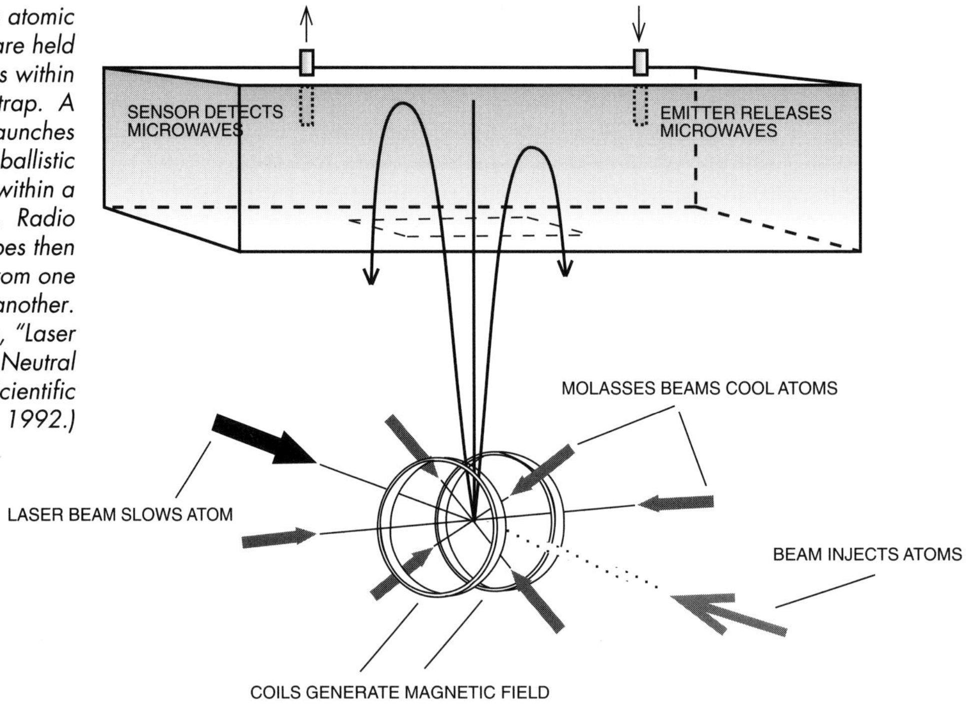

FIGURE 2.8 *An atomic fountain. Atoms are held in optical molasses within a magnetic trap. A pushing beam launches them onto ballistic trajectories within a vacuum chamber. Radio frequency probes then excite the atoms from one energy state to another. (Adapted from Chu, "Laser Trapping of Neutral Particles," Scientific American, Feb. 1992.)*

relatively dense beam of very slow atoms that can feed a continuous atomic fountain. "The Stanford funnel accepts atoms with a large velocity spread and cools them into a localized and collimated beam," he writes. "With our funnel, a beam of atoms was produced with a flux of 10^9 atoms per second at a velocity of 270 centimeters per second and a temperature of 2×10^{-4} Kelvin."

The pursuit of ever-lower temperatures represents another area of current activity. This work relies on a subtle point: It is not the absolute values of atomic velocity that define their temperature but rather the velocity spread or range of variation about an average. Hence, to achieve ultralow temperatures, it suffices to greatly reduce this spread, even if the atoms continue to show a well-defined (and uniform) velocity. This work takes advantage of today's relatively long trapping times for atoms, which makes it possible to drive their velocity spread close to zero. Working with a colleague, Mark Kasevich, Chu reports producing a collection of sodium atoms with a velocity spread of only 0.027 centimeters per second, corresponding to an effective temperature of 2.4×10^{-11} Kelvin.

SOME USES OF PRECISION

What do such measurements offer to the working physicist? Their great precision offers the prospect of new time and frequency standards that have considerable improvements in accuracy. The present time standard is an atomic clock that relies on cesium atoms. It keeps time with an error of one part in 10^{13}, corresponding to a few ten-thousandths of a second over a human lifetime. That certainly is far better accuracy than anyone needs for ordinary purposes, but there are situations where even this is barely adequate. Tests of Einstein's relativity, for instance, demand all the precision a physicist can achieve. Now the atomic fountain offers the prospect of a thousandfold improvement, reducing the error to one part in 10^{16}.

Chu notes that this would make it possible to determine whether the constants of physics may be changing slowly with time. These constants include the speed of light, the charge on the electron, and Planck's constant in quantum mechanics. Standard physical theories hold that they all should indeed remain invariant, but it would be useful to check. As with Dehmelt's study of the radius of the electron, such a check might show that some of our basic ideas are wrong. Chu notes that such an experiment could feature atomic clocks of very great precision that would rely on distinctly different physical principles. One clock, based on optical transitions, would have its frequency determined primarily by quantum electrodynamics (QED), while another, based on the hyperfine splitting of certain spectral lines of atoms, would be determined by a combination of QED and nuclear forces. If the strength of the nuclear forces changed relative to the electromagnetic forces, the two very precise clocks would begin to "tick" at different rates.

Other types of observation would involve novel aspects of quantum mechanics. This branch of physics predicts that an atom has wave properties similar to those of photons of light. This is different from saying that the atom absorbs photons having the right wavelength; it means that the atom itself behaves like a wave rather than as a particle. Further, its wavelength increases as the atom cools to ultralow temperatures. This opens the prospect of carrying out experiments with atoms that are done ordinarily with light.

Among these is interferometry, which demonstrates that light consists of waves. Such an experiment conventionally features two thin slits set close together, with a screen behind them. A laser shines through the slits. It does not produce shadows but yields a pattern of alternating light and dark bands on the screen, as a result of interference effects. One can also carry out a similar experiment using a beam of electrons, which have their own wave properties. Here, too, the result is a pattern of alternating bands. Moreover, it is possible to reduce the intensity of the source of the light, or of the electron beam, so that at any moment only one photon or electron is in flight. Given time, though, the same pattern will form. It would also form using a beam of slow, and ultracold atoms.

There are several advantages to constructing an interferometer based on slowly moving atoms. For example, such an instrument can be a very sensitive inertial sensor. Chu and Kasevich already have built an interferometer that uses slow atoms and that measures the acceleration of gravity with an accuracy of at least three parts in 10^8. Moreover, Chu expects to achieve a further improvement of a thousandfold. The result could be an atomic standard for gravity measurements that has greatly increased precision. In turn, that could influence the highly practical matter of searching for oil. Changes in the local gravity sometimes point out oil-bearing formations to geologists, and improvements in such gravity measurements might aid the mapping of such deposits.

Ultracold atoms can also assist in opening up other fundamental topics in quantum mechanics. At sufficiently low temperatures, such atoms should undergo a phenomenon called Bose condensation. This is not the condensation that occurs when a gas freezes into a solid. Rather, it is a quantum effect whereby all the atoms would take on the same ground state. Analogs exist in a laser or in superfluid helium, where many of the particles—photons or helium atoms—are considered to be in the same quantum state. Because ultracold atoms take on wave properties akin to those of photons and electrons, they should also undergo Bose condensation. The difference is that the ultracold atoms will form a dilute Bose gas, a new state of matter that has never before been observed or studied.

APPLICATIONS IN BIOLOGY

The direct manipulation of atoms and other elementary particles, with the scanning tunneling microscope, with lasers, and by using electric and magnetic fields, thus opens a host of prospects. These include fundamental tests of basic features in physical theories, new states of matter, studies of chemical reactions at the atomic level, new instruments for the precise measurement of time and of gravity fields, and even a new approach in searching for oil. Moreover, some of the basic laser techniques used in slowing and cooling atoms are also finding adaptation for studies of the DNA molecule, and even of bacteria and other cells.

A point of departure for this work lies in the fact that micron-size polystyrene spheres can respond to a single focused laser in the same fashion as atoms. These spheres also redistribute their internal electrons in response to the rapidly varying electric field of the laser and then move in the direction of the focus of its light, where its intensity is the strongest. As with atoms, the rule in choosing a laser frequency is simply to avoid one at which the spheres absorb light strongly. But these spheres, unlike atoms, are large enough to be visible through a light microscope. This makes it possible to integrate such a laser with this microscope and to observe the particles while manipulating them.

The basic idea, that of using a single laser beam for this purpose, is that of Arthur Ashkin, of AT&T Bell Laboratories. Chu and his colleagues call it "optical tweezers" and have been using it in studies of DNA. The DNA in a human cell is a meter in total length, yet it coils up to fit neatly within the cell's nucleus. Chu's group has begun its studies of DNA by attaching tiny spheres to each end of a strand of DNA, stretching it out to full length using a pair of optical tweezers. By measuring the force required to stretch the DNA out to a given length, they are able to compare their results to basic models of polymer elasticity. In addition, with this new-found method of manipulating the molecule, they intend to study the function of enzymes that act on the DNA.

They also have spot welded such spheres to a microscope slide, by increasing the laser power, leaving the DNA fixed to the slide and available for further research. Chu notes that these could include studies of the interaction of enzymes with DNA, including those involved with the expression of genes and with the editing out of DNA errors, which could permit the DNA to mutate into genetic nonsense. In another study of the mechanical properties of large biological molecules, Chu's group has been studying the contraction of muscle at a fundamental level. The pertinent molecule is called myosin, and these investigators have been using optical tweezers to study the force of its contraction.

Other investigators have used such tweezers to manipulate cellular organelles. Ashkin and his colleague J. B. Dziedzic have made the surprising discovery that optical tweezers can handle live bacteria and other cells without damaging them. This is possible because the cells are nearly transparent to the laser and are immersed in water, which offers an effective coolant. Indeed, Ashkin has manipulated objects within a cell without puncturing the cell wall. In a potentially important application, Michael Berns and colleagues, at the University of California at Irvine, have manipulated chromosomes within a cell's nucleus.

WHERE DO WE STAND?

At present, the world of atomic and ion manipulations is largely one of laboratory experimentation. It is not yet possible to consult a Hewlett-Packard catalog for an atomic clock based on these principles or a gravity meter based on interferometry of ultracold atoms. Still less have such instruments served to make fundamentally new measurements or observations, which would stand as important contributions in their own right. Such an assessment even includes Dehmelt's proposal that the electron has a physical diameter and hence may possess a subquark structure. Though intriguing and potentially of great significance, such a suggestion leaves us with no way to study or examine that structure, for its energies are beyond our reach.

As the pertinent instruments and techniques enter widespread use, however, they indeed may bring forth a solid stream of significant findings. Certainly, there already is a considerable breadth to this field, both in methods and in scope for their applications. Significantly, the pertinent equipment is likely to accommodate well to working scientists with respect to size, price, and ease of use. Rather than finding concentration in some "National Center for Atomic Manipulation," such apparatus is likely to find space on the laboratory benches of physicists, chemists, and biologists. Amid such convenience and low cost, then, these instruments and methods may cease to dazzle. Instead, they will fold into the workday world of investigators, as these people focus instead on the advances they can pursue.

BIBLIOGRAPHY

Binnig, G., and H. Rohrer. 1985. The scanning tunneling microscope. Scientific American 253(Aug.):50–56.

Bollinger, J. J., and D. J. Wineland. 1990. Microplasmas. Scientific American 262(Jan.):124–130.

Chu, S. 1991. Laser manipulation of atoms and particles. Science 253(Aug. 23): 861–866.

Chu, S. 1992. Laser trapping of neutral particles. Scientific American 266(Feb.):70–76.

Dehmelt, H. 1990. Experiments on the structure of an individual elementary particle. Science 247(Feb. 2):539–545.

Feynman, R. 1960. There's plenty of room at the bottom. Engineering and Science 22(Feb.) (Also reprinted in Miniaturization, 1961, H.D. Gilmore, ed., Reinhold Press, New York.)

Itano, W. M., J.C. Bergquist, and D. J. Wineland. 1987. Laser spectroscopy of trapped atomic ions. Science 237(Aug. 7):612–617.

Petit, C. 1989. Beyond the cutting edge. Mosaic 20(Summer):24–35.

Phillips, W. D., and H. J. Metcalf. 1987. Cooling and trapping atoms. Scientific American 256(March):50–56.

Phillips, W. D., P. L. Gould, and P. D. Lett. 1988. Cooling, stopping, and trapping atoms. Science 239(Feb. 19):877–882.

AIDS
Solving the Molecular Puzzle

by Michelle Hoffman

In his 1993 inaugural address, President Clinton warned about the twin enemies of "ancient hatreds and new plagues" that continue to threaten and shape the world. There can be no doubt that among those plagues the new president was thinking of AIDS.

ANCIENT HATREDS AND NEW PLAGUES

Acquired immunodeficiency syndrome, or AIDS, is very much a modern plague, a reflection of the times in which it emerged. In a world where cultural and economic boundaries between countries are less firm than geographic ones, the grim reality is that epidemics respect no borders. AIDS is a plague that could emerge only in a modern world where, for better or worse, rapid mass transportation links materials, people, and pathogens from all over. And so it is that the human immunodeficiency virus (HIV), the agent that causes AIDS, could first evolve in a remote corner of the world and within little more than a few decades become a dread fear in every corner of the world. Today, more than 600,000 people worldwide have been diagnosed with AIDS (see Figure 3.1), and in the United States alone at least 1 million people are infected with HIV (see Figure 3.2). Throughout the world, still more are infected and have yet to know or have yet to develop the symptoms of AIDS .

But numbers alone do not reflect the dimensions of the "world AIDS crisis," as President Clinton called it. Certainly, more people have had flu from influenza virus than have suffered from AIDS. What distinguishes AIDS from every infectious disease to come before it is the scientific challenge it represents to a research community that has successfully controlled all of the plagues of the past. Hardly anyone living in the United States today can remember a time when they were concerned about typhoid or tetanus. Few children today will grow up with the threat of polio or smallpox. Improved public health measures and potent vaccines have made these diseases the exception rather than the rule in the United States and in most of the industrialized world. In more recent memory, both toxic shock syndrome and Legionnaire's disease emerged and were vanquished by science within a few short years.

Against this backdrop of success fighting infectious diseases, comes HIV, a virus whose natural history is so unusual that no prior experience with any infectious agent has appropriately prepared modern science to tackle it. And yet no previous era has been better prepared to take on the challenge. Only now does science have the experimental tools to unravel the details of the peculiar life style of this virus. It is one of the greatest ironies of this epidemic that stopping it will owe as much to the technology of the era as its spread.

The hallmark of AIDS is the slow but complete erosion of the immune system. Years, sometimes many years, after a person becomes infected with the virus, his or her immune system is so weakened that it is unable to fight off the bacterial and viral assaults that a strong immune system could easily defeat. For that reason people with AIDS are vulnerable to a host of diseases rarely found in the general population. These are called opportunistic infections because they take the opportunity to strike someone with depleted immune defenses. Ultimately, the patient dies from the effects of one of these diseases.

The list of opportunistic infections that most frequently affect AIDS patients itself reads like a catalog of plagues. Tuberculosis, recurrent

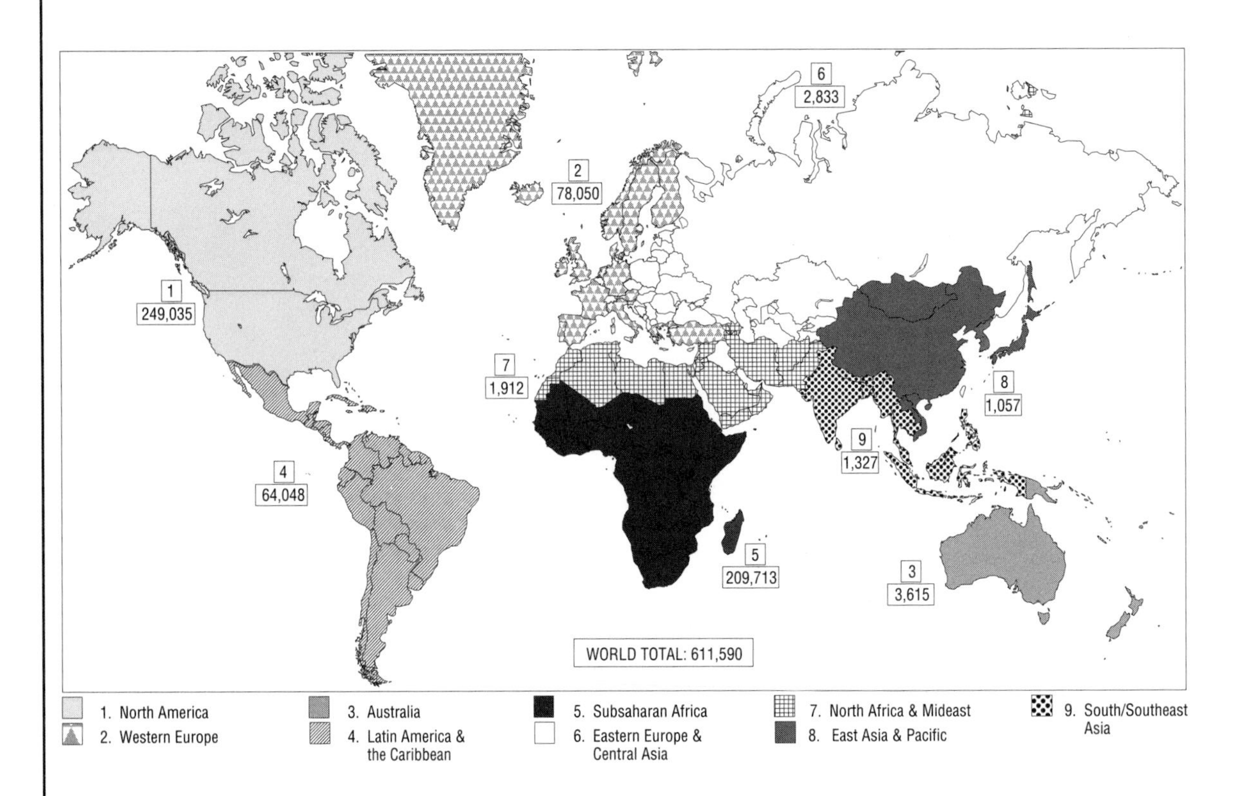

FIGURE 3.1 *Cumulative number of reported adult AIDS cases worldwide through late 1992. Map derived from data provided by the World Health Organization, 1993.*

pneumonia (caused by a parasite called *Pneumocystis carinii*), and a rare skin cancer called Kaposi's sarcoma are some of the most common causes of death for people with AIDS in the United States. Other AIDS-related conditions include lymphoma and other cancers, systemic yeast infections, toxoplasmosis and other degenerative disorders of the nervous system, and an unexplainable wasting syndrome (see Table 3.1). If any one of these conditions in itself is not fatal, the cumulative effect of fighting off so many infections with an increasingly damaged immune system often is. In one sense the solution to the problem is clear. If science could find a way to lessen the immune-depleting effects of the virus, people with AIDS would not be so defenseless against opportunistic infections.

But this strategy is not as simple as it sounds. Once the virus infects an individual, the virus and immune system are locked in an intimate and paradoxical relationship. Following infection the virus becomes part of the immune system and in so doing undermines the very defenses that are supposed to fight it. Eliminating the virus therefore means turning the immune system against itself. And there lies another of the disease's ironies and scientific challenges.

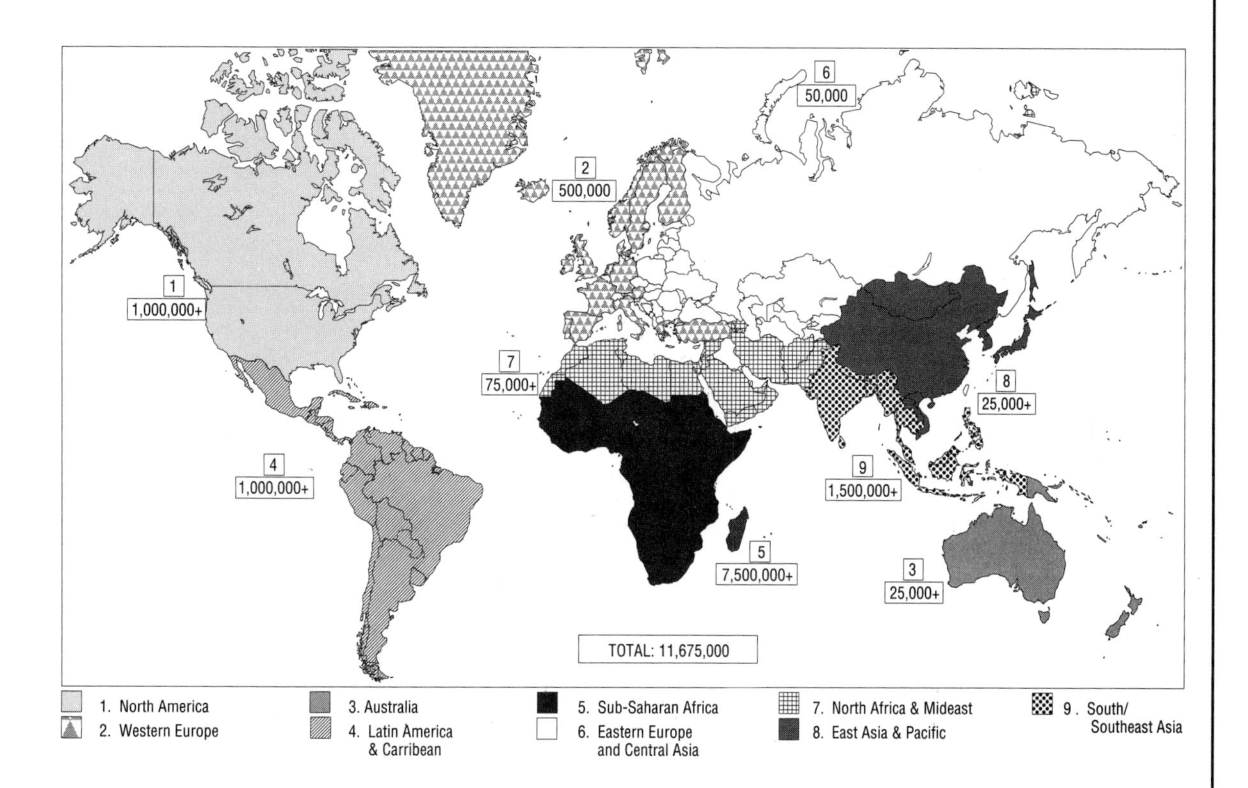

FIGURE 3.2 *Estimate of HIV infections in adults as of late 1992. Source: World Health Organization, 1993.*

The ultimate goal of current AIDS research is to extricate the virus from the immune system without causing further damage. To do that scientists over the past decade have focused on the complex relationship between the virus and the cells of the immune system as well as relationships of the different immune cells to each other.

VIRUSES

A famous scientist once defined a virus as "a bit of bad news wrapped in protein." Indeed, while the experimental study of viruses can unlock a wealth of scientific information, viruses offer little else of value to human beings. Unlike the occasional bacterial species that can provide a beneficial service to humans (e.g., bacteria found lining the human intestines that aid digestion), viruses are never Good Samaritans. They seem perfectly willing to take up residence in the human body without doing anything to earn their keep.

But the scientist's comment also touched on the very simple structure of a virus. Compared with the complexity of the mammalian or bacterial cell, viruses are remarkably minimal. Most viruses are

TABLE 3.1 *Conditions Included in the 1993 AIDS Surveillance Case Definition*

- Candidiasis of bronchi, trachea, or lungs
- Candidiasis, esophageal
- Cervical cancer, invasive*
- Coccidioidomycosis, disseminated or extrapulmonary
- Cryptococcosis, extrapulmonary
- Cryptosporidiosis, chronic intestinal (>1 month's duration)
- Cytomegalovirus disease (other than liver, spleen, or nodes)
- Cytomegalovirus retinitis (with loss of vision)
- HIV encephalopathy
- Herpes simplex: chronic ulcer(s) (>1 month's duration); or bronchitis, pneumonitis, or esophagitis
- Histoplasmosis, disseminated or extrapulmonary
- Isosporiasis, chronic intestinal (>1 month's duration)
- Kaposi's sarcoma
- Lymphoma, Burkitt's (or equivalent term)
- Lymphoma, immunoblastic (or equivalent term)
- Lymphoma, primary in brain
- *Mycobacterium avium* complex or *M. kansasii,* disseminated or extrapulmonary
- *Mycobacterium tuberculosis,* any site (pulmonary* or extrapulmonary)
- *Mycobacterium,* other species or unidentified species, disseminated or extrapulmonary
- *Pneumocystis carinii* pneumonia
- Pneumonia, recurrent*
- Progressive multifocal leukoencephalopathy
- *Salmonella* septicemia, recurrent
- Toxoplasmosis of brain
- Wasting syndrome due to HIV

*Added in the 1993 expansion of the AIDS surveillance case definition.

SOURCE: Centers for Disease Control, January, 1993

nothing more than a shell of protein that contains a packet of genes on the inside.

Viewed externally, each HIV particle is shaped like a 20-sided soccer ball (see Figure 3.3). The virus takes its form from the protein making up its outer shell, which is sometimes called the capsid. Overlying the outer capsid is a lipid membrane, pilfered from the human host cells that the virus infects. Poking through the capsid and the membrane are proteins studded with sugars, called glycoproteins. The glycoproteins on HIV are often depicted as lollipop shaped. The stick of the lollipop is made of a glycoprotein with a molecular weight of 41 kilodaltons, and, according to the convention of referring to proteins and glycoproteins by their molecular weights, it is named gp41. The "candy" of the lollipop is a glycoprotein of 120 kilodaltons, called gp120. Because gp41 and gp120 are initially associated in a single protein, scientists refer to this precursor form as gp160.

Contained within the outer shell is another protein capsule. This cone-shaped capsule, called the core, is made up of many protein molecules, each with a molecular weight of 24 kilodaltons, named, appropriately, p24.

Each component of the structure serves a distinct purpose. The glycoproteins gp41 and gp120 are known to be crucial in anchoring the virus to the cell it will infect, while the lipid membrane helps the virus enter that cell. The capsid protects the virus and gives it shape. And all of this protective packaging is designed to safeguard the precious cargo found within the viral core. There the virus places its hope for the future: Its genes and a few proteins that will help it carry out its reproductive mission.

Because viral structure is so pared down, many scientists have argued that viruses cannot even be considered living organisms. To be alive in the biological sense, an organism must have the capacity to feed itself and replicate itself. Viruses can do neither on their own. The viral genes provide the instructions for making more viruses. They are the guidelines for the manufacture of structural proteins that make up the

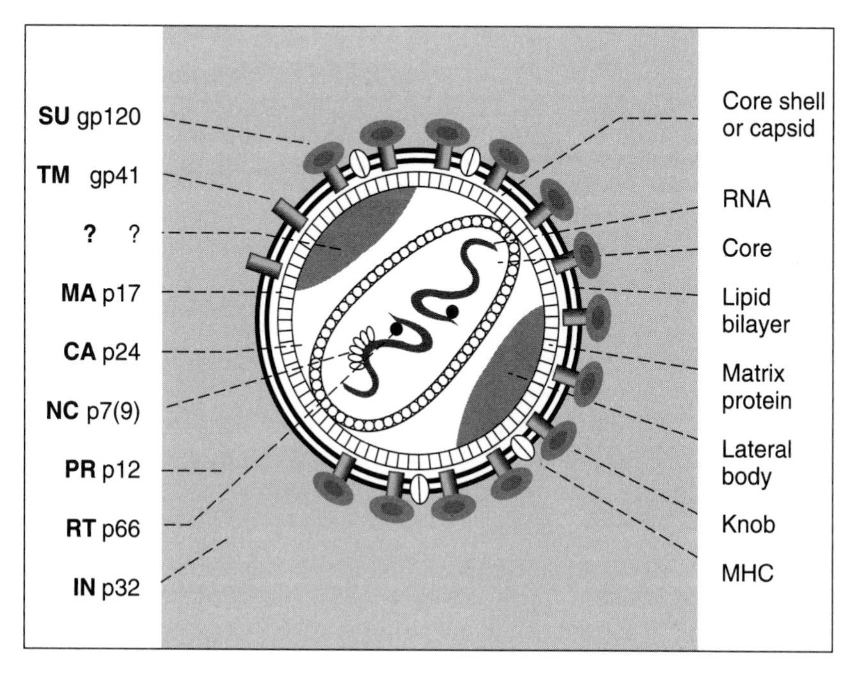

FIGURE 3.3 *The structure of the Human Immunodeficiency Virus (HIV). At the center of the virus is a cone-shaped core that contains the viral genetic material and auxiliary proteins required for viral replication. The cone sits within a protein shell, called the capsid, and the entire assembly is surrounded by a membrane, pilfered from the human cell in which that particular virus particle was replicated. Sticking through the membrane are the viral proteins that allow the virus to anchor onto and then infect a new cell. (Adapted from Gelderbloom et al., 1987. Virology, Volume 156, pp. 171–176.)*

virus's capsid and core and enzymatic proteins that reproduce the viral genes and help package them into the shell. But the actual work of manufacturing the genes and proteins needed to produce more viral particles is achieved within the infected host cell, using host cellular machinery, energy, and the raw materials—the chemical subunits—found within the cell. The production of new viral particles often means that the host cell must turn away from its ordinary business and devote itself entirely to the assembly of new viral particles.

Viral dependence on the host cell requires that at some stage in the virus's life cycle it must enter the host cell and there fulfill its biological mission to reproduce. And so it is that newly manufactured viral particles released into the blood must find new host cells to infect, so that they too can start to reproduce. Viruses must spread from cell to cell within the infected individual and to new individuals, in search of new cellular hosts that provide the facilities for them to meet their reproductive needs.

THE IMMUNE SYSTEM

Healthy people are anything but defenseless against microbial invaders. On the contrary, the human body can be viewed as a veritable fortress. The skin provides a strong and almost impenetrable barrier to infection, and saliva is full of degradative enzymes. If, however, a microbe should get past either of these barriers, it could well end up in the lungs, where it might induce the infected person to cough. Coughing tosses the microbe up the windpipe and down the esophagus where it will land in the stomach. From the point of view of a microbe unfortunate enough to end up there, the stomach is nothing more than a vat of harsh acids and more degradative enzymes. Needless to say, this points to sure destruction for the microbe.

As effective as these barriers usually are, the human body is not completely impregnable to pathogens, and people do get sick. In the event a pathogen does get past the body's defensive barriers, humans also have at their disposal an internal system for removing invasive pathogens. It is this internal system—composed of cells and the chemicals they secrete—that is referred to as the immune system.

The immune system is often compared to the military, and rightly so. Like soldiers, the cells of the immune system are designed to recognize foreign invaders and attack them. But the analogy extends even farther. In both systems, protective forces are deployed to the site of attack to fight in the same mode as the attackers. For example, military forces are prepared to respond to attacks in the air, sea, or on land, depending on where the enemy threatens.

Pathogens, like enemy armies, have different modes of attack, and the immune system, like the military, has specialized divisions to fight back in the relevant mode.

Broadly speaking, pathogens invade the body in two modes. Some pathogens live in the body's fluids but remain outside the host cells. These are called extracellular pathogens. Other pathogens are intracellular. Intracellular pathogens spend at least part of their life cycles inside host cells. As a general rule, these two modes of attack correspond to the two major classes of infectious disease-causing agents: bacteria and viruses.

Most bacteria are extracellular pathogens. As such, they live and thrive outside the host cells. For the bacteria the host is a sort of incubator and a source of nutrients, providing a warm and fertile environment for the bacteria to grow and multiply. But most bacteria are self-sufficient in at least one regard. They contain within themselves all of the cellular machinery required to multiply. Reproductively independent as they are, bacteria can in some but not all cases be less than benevolent to their human hosts. During the course of their life cycles, bacteria can manufacture and secrete biochemicals that are toxic to their hosts. Even though they remain outside host cells, bacteria can still cause disease via their toxins.

In contrast to the relatively independent bacteria, viruses spend only part of the time outside the host cells. When it comes time to multiply, they must enter the host cells. While viruses contain a complete set of their own genes, they carry little else inside their protein shells. Without any means to replicate their own genes or manufacture the proteins that encase those genes, viruses are compelled to enter a person's cells, making the infected person's cells an unwilling host to an all too often pernicious viral guest. The unwanted viral visitor co-opts host cellular machinery to carry out its own gene and protein synthesis. Viruses are therefore intracellular pathogens.

Since pathogens can exist either intracellularly or extracellularly, the immune system must be ready to respond in either context. Or, returning to the military analogy, the immune system must have at least two branches: one that is prepared to fight extracellular pathogens and another to fight intracellular ones—and indeed it does.

To fight extracellular pathogens, the body produces antibodies. These highly specific molecules are manufactured and secreted into the blood by a type of white blood cell called a B-cell. Once in the blood the antibodies can fight infectious agents that are swimming there. The blood can also help circulate the antibodies to other tissues in which bacteria and newly released viral particles might be found.

Antibodies are extremely specific in what they recognize. A healthy person manufactures antibodies that recognize substances foreign to the body but not the body's own chemicals. Each antibody picks out only a small portion of a single molecule on the surface of a bacteria or a virus. But that is sufficient for the antibody to bind to that molecule, and eventually many antibody molecules coat the pathogen's surface. The antibody-coated pathogen then becomes a target for destruction by macrophages, another cell type in the immune system.

Fighting intracellular pathogens is a trickier business. Because the pathogens gain entry to and are hidden within the body's own cells, infected cells look to the immune system just like every other host cell. But infected cells are not without their defenses. All cells have a kind of housekeeping system that digests particular proteins when they are no longer needed. In some instances, protein fragments, called peptides, made during this digestion process are carried out to the cell's surface and held there by surface proteins. When peptides derived from the host's own proteins are displayed on the cell surface, they are ignored by the immune system, which has a mechanism for distinguishing proteins of the host from foreign proteins. In an infected cell, peptides from the pathogen are also transported to and displayed on the cell's surface. These foreign peptides serve as a distress signal that tells the immune system that the host cell in question is harboring a pathogen.

The immune system responds to the distress call by deploying another type of white blood cell, a T-cell, which is specially equipped to detect the foreign peptides on the surface of the infected cells. Once the infected host cell has been identified, the T-cell destroys the host cell along with its infectious cargo. Because of its activity, this type of T-cell is called a cytotoxic T-lymphocyte (CTL) or, more colloquially, a "killer" T-cell.

It is interesting to note that under a microscope the antibody-secreting B-cells and killer T-cells are almost indistinguishable. B-cells are distinguished from T-cells first by their site of maturation. B-cells mature in the bone marrow, T-cells in the thymus. But beyond that, B-cells and T-cells carry different proteins on their surfaces and manufacture and secrete different sets of biochemicals that help them carry out their different functions.

Like any well-organized military system, the immune system is tightly coordinated. In this defense network the generals are another type of T-cell, called a "helper" T-cell (see Figure 3.4). By secreting the

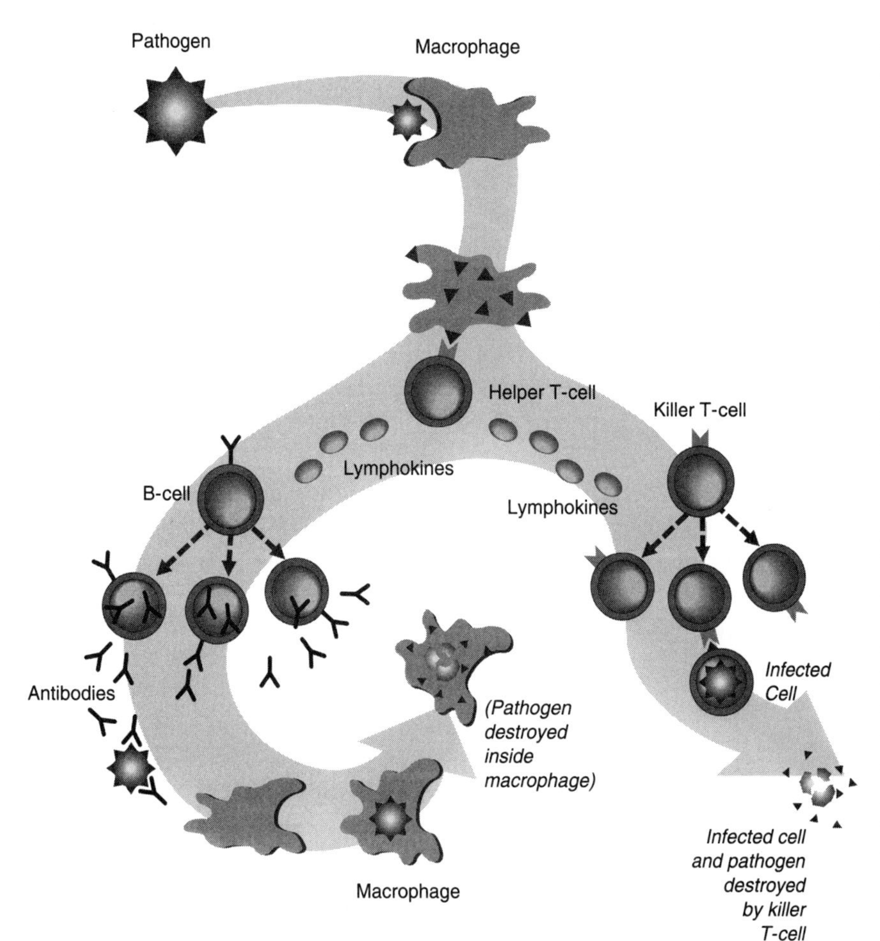

FIGURE 3.4 *The immune system relies on the concerted activities of a number of different cell types. At the center of this activity is the helper T-cell, which receives information about invading pathogens from macrophages and stimulates pathogen-fighting activities in other immune cell types. The helper T-cell will trigger antibody production in B-cells and will activate "killer" T-cells, which destroy other cells in the body that are infected with the pathogen. Unfortunately, the helper T-cell is the target for infection with and destruction by HIV. As the number of helper T-cells decreases throughout the progression of AIDS, the person's immune system becomes depleted and unable to fight off pathogens effectively.*

proper chemical signals at the appropriate time, helper T-cells can boost the populations of B-cells that secrete a particular antibody or they can increase the number of killer T-cells that recognize and eliminate virus-infected host cells. As such, helper T-cells are pivotal in plotting, coordinating, and implementing the defense strategy against both intra- and extracellular pathogens.

Viruses and bacteria have evolved many complex mechanisms to dodge immune detection and destruction. Pathogens have been known, for example, to use camouflage, cloaking themselves in proteins that resemble the host's own. Other pathogens use changeable disguises. As soon as the immune system learns what the invader looks like, the pathogen has changed its chemical "face," which the immune system has to learn to recognize anew and which leaves it continually trying to play "catch up" with the pathogen. To be sure, HIV employs some of these guerilla tactics in its strategy to avoid the immune system, too. But this virus goes beyond that. HIV, it seems, does not merely dodge the immune system, it destroys it.

An invader wishing to destroy host immune defenses can choose one of two ways to do it. The invader can kill each of the soldiers in the military or it can take hold of the command center. An immune system may have a complete set of soldiers—in this case B-cells and killer T-cells—but still remain immobile if the appropriate command never goes out to activate it. And the proper command might be impaired if something were to happen to the helper T-cells.

As it turns out, HIV has a special affinity for and may be particularly harmful to the helper T-cells. While many of the interactions between HIV and the immune system are being worked out, and scientists are still debating how each of these interactions contributes to AIDS, most researchers will agree that part of the answer lies in the special relationship between HIV and the helper T-cell.

LIFE CYCLE OF HIV

Initial Contact, the Receptor

The association between viruses and host cells begins when the virus tries to enter the cell. Because this is a crucial step in viral replication and it starts an infected individual on the road to developing a disease, a great deal of scientific investigation has focused on this process.

Obviously, no host cell willingly invites a virus inside it. And yet viral access to host cells by no means constitutes a forced entry. Viruses use a kind of chemical subterfuge to gain access to their hosts. In the most common scenario, a virus poses as something that would normally be admitted to the cell and uses the normal ports of entry. This method is not unlike a thief who pretends to be a locksmith or an appliance-repair person to enter a house targeted for robbery.

The cellular equivalent of a door is a surface protein called a receptor. A host cell has on its surface a great number of receptors that help the cell sense and interact with its external environment. To be sure, not all receptors are doorways. Most receptors receive external chemical messages and then initiate some sort of internal cellular response. Locomotion, gene expression, or protein synthesis and secretion are often touched off when a receptor on the cell's surface receives the appropriate chemical signal. Other receptors do help import substances from the outside to the cell's interior. The external substance is held by the receptor while the entire complex is brought into the cell in a process called endocytosis.

Whatever the outcome of the interaction, the association between a substance and a receptor is extremely specific. And each receptor is designed to recognize one or, at most, a small number of closely related molecules. The kinds of receptors on a cell's surface therefore determine the kinds of messages it will respond to. Nerve cells, for example, have receptors for the special chemical messengers found in the brain called neurotransmitters. Red blood cells do not have receptors for neurotransmitters and are therefore insensitive to them when they encounter them.

Rather than being excluded by the specificity of the receptor then, viruses exploit this. On the viral surface are molecules that can interact with host cell receptors in much the same way that native substances can. This interaction leads to viral entry into the host cell.

In the case of HIV the doorway to host cells is a cell surface molecule called CD4. This means that all cells that carry CD4 are potential targets for HIV infection. But very few human cell types have CD4. Those that do play a crucial role in maintaining an individual's health. To date, all known CD4-bearing cells are immune-related cells. And there lies one of the secrets to the devastation wrought by HIV: The virus infects the very cells that are supposed to fight infection.

Not all cellular components of the immune system display CD4 on their surfaces. For example, the killer T-cells do not. Scientists hope to learn more about the way HIV spreads in an infected individual by tracing the interactions between CD4-bearing cells.

The "Pick-up"

One of the most common routes of HIV infection is through sexual contact, so one would assume that some receptive cell types are found in the blood going to the skin lining the vagina, rectum, and other body orifices. Indeed, CD4-bearing cells of the immune system are posted in exactly those sites. These cells, called macrophages, are like sentinels and have the job of roaming areas exposed to the environment in search of foreign pathogens. Normally, macrophages pick up the pathogens on surface molecules designed for the purpose and "present" the pathogens to the commander of the immune system, the helper T-cell. Presentation alerts the helper cell to the presence of the pathogen. The helper cell then gives the command to the specific B-cells and killer T-cells that are equipped to recognize that particular pathogen. The B-cells respond by secreting antibodies into the blood to help remove any circulating pathogen, and the killer cells destroy any host cells infected by the pathogen.

In an HIV infection, however, this scenario becomes a little more ambiguous. The cell-to-cell contact between macrophage and helper cell that serves as such an effective alert system against most pathogens activates the immune response against HIV. At the same time, it spreads the virus to more host cells. The presenting macrophage has CD4 molecules on its surface, so more than likely the presenting macrophage is actually infected with the virus. And, as fate would have it, the helper T-cell also has CD4 molecules on its surface and is a prime target for HIV infection. Many scientists now believe that the presentation process has been subverted by the virus to become a means of infecting the helper T-cells. In other words, the macrophages help spread the virus from the skin lining the body's orifices to the T-cells circulating in the blood as well as to those in the spleen, lymph nodes, and other tissues. Without a doubt, the infection of a cell as important to immune functions as the helper T-cell can deal a profound blow to the integrity of the immune system. For that reason, immunologists, virologists, and molecular biologists have all contributed their expertise in deciphering the events that follow the initial contact between the virus and the CD4 molecule.

Uncoating

We have already seen that a virus is one of the most minimalist forms of life, if it can be considered living at all. Basically, it is a protein-wrapped package of genes that has no other purpose than to create more packages like itself. The information for this replication is contained entirely within the viral genes, the protein coating being nothing more than protective packaging. As such, the mission of the virus infecting a cell is to inject its genes inside the cell. But the protein coat does not necessarily have to enter, and in the case of HIV some of that protein shell is left outside.

The specifics of this process, called uncoating, have not been worked out in their entirety for HIV. What is known, however, is that the glycoprotein gp120 on the surface of HIV contacts and binds to the CD4 receptor on the immune cell. Sometime afterwards the lipid membrane surrounding HIV fuses with the lipid membrane of the cell, an action that effectively opens the outer shell of the virus and juxtaposes the viral contents with the cell's interior. From that point it is an easy matter for the virus to inject its core, the conical inner protein shell that carries the virus's genes, into the cell.

Many of the events in this sequence are still subjects for investigation. For example, important questions still surround the nature of the initial contact between the virus and the cell. While the interaction between the surface molecules—CD4 on the cell and gp120 on the virus—is crucial in binding virus to cell, scientists have yet to determine whether other surface molecules are involved as well. Some studies indicate that this interaction is itself sufficient for both binding and the subsequent fusion of the viral and cellular membranes. Other studies suggest that CD4-gp120 interactions allow binding only. Additional surface molecules on the cell and possibly on the virus are needed for fusion. Some studies, for example, point to the viral gp41 as a facilitator of viral fusion with the cell. As we will see later, questions such as these are important in designing drug strategies that interrupt viral binding, fusion, and entry into the target cell. The hope would be to disrupt these activities since they are the prelude to viral replication.

Replication

The business of viruses is to make more viruses. And the directions for making more viruses are contained within the core protein. There lie two complete copies of the virus's genes, the unabridged instructions for making more virus particles. Viral uncoating paves the way for what is truly the sine qua non of the replication process—the injection of the virus's genes into the host cell. Once inside the cells, viruses are expert at redirecting cellular activity toward viral reproduction and away from the normal order of cellular business (see Figure 3.5).

After infection, most host cells stop reproducing their own genetic material and start reproducing the virus's. But the virus does not stop there. It also needs viral protein shells to be made. So the virus co-opts the host cellular machinery and uses it to manufacture shell proteins. The directions for viral protein synthesis are contained within the viral genes. In addition to the shell proteins, many of the viral proteins that will help assemble the new virus particles are also synthesized by the infected host cell. Once all of the viral proteins and genes have been synthesized, the viral shells are assembled, and the viral genes and assembly proteins are packaged inside. Viral assembly is also accomplished within the host cell.

Each virus has its own method for seizing host cellular machinery, but in most cases viruses use the conventional systems of genetics and

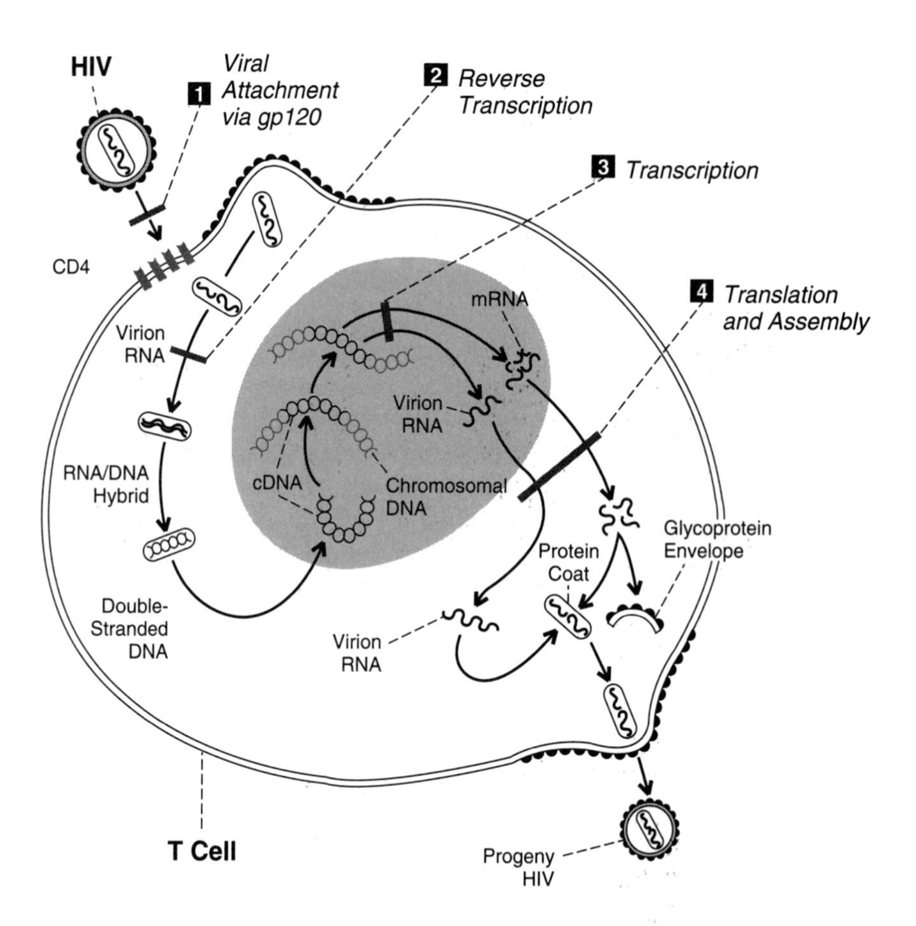

FIGURE 3.5 *HIV starts its replicative cycle by binding to a host cell. The gp120 molecule on the viral surface contacts the CD4 receptor on the cell of the human host. The viral core is injected into the cell and then disintegrates enough to release the viral genetic material in the form of RNA. But the cell's genetic material is in the form of DNA, so the viral RNA must be copied into DNA, a task done by the enzyme reverse transcriptase, which is also contained in the viral core. Viral DNA then enters the cell's nucleus and becomes integrated into the host cell's chromosomes, where it may sit quietly and undetected for many years. At some point, however, the virus will enter an actively reproducing stage. Inside the cell's nucleus, copies of messenger RNA will then be made from the viral DNA. Messenger RNA will leave the nucleus and will direct the synthesis of proteins for the virus's core and capsid, as well as proteins, such as reverse transcriptase that facilitate viral replication. In addition, RNA molecules that contain all of the viral genetic program will be copied from viral DNA. Viral proteins and genetic material will be assembled into a new viral particle outside of the cell's nucleus, very near the cell's membrane. In fact, the final particle is made as the virus is extruded through the membrane. The new viral particle is free to infect another cell and start the cycle over again. (Adapted from Hospital Practice, Sept. 15, 1992, p. 147. Reprinted by permission of the artist, Alan Iselin.)*

information transfer that cells use. HIV does not. The very chemical it uses as its genetic material is different from that used by all cells and most viruses.

CENTRAL DOGMA

The genes of most of the world's creatures are stored in the form of a large molecule called deoxyribonucleic acid, or DNA. An individual's (or a cell's) DNA contains all of the information that the cell will require to produce proteins, the real stuff of which the cells are made, and the molecules that accomplish all of the cellular work. Proteins are made of smaller molecules called amino acids. Twenty different amino acids go into the construction of most human proteins. DNA is often called a blueprint, and it tells the cell the correct order in which to string the amino acid building blocks of a particular protein.

In a way, DNA can be considered an archive, a catalog of all the proteins a cell will ever need to carry out its replication, daily maintenance, and the specialized functions it performs. For that reason preservation of a cell's DNA is paramount. Within the cells of all plants and animals the DNA is sequestered in a special compartment called the nucleus, the way a precious book is kept under glass in the library.

Of course, the DNA must be consulted at times when particular proteins must be manufactured. A system has evolved in which the archival DNA is maintained in the nucleus but where the necessary information can be exported outside the nucleus into the large cellular compartment known as the cytoplasm where the proteins are made. Since it would be too risky for the cell to allow its DNA to be transported into the cytoplasm, where it could possibly be damaged, copies of the relevant portions of the DNA are made as the proteins they code for are needed by the cell. The process again is analogous to the library where photocopies of particular pages of the precious book can be made as they need to be read. In the context of the cell, the photocopy is another molecule, chemically very similar to DNA, called ribonucleic acid, or RNA.

When a cell needs to manufacture a particular protein, an RNA duplicate of the DNA coding for that protein is copied, or transcribed, in the nucleus. This RNA copy leaves the nucleus and enters the cytoplasm, where it is "translated" into the proper sequence of amino acids required for the final protein product. When it is no longer necessary to produce the protein, the RNA copy in the cytoplasm is destroyed. Thus, the RNA transcript is as temporary as the DNA is permanent. The DNA archive remains safe in the nucleus, where it can be consulted again when necessary.

The scenario described above is so ubiquitous among living organisms that it has come to be called the Central Dogma. Scientists often summarize the principle of this dogma, which describes the flow of information in a cell, as DNA to RNA to protein.

UNORTHODOX INFORMATION

One of the rare examples of a violation of the central dogma is presented by HIV and related viruses. For these viruses the genetic archive is not in the form of DNA but rather RNA. This poses an obvious dilemma for the virus. The virus's genetic program is carried out by the human cell it infects, but that cell obeys the central dogma, while the virus does not.

For HIV, the solution lies in converting its genetic information into the same form used by the cell. In short, the virus must change its archive from RNA to DNA. In fact, this is one of the first things that happens when the viral core is injected into the cell. At that point the virus, with the help of an enzyme packaged in the core, changes its genetic information from an RNA to a DNA archive.

This conversion stunned the researchers who first observed it in the viruses they were working with. The phenomenon defied the central dogma in that information flowed in the reverse direction from that predicted. These viruses are now known as retroviruses, to reflect the backward flow of information. The conversion of RNA to DNA is called reverse transcription, a process carried out by an enzyme called, appropriately enough, reverse transcriptase.

Once the viral genes have been converted to DNA, they are ready to interact with the host cell's genes. The viral DNA enters the cell's nucleus, where it is integrated into the host's DNA, sewn seamlessly into host chromosomes by another viral enzyme, called integrase. Like reverse transcriptase and the virus's genes, integrase is also carried into the cell in the viral core.

Now that the viral genes have been changed into the form of DNA and integrated into the host's chromosomes, they are permanent residents of that cell and all of that cell's progeny. Integrated viral DNA is treated by the cell just as native DNA is. Whenever the cell divides and reproduces its own genes, the viral genes are also reproduced. When host cells pass their own genes on to daughter cells, the viral genes will be passed on too. In effect, this constitutes another mechanism for viral spread. We have already seen that HIV can enter a cell actively via the CD4 receptor, and now we see that the virus can, in effect, gain access to other cells as part of their inherited legacy.

Once viral genes are integrated into the host cell, it is possible that nothing more will happen in that cell for a long time, maybe years. A cell harboring quiescent viral genes looks and acts perfectly normal. Cellular gene expression and protein production go on as if nothing were unusual. Unlike an infected cell that is actively producing new viral particles, a so-called latently infected cell is not manufacturing viral components. It therefore cannot display fragments of these components on its surface, which serves as a signal to the immune system that all is not well. The immune system cannot fight what it does not see. So, like a clever spy, the latently infected cell slips past host defenses. This is one of the ways in which HIV evades the immune system, and, as we shall discuss later, it is one of the primary challenges in developing therapeutics that eliminate HIV-infected cells.

But the viral genes do not remain inactive indefinitely. Sooner or later, something will trigger a switch, and the cell will move from latent infection to an active producer of viral particles. During this time, the cell will be given over almost entirely to the production of components—the genes, shell proteins, and enzymes—that will go into the new viral particles. No one is absolutely certain how the switch is tripped, but once it is life will never again be the same for the infected cell.

THE BIRTH OF NEW PARTICLES

Active production of viral particles means that the genes for the viral components are expressed—that is, the proteins they encode are manufactured in the cell's cytoplasm. Now that the viral genes have been converted into DNA, the expression of viral genes can be carried out as dictated by the central dogma. Segments of the viral DNA are copied, or transcribed, into the more transient form, RNA. RNA carries the message encoded in the DNA out of the nucleus and into the cytoplasm, where protein synthesis is carried out using the messenger RNA as a guide.

Because the virus itself is not very complicated, one would expect that it would not contain a vast number of genes—and in relative terms that is true. Compared with the genetic complexity of its host cell, HIV is genetically simple. Yet many virologists have been struck by the virus's complexity relative to other viruses in general and to its closest kin, in particular.

Most retroviruses, HIV included, have three large genes. These are called *env*, *gag*, and *pol*. The *env* gene provides the instructions necessary to make the protein in the glycoproteins gp41 and gp120, which stick out from the viral surface. (The sugar molecules are attached shortly after that part of the protein synthesis has been completed.) The *gag* gene encodes p17, the protein used in constructing the outer shell; p24, the protein making up the inner core; and p9, a small protein that helps wrap up the viral RNA into the core. Finally, the *pol* gene encodes the reverse transcriptase enzyme that converts viral RNA into viral DNA, the integrase enzyme that integrates the viral DNA into the host chromosome, and a third enzyme called a protease. The protease cuts proteins, which is necessary because a single viral gene often encodes more than one protein. This means that the product of one of these genes—*gag*, for example—will be one very long protein, from which the three functional proteins p17, p24, and p9 must be excised.

In addition to the genes encoding the structural proteins that constitute the actual components of the mature virus are viral genes whose protein products regulate the expression of the other genes. These are called regulatory genes, and they will, in effect, help determine when viral genes are expressed and in what quantities. HIV has more regulatory genes than most other retroviruses and is therefore more complex. The roles of the additional regulatory genes and proteins are only now being elucidated.

However, among the more than six regulatory genes, three in particular have attracted the most attention. Many scientists have been especially interested in the gene whose protein product is referred to as Tat. If any of the known viral proteins is part of the switch that causes cells to produce viral proteins instead of cellular ones, Tat is it.

Tat seems to be required to initiate expression of the viral genes. And while host proteins do most of the work of making RNA transcripts from viral DNA, Tat may help recruit the transcriptional machinery so that it preferentially transcribes viral DNA over that of the host. Having said that, one is also faced with a sort of chicken or egg problem. If viral DNA requires Tat protein in order to be transcribed, how does Tat get made initially, since it is one of the proteins encoded in that viral DNA? When scientists have the answer to that question, they will have significant insight into how the switch is made from a latent infection, one in which viral DNA sits quietly and unobtrusively in the host, and active infection, where viral components are being made.

One possible solution to the switch conundrum suggests that an external agent activates Tat expression and production. In some scenarios this external agent can act indirectly, stimulating the host cell in some as yet undefined way. Since the host cell is an immune cell, some evidence exists that when the host is stimulated to fight some other infection, the HIV genes become available for transcription in the process. It is also possible that the gene for Tat becomes directly activated by proteins coming from other viruses that infect the cells.

Yet another theory rejects the notion that the genetic material of HIV is at any time completely inactive or completely active. Rather, it is possible that a low level of viral transcription is going on even during the seemingly latent periods. According to this theory, levels of Tat slowly escalate until they reach a critical point where the protein can initiate enough transcription of the viral genes to make detectable amounts of viral protein and from there large numbers of new viral particles.

A second important regulatory protein is the one called Rev. Like Tat, Rev helps determine when and how viral proteins are made. However, relative to the role of Tat, the role of Rev is somewhat more limited. While Tat is required to initiate transcription of all the viral genes, both structural and regulatory, Rev's presence ensures the synthesis of the structural proteins only. It is likely that together Tat and Rev proteins regulate the relationship between regulatory and structural proteins and as such regulate the course of viral replication, its initiation and magnitude. Scientists are therefore very eager to learn more about the relationships between these two proteins with the hope that they will someday be able to manipulate them to reduce or completely inhibit viral replication. Added to these two proteins is a third one called Nef. The reproduction of viral particles is not dependent on Nef, but the rate of reproduction seems, at least in part, to be determined by this protein. Several other viral genes have been identified, making HIV one of the most genetically complex retroviruses. Some scientists believe that these additional genes determine the pathogenicity of the virus—that is, that these genes in some as yet undefined way determine the ability of HIV to cause disease.

VIRAL ASSEMBLY

However the viral genetic program is activated, viral components will eventually start to accumulate in the infected cell. To complete the cycle, these components must be assembled, after which the newly assembled particles can be released from one cell and can find other host cells to infect.

Some retroviruses are assembled within the cell's interior and then make use of cellular transportation mechanisms to reach the cell's membrane before making their exits. In contrast, components of HIV are shipped to the cell's periphery and are assembled just before or even as the virus is being extruded from the host cell.

The capsid and core proteins are products of the *gag* gene and are manufactured as one long protein string. Approximately 2000 of these protein strings are shipped out to the internal face of the cell's membrane. There the strings aggregate into a ball of concentric protein shells. Proteins destined to form the outer capsid are on the outside of the ball, and those destined to become core proteins are on the inside. Late in the assembly process the protease clips the various proteins and resolves the sphere into the capsid and core.

But the virus would be nothing more than a hollow shell if the viral genes, in the form of RNA, were not enclosed within. Not much is known about how viral RNA becomes associated with the shell proteins. The current belief is that the RNA is attached to the segment of the long protein string that will ultimately form the core protein. Scientists have identified a segment on the viral RNA that acts as a handle for the protein to grab on to. More than likely, the p9 protein is somehow involved in attaching the viral RNA to the core protein. Other proteins, namely integrase and reverse transcriptase, are also packaged within the core by a mechanism that is still undetermined.

All that remains now is for the final touches to be put on the virus. At this stage the viral particles are lacking the lollipop-shaped glycoprotein gp160, made up of gp41 and gp120. Without these the new viral particles would be unable to attach to CD4 receptors on new hosts, and the whole exercise would have been wasted. The glycoproteins are efficiently and independently transported to the cell membrane, awaiting the formation of the protein shells.

As a matter of fact, the glycoproteins are actually stuck in the cell membrane, where they can migrate rather freely. The incomplete virus pushes against the membrane in its initial attempts to exit the cell, and through this action the glycoproteins may become associated with the virus. It is also likely that one end of the glycoprotein in some way hooks on to the capsid protein and permanently attaches.

Not only do the viral glycoproteins associate with the capsid as the virus exits the cell, but the exiting viral particle also takes with it a piece of the host cell membrane. Because the virus is cloaked in this membrane, it is vulnerable to any agents that dry or otherwise destroy cell membranes. The virus is therefore unable to survive exposure to air, which can desiccate the membrane, or to harsh detergents, which can destroy the membrane's structural integrity. HIV's fragility in air and in the environment in general means that it cannot be passed between individuals by touching them or by touching surfaces they have touched.

The virus that leaves the cell looks pretty much like the one that entered it. The mature virus has an outer protein shell and an inner one that carries viral RNA, integrase, and reverse transcriptase. On its exterior the 20-sided viral particle is wrapped in a cellular membrane. The viral surface is studded with the lollipop-shaped glycoproteins that stand poised and ready to gain entry to another cell via the CD4 receptor. For a time the newly manufactured viral particle circulates freely in the blood, or in whatever body fluid its host cell is floating, until it encounters another CD4-bearing cell, infects it, and the whole cycle starts over until more and more cells become infected.

HOW HIV CAUSES AIDS

Since the mid-1980s when HIV was identified as the cause of AIDS, scientists have learned a tremendous amount about the natural history of the virus and the symptomatology of the disease. Yet putting these things together has proved more difficult than anyone would have imagined. As the virus's life style is understood in greater detail, it becomes less clear exactly how it decimates the immune system and causes AIDS.

The problem is not in identifying ways in which the virus's behavior is destructive to the immune system. On the contrary. The problem is that the virus is destructive to the immune system in so many ways that it is difficult to identify the one activity, or constellation of activities, that constitutes the primary cause of the destruction. The hope in identifying the primary cause is to develop therapeutics that will block that avenue of destruction and reduce the toll the virus takes on a person with AIDS.

Infection

Technically, it would be wrong to say that the disease starts at the moment of infection. There is a sharp distinction between infection and disease. Yet it is clear that infection with HIV is a necessary precondition for later developing AIDS. It is at the moment of infection that the virus has the potential of embarking on its life cycle, multiplying, and causing disease. It is also currently believed that the destruction of the immune system that ultimately leads to AIDS begins soon after infection.

But infection requires a very particular kind of contact between individuals. We have already seen that the virus cannot withstand the environment for very long because its membrane will quickly deteriorate. For that reason, HIV, unlike many other viruses, cannot be passed along by touching a surface that has been touched by an infected person.

On the other hand, any circumstances that protect the virus from the environment carry the potential for transmitting the virus to another person. Scientists know, for example, that the virus continues to thrive when it is inside host cells. So one mode of infection involves passing an infected cell from one individual to another. HIV particles can also be found circulating freely in the body fluids of an infected individual, so another mode of transmission involves the exchange of body fluids containing free viral particles from an infected to an uninfected person.

Infection is a two-way street and requires not only a donor but also a recipient. Viral transmission also requires that the virus be delivered to a receptive environment, one replete with CD4-bearing cells. Given these facts, it becomes more clear why transmission often involves blood contact between individuals. Of all body fluids that can support either free virus particles or virally infected cells, blood is the fluid with the highest concentration of both (see Table 3.2). The blood of an infected individual provides a milieu that supports free virus and infected cells, while the blood of an uninfected individual contains many potentially infectable new cellular hosts.

Transfusion of infected blood into an uninfected recipient is the equivalent of injecting live virus into the person and is an almost certain route of infection. Needle sharing among intravenous drug users is not much different. The virus can also be effectively spread by sexual intercourse, and viral spread can be further enhanced by sexual practices that can tear skin and expose the blood of an uninfected sexual partner to the body fluids of an infected partner. In addition to blood, semen can carry both free virus and virally infected T-helper cells and infected macrophages. So any sexual act that puts an uninfected partner's genital or rectal tissues in contact with infected semen puts the uninfected partner at risk of infection. An additional factor that may explain the higher frequency of HIV infection from anal intercourse is the high density of Peyers patches—regions containing white blood cells—in the rectum.

In addition to semen, other fluids contain either free virus or virally infected cells. These include tears, ear secretions, saliva, urine, vaginal or cervical fluids, breast milk, bronchial fluid, and cerebrospinal fluid (CSF). Of these, the highest concentrations of free virus, virally infected cells, or both are found in blood, semen, and CSF. Activities involving

TABLE 3.2 *Body Fluids and Cells from which HIV Can Be Isolated*

	Isolations/Attempts	Estimated Quantity
Fluid		
Plasma/serum	45/46	10-1,000 infectious particles (IP)/ml*
Tears	2/5	<1 IP/ml
Ear secretions	1/8	5-10 IP/ml
Saliva	3/55	<1 IP/ml
Urine	1/5	<1 IP/ml
Vaginal or cervical	6/16	<1 IP/ml
Semen	5/15	10-50 IP/ml
Milk	1/5	<1 IP/ml
Cerebrospinal fluid	21/40	10-1,000 IP/ml
Cells		
Peripheral blood mononuclear cells	89/92	0.001%-1.0% of infected cells
Saliva	4/11	<0.01%
Bronchial fluid	3/24	Unknown
Vaginal or cervical fluid	7/16	Unknown
Semen	11/28	0.01%-5.0%

*For comparison, note that for hepatitis B infection, particle counts may range from 10^6 to 10^9 IP/ml.

SOURCE: *Hospital Practice*, November 15, 1990, p. 45.

exchange of these fluids between individuals also carry the potential for spreading HIV (see Table 3.3).

The last mode of transmission occurs when infected mothers pass the virus on to their newborns before or around the time of birth. The presence of HIV in aborted fetuses indicates that a woman can transmit the virus to her unborn child in utero, but the mechanism for prenatal transmission is not entirely known. The act of birth itself can facilitate transmission since the neonate is exposed to the mother's infected blood and amniotic fluid. HIV can also be passed on to a newborn through mother's breast milk.

The events that follow HIV infection are fairly standard regardless of the mode of infection. The early stages of infection resemble those of almost any other viral infection. Within 1 to 2 months of becoming infected, many people feel sick and experience severe flu-like symptoms similar to those of mononucleosis, with fatigue, fever, and muscle aches. Some report blinding headaches as well.

It is during this time, called the acute phase, that scientists can detect

TABLE 3.3 *Cumulative Number of Adult/Adolescent AIDS Cases in the United States by Exposure Group, Through June, 1993*

	Cumulative Total	
	Number	%
Men who have sex with men	172,085	55
Injecting drug use	73,610	24
Men who have sex with men and inject drugs	18,557	6
Hemophilia/coagulation disorder	2,782	1
Heterosexual contact:		
Sex with injecting drug user	10,800	
Sex with bisexual male	1,115	
Sex with person with hemophilia	172	
Born in Pattern-II* country	3,557	
Sex with person born in Pattern-II country	261	
Sex with transfusion recipient with HIV infection	420	
Sex with HIV-infected person, risk not specified	5,548	
Total heterosexual contact	21,873	7
Recipient of blood transfusion, blood components or tissue	5,733	2
Other risk, not identified	15,080	5
Total	309,720	100

*Pattern II transmission is observed in areas of sub-Saharan Africa and in some Caribbean countries. In these countries, most of the reported cases occur in heterosexuals, and the male-to-female ratio is approximately 1 to 1. Injecting drug use and homosexual transmission either do not occur or occur rarely.

SOURCE: Centers for Disease Control, July 1993.

large numbers of free virus particles in the bloodstream. A month later the number of particles in the blood is on the decline. As the virus becomes less detectable in the blood, the protective components of the immune system become more evident, indicating that the immune system may be effectively keeping the virus at bay. Large amounts of antibodies and killer T-cells are specifically deployed to fight free virus and virally infected cells. In fact, these are the antibodies detected by standard blood tests that determine a person's HIV status. Individuals in whom antibodies are detected are said to be seropositive—that is, their blood serum tests positive for the presence of HIV. The absence of HIV in the blood means an individual is seronegative.

Following these initial events, the flu-like illness abates, and the seropositive individual enters a phase during which no obvious manifestations of infection or disease symptoms are evident. This period lasts, on average, from 8 to 10 years and is referred to as the asymptomatic phase. For a long time scientists believed that viral genes were not being expressed during this phase and that little or no viral replication was taking place—and to some degree that may still be true.

More recent data indicate a slow and gradual decline in the number of helper T-cells during the asymptomatic period. Currently, many researchers think that while most virally infected cells may not be actively producing new virus, some proportion of infected cells are. But at this stage an individual's immune system is still strong enough to eliminate a sufficient amount of the free virus and infected cells, so that the individual remains healthy.

The fact that most individuals can remain healthy for 10 years suggests that the immune system has a tremendous excess capacity, and significant numbers of T-cells can be eliminated without causing the host any severe difficulties. Eventually, however, the number of helper T-cells killed brings the HIV seropositive individual to a point where he or she can no longer handle the viral load, and symptoms of disease start to appear. Since the helper T-cells are the commander of the immune army, killing them can potentially inactivate the entire immune defense system.

T-cell Mystery

While the scenario depicted above seems logical, one great mystery remains unanswered. Scientists still cannot explain why so many helper T-cells are killed. They still do not know whether the virus kills host cells directly or whether some indirect viral activity kills the cells. Even if they did know the mechanism for viral killing, there is one nagging statistic they still must explain.

Healthy individuals have 550 to 1200 helper T-cells in each microliter (one-millionth of a liter) of blood, with the average falling around 800 cells per microliter. Individuals are defined as having AIDS if they are HIV seropositive and have fewer than 200 helper T-cells per microliter of blood. That means that a person with AIDS has lost 75 percent of his or her helper T-cells. Yet when scientists try to count the number of helper cells in which they can actually detect the virus, they find HIV in less than 1 percent of cells. How can so many cells be killed when so few seem to be infected? To solve this puzzle, scientists have proposed that HIV has some indirect means of affecting healthy immune cells in such a way that they become inactive or are targeted for immune destruction.

Shedding

So many things are going on between the immune cells and the virus that infects them that it is difficult to sort out even the most direct means of helper T-cell killing. Many scientists believe that the host cell is worn down from reproducing so many virus particles and releasing them to the environment. Eventually the host cell simply dies. Added to this very direct interaction is a wealth of indirect interactions between pathogen and host.

One theory that has received much attention concerns the role of the lollipop-shaped glycoproteins. Gp120, the candy part of the lollipop-shaped gp160 complex, sticks out of the surface of the virus and facilitates viral binding to the CD4 receptor on host T-helper cells and macrophages. Fragments of these and other viral proteins are displayed by the host cell on its surface. The foreign peptide sticking out of the infected host cell serves as a red flag to the immune system, alerting the immune cells that a host cell has been invaded by a foreign pathogen. The infected cell is now targeted for destruction by the killer T-cells.

But the killing might not stop there. Research has shown that infected cells shed gp120 molecules. These free-floating molecules can be captured by uninfected helper T-cells. In some cases the healthy helper T-cell may display the shed gp120 molecules on its surface, so now it looks to the immune system just like an infected cell. Even though the healthy cell has no virus inside it, it becomes a target for immune destruction and is eliminated by killer T-cells as though it were infected. Such a mechanism would raise the number of helper cells killed without necessarily increasing the number of cells infected.

Shedding of molecules is not the only indirect effect of HIV infection. The gp120 molecules sticking out of an infected cell may also have direct contact with other uninfected cells. Since gp120 can bind to CD4 molecules, it is possible that infected cells actually bind to and fuse with uninfected cells via the gp120-CD4 interaction. The result of this fusion is the formation of giant masses of infected and uninfected cells in what is technically called a syncytium. Syncytia have been observed in the test tube when infected and uninfected cells are mixed together. Some scientists believe that syncytia may form in people with AIDS as well, but no such syncytium has ever been isolated from an AIDS patient. Other scientists interpret this discrepancy to mean that syncytia are only a test tube phenomenon but that it might hint at some other harmful interaction between infected and uninfected cells in the HIV seropositive individual.

Gp120 is not the only molecule that can pass from infected to uninfected cells. The Tat protein also may diffuse from cell to cell. If that is the case, scientists have identified several potential consequences. For one thing Tat is a potent activator of viral gene expression, so it is possible that Tat secretion initiates a spate of viral reproduction in previously latently infected cells. In addition, there are some indications that Tat may directly kill uninfected cells. Alternatively, Tat may also alter the chemical signals that are sent out from a helper T-cell. According to this theory, the altered signals result in an immune response that is ineffective or inappropriate. For example, there may be too many signals stimulating antibody production and not enough stimulating killer T-cell activity, so that while there is an immune response it is ineffectual and does not eliminate the virus.

New evidence has come to light to suggest that the Nef protein may also be secreted. Some scientists think that Nef may act as a so-called superantigen. Superantigens are thought not to depress immune activity but to overstimulate it. As a consequence, the immune system starts to look like a "Keystone Cops" episode where the perpetrator slips past a frenzied and disorganized immune system.

It is important to note that many of these phenomena are studied in the laboratory using viruses and immune cells that have been cultured outside the human hosts from which they were taken. The artificial circumstances of culture conditions may well alter some of the biochemistry of these entities, and scientists caution that the behavior of cells and viruses may be different in intact living systems.

So the question of whether the laboratory situation realistically reproduces real life must be addressed by scientists hoping to understand the relationship between HIV and AIDS. But even if all of these phenomena occur in a person with AIDS, scientists must also assess which mechanisms are of primary importance in helper T-cell death and which are fairly minor contributors. New evidence comes to light almost daily that makes the research community evaluate these questions in a new way. Some scientists believe that as the virus replicates within a host new strains emerge that are more efficient killers, more aggressive at reproducing, and more effective at infecting cells than its progenitors. Those who believe this suggest that the emergence of these potent new strains accounts for the sudden and rapid decline of the immune system after 10 or more years of calm.

As research techniques improve, scientists discover aspects of the interaction they could not previously discern. For example, using higher-resolution techniques, some research groups are finding that many more than 1 percent of helper T-cells are actually infected with HIV. New estimates suggest that 20 to 30 percent may be infected at any one time. Some researchers think these numbers are adequate to explain the observed loss of helper cells. In that case the indirect mechanisms of killing may add to immune destruction but may play a relatively minor role. Although not all scientists agree on the relative contributions of these various mechanisms to the development of AIDS, they are in agreement that there may yet be facets of HIV that remain to be uncovered. Some hope that a definitive answer might lie in these still undiscovered interactions.

VACCINES AND THERAPEUTICS

Scientists are interested in the natural history of the virus in part because it is so unusual and its study promises to unlock many mysteries about the different ways that evolution has invented to do things. Take, for example, the unexpected finding of reverse transcription, which taught scientists that information could flow in more than one direction.

Beyond satisfying intellectual curiosity, the intense interest that HIV and AIDS have received is motivated by a deep desire by the scientific community to eradicate HIV and AIDS. In the past, science has eliminated plagues with vaccines and drugs, and it is looking toward these to deal with HIV and AIDS. While people can try to avoid contact with infectious agents, some infection is inevitable. Most encounters with pathogens are completely random, and while no one can control when they might run across a virus or a bacterium that will cause them to get sick, vaccines can help them prepare in advance to fight the infection before it gains access to their system.

The immune system is a defense network. Like any defense system, it needs to know what its enemy looks like, so it knows what to attack. Of course, after infection, once an invader has made its way into a person's system, the immune system will learn to recognize it and will respond by producing antibodies and deploying killer T-cells. But mobilizing such an attack takes time. The immune system needs to decipher the features of the foreigner and discern that the pathogen is in fact foreign. While the immune system is doing that, the pathogen is multiplying and causing disease.

With vaccines, physicians try to eliminate or at least reduce the lag time so that the immune system is ready to kill an invading pathogen at the moment it enters the body, before it can infect cells, multiply, or cause disease. The vaccine is a device to teach the immune system of an uninfected person the features of a pathogen before that person ever comes in contact with it. The concept is identical with hanging a "Wanted" poster in public places, so criminals will be recognized and apprehended before they commit another crime.

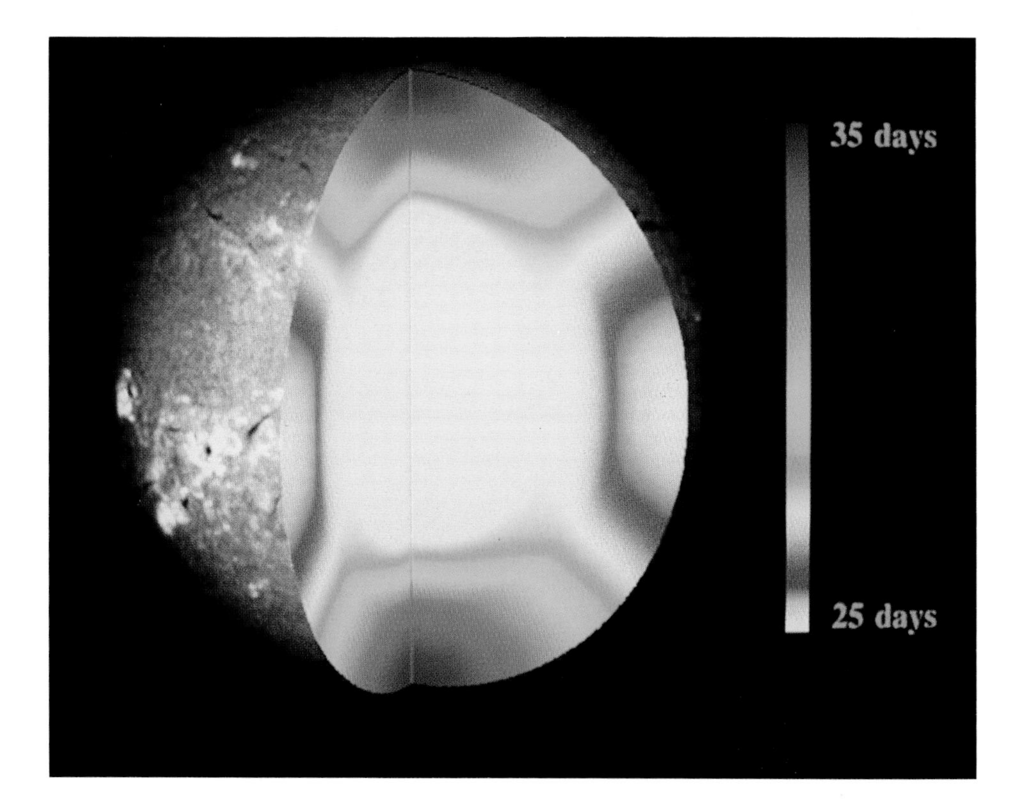

(Above) *A cutaway picture of the sun showing how solar rotation varies with depth and latitude. The period of rotation ranges from 36 days at the poles (blue) to 25 days at the equator (red). Starting near the base of the convection zone, the sun begins to rotate as a solid body with a period of 27 days (yellow). (Courtesy of Ken Libbrecht, Caltech.)*

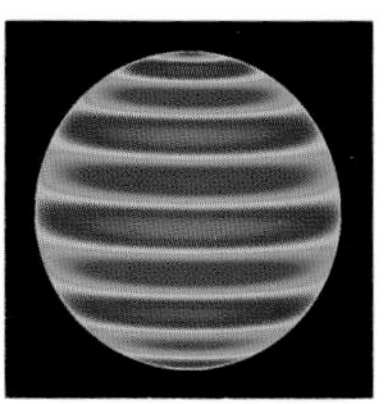

l = 10, m = 0

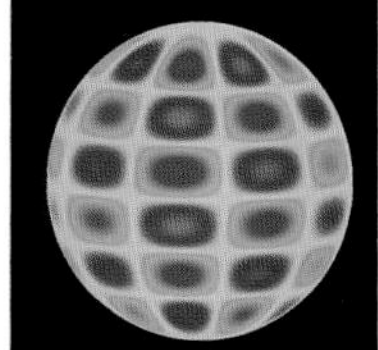

l = 10, m = 5

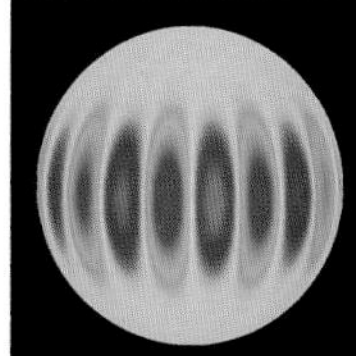

l = 10, m = 10

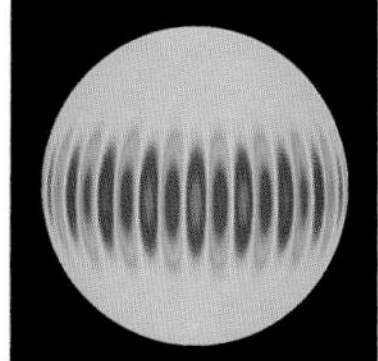

l = 20, m = 20

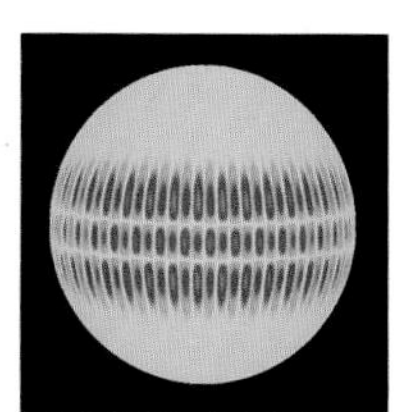

l = 40, m = 38

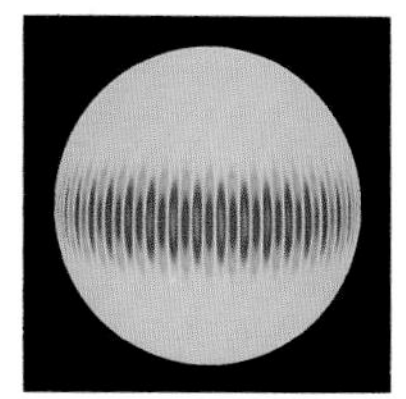

l = 40, m = 40

(Right) *These computer-generated images show a few of the ways in which the sun can oscillate. The degree of the acoustic mode, the parameter l, represents the total number of nodes (the stationary planes both perpendicular and parallel to the equator) belonging to each oscillation. The m value specifically refers to the number of nodal planes cutting perpendicular to the equator (with m = 0, the cuts are all horizontal). The yellow regions in these pictures depict the nodes, while the blue areas are moving toward us and the red areas moving away. Note that as the degree of the acoustic mode increases, smaller and smaller regions of the sun are separately oscillating. (Courtesy of Douglas Gough, Cambridge University.)*

Letters "I B M" spelled out with xenon atoms. (Courtesy of Donald Eigler, IBM.)

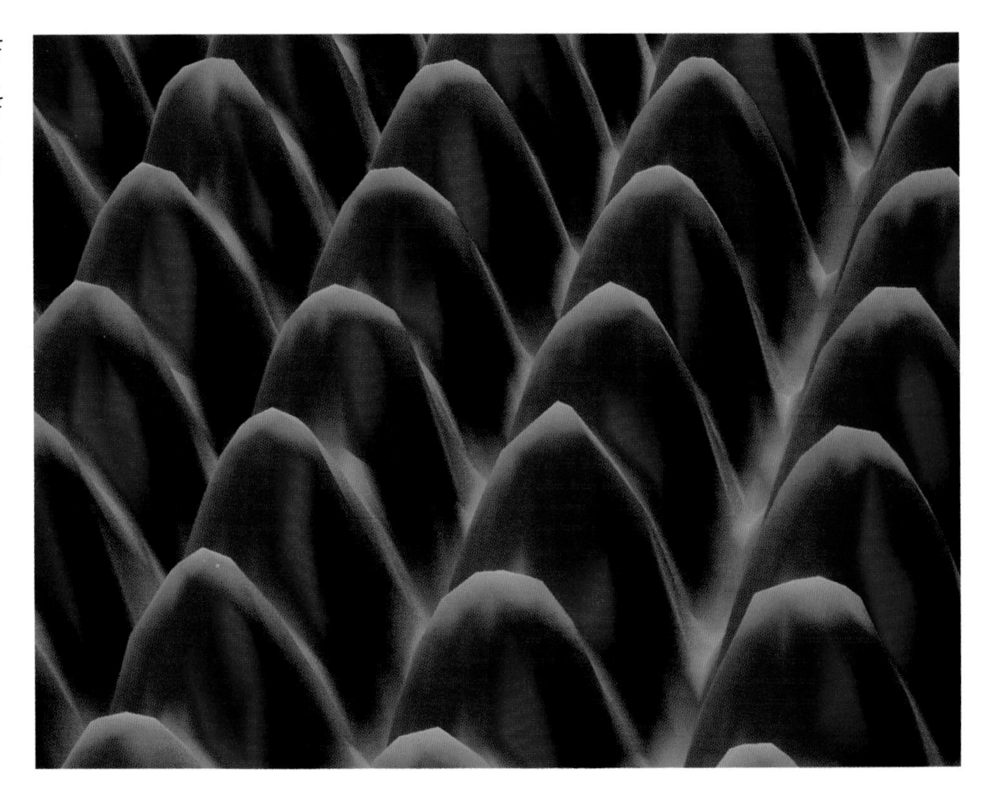

Detail of image of nickel surface. (Courtesy of Donald Eigler, IBM.)

(Opposite) *Unveiled by Magellan's radar altimeter, topographic features of Venus are displayed in colors ranging from red, for greatest elevations, to blue, for lowest. Gray areas represent gaps in the data. This map was assembled from images acquired over one Venusian day, a period of 243 earth days. Note the Australia-size highland region called Ishtar Terra, which contains the planet's highest mountains, the towering Maxwell Montes. South of Ishtar lies Eistla Regio, whose tectonic features were studied by geophysicists Robert Grimm and Roger Phillips. (Courtesy of NASA and the Massachusetts Institute of Technology.)*

Magellan spacecraft (Courtesy of the Jet Propulsion Laboratory, NASA.)

The face of Venus: An entire hemisphere of the planet, centered at 180 degrees longitude, is shown in this computer-generated mosaic of Magellan images. Bright areas (i.e., highly reflective to radar waves) are thought to be rough, while dark regions (poorer reflectors) are interpreted as relatively smooth. The orange tint has been added to simulate the appearance of the surface as indicated by Soviet landers; a ruddy cast would result from absorption of blue light by the planet's dense, cloudy atmosphere. Near the left-hand edge the Africa-size highlands of Aphrodite Terra show up as a bright area, while right of center lies a smaller bright patch called Atla Regio. Connecting them is a belt of bright rifts and ridges, part of the so-called volcano-tectonic zone. Along the zone's bottom margin, several ring-shaped features called coronae are visible. Much of the surface is covered by relatively smooth volcanic plains. (Courtesy of the Jet Propulsion Laboratory, NASA.)

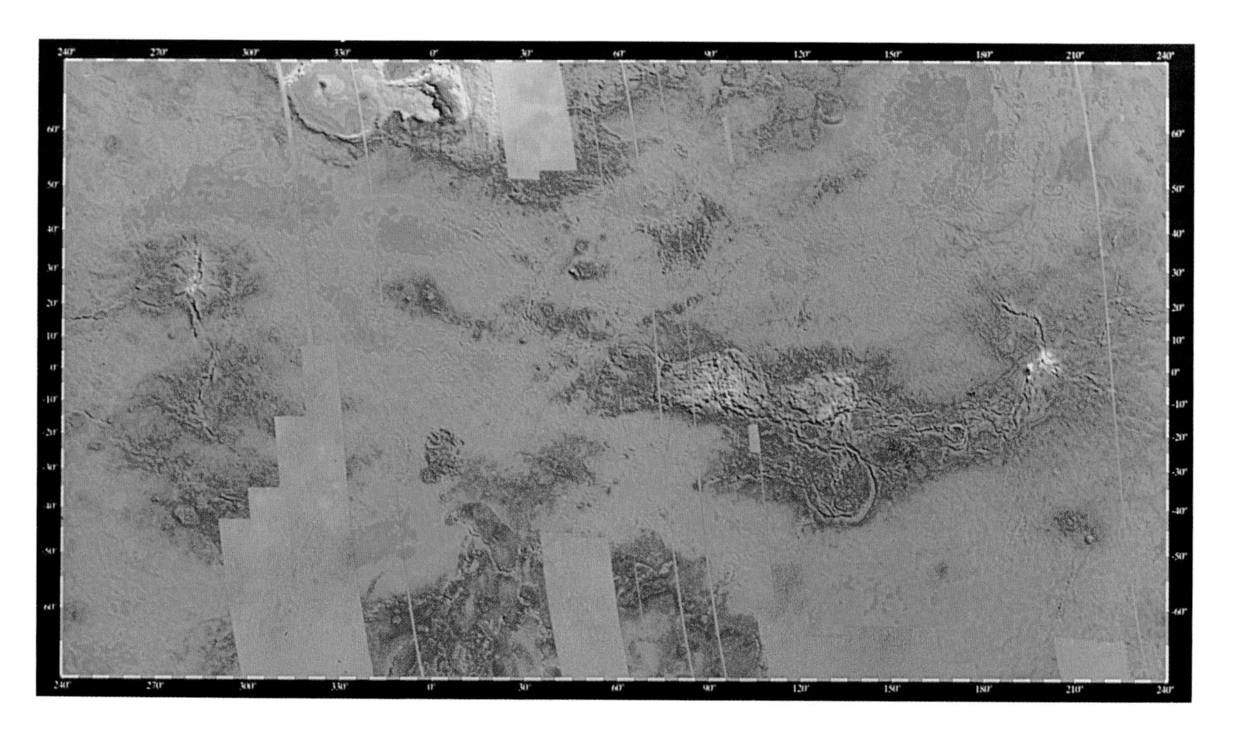

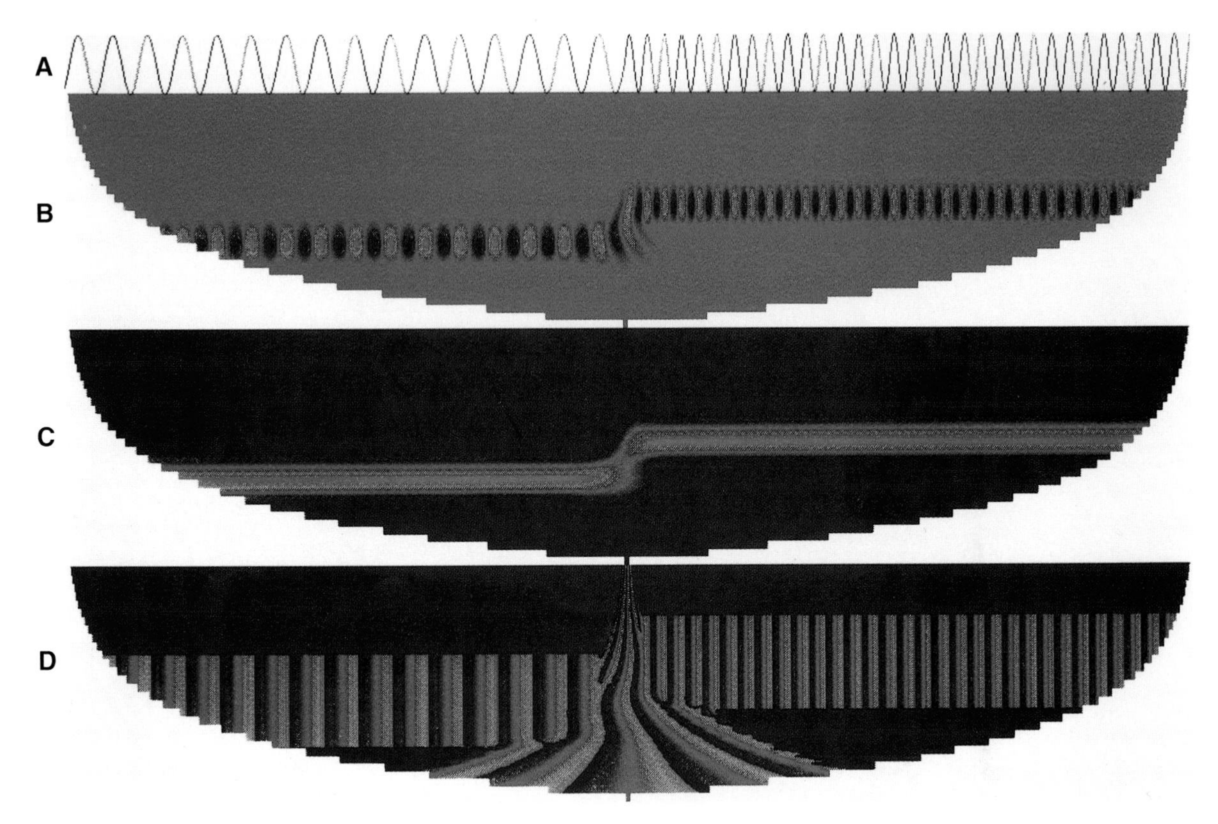

Analysis of a signal showing a doubling of frequency, using Morlet complex-valued wavelets. The signal (a) is at top. The analysis is shown in three parts; in each the horizontal axis represents time or space, and the vertical axis represents the logarithm of scale (i.e., the doubling of scale). From top to bottom: (b) The real part of the wavelet coefficients, indicating the content of the signal at different scales. Red indicates positive values, blue negative values, green zero. (c) The modules of the wavelet coefficients, showing energy density. Red is high energy, green is average, and blue is low. (d) The phase of the wavelet coefficients, showing the change in frequency. (Software TecLetD, Courtesy of Science & Tec, Les Algorithmes, 91194 Saint-Aubin, France.)

Wavelet transform of the sound of a saxophone (32 ms). Complex-valued wavelets were used (each wavelet consist of a "real" wavelet and an "imaginary" wavelet shifted in phase by 90 degrees). The upper portion shows the modulus, or energy density. The colors indicate how the energy of the signal varies with time (horizontal axis) and frequency (vertical axis). The lower portion is the phase; it indicates how the frequencies of the signal (again, the vertical axis) vary with time. (Produced at the CNRS Laboratory of Mechanics and Acoustics-Marseille, courtesy of R. Kronland-Martinet and P. Guillemain.)

Vaccines

In immunological terms there are several ways of hanging a "Wanted" poster. But the basic idea in all vaccine strategies is to inject some portion of the pathogen into the blood where immune cells will encounter it and learn to recognize it. Some small portion of these immune cells will always remain in the blood whether or not the pathogen is ever encountered. These few cells constitute what is called immunological memory, because they remember what the pathogen looks like. If a vaccinated person is challenged by the real pathogen, these few cells will proliferate, producing many more cells like them that can either kill infected host cells or secrete antibodies against the pathogen. In effect, the vaccine allows a person to respond to a pathogen much more quickly than he or she could if their immune system had no prior experience with the pathogen.

The trick in developing a vaccine is to choose the correct portion of the pathogen, one that will elicit an appropriate immune response. For example, if one were to choose a protein fragment from the pathogen that was identical with a protein fragment found in the host, the immune system would not identify the fragment as foreign and would not bother to mount an immune response against it. Since it is difficult to know in advance what the immune system will see as foreign, the traditional approach to antiviral vaccines has been to expose the immune system to the entire virus. Naturally, viruses used in vaccines must be modified so that they do not cause disease. One form of polio vaccine, for example, makes use of viral particles that have been killed. Injecting these killed polio virus particles helps a person's immune system to become familiar with the features of the polio virus and develop immunological memory to it. But the killed vaccine itself cannot infect cells, reproduce, or cause disease. As an alternative to using killed virus, some vaccines are made of live virus. Even if the virus is live, however, it has been modified in such a way that its potency is attenuated. It cannot multiply or cause disease. A second form of the polio vaccine uses live attenuated virus.

Both approaches are being explored experimentally against HIV, using animal models. While some research groups have reported limited success, it must be stressed that they have employed very special conditions, and critics of these experiments point out that the experimental conditions are a far cry from those presented in the real world. Many scientists are also concerned about the advisability of using either approach, but especially live attenuated virus, when dealing with a virus as harmful as HIV. The attenuated virus is genetically disabled so that theoretically it cannot reproduce, but skeptics question whether it might spontaneously mutate back to a form that can reproduce, which would make a vaccinated person sick rather than immunologically prepared to fight disease. The slight chance of such a mutation makes most people very cautious when proposing the use of live attenuated vaccines against HIV.

Newer approaches to vaccine development aim at skirting the safety issue. Because some researchers are uncomfortable with the idea of using either dead or attenuated whole virus, creative new vaccines, which are as yet untried in humans, offer protein fragments or peptides derived from the virus. Since there is no chance that new viral particles can be generated from peptides, they are deemed safer, at least in that respect, than the classical whole-virus vaccines. Candidate peptides include segments from the surface glycoproteins, but scientists are experimenting with a number of peptides derived from other viral proteins, like the reverse transcriptase and the core proteins. Even though these proteins are found on the inside of the virus, they are chewed up by the host cell following infection, and some of the resulting peptides are displayed on the infected cell surface, detectable by the immune system. As such, it may be worthwhile to train the immune system to recognize them as well as proteins on the viral exterior.

In the case of HIV, however, safety is only one of the concerns of vaccine developers. The concept of a vaccine assumes that the virus used to make it looks very much like the real virus that a vaccinated person might encounter. But scientists have learned not to make that assumption with HIV because it is highly mutable. Some portions of the gp120 molecule, for example, can change over just one reproductive cycle, so in the course of a year a virus may not at all resemble the original infecting virus, at least not in one of these highly changeable segments of gp120.

Viral mutability poses a frustrating roadblock to vaccine development. It seems that the variable regions of gp120 are highly effective in activating an immune response, so scientists want to make use of them in vaccines. But because they change so rapidly, the immune system is always lagging behind the viral disguise, looking for a face that has already changed. Sadly, the regions of gp120 that cannot change because they are required for the interaction with the CD4 receptor are not as good at stimulating the immune response. A truly effective viral vaccine would therefore have to anticipate all of the viral variants a person is likely to encounter and include a mixture of them so that the immune system can get to know all of them. That, scientists say, is simply not yet possible.

Even if both the safety and the viral variability issues could be overcome, HIV presents one final challenge to vaccine developers. The immune system can fight only enemies that it can see, but HIV creates many unseen enemies. The virus has a latency period, where the host cell is infected, harbors viral genes and has the potential to reproduce new viral particles at any moment but does not. In this quiescent state the infected cell displays no viral peptides on its surface and has no way of signaling its distress. Developing a vaccine that will distinguish infected but quiescent cells from uninfected cells is an intellectually imposing task.

In one particularly innovative approach, peptides are administered consistently over time, so that the immune system remains at a high state of readiness against HIV. The hope here is that a highly alert immune system might be able to kill off the virus swiftly, either to avert initial infection or to quell reproduction of new viral particles that emerge as latent cells become activated. The strategy taken in this approach is to package the genes for the immunologically stimulatory peptides inside another organism that will continuously produce them and "remind" the immune system what the HIV perpetrator looks like. The way this vaccine works is that genes for some HIV proteins are placed inside a single-celled organism called BCG, a harmless derivative of the bacillus that causes tuberculosis. As a vaccine against tuberculosis, BCG has already been widely used and proven safe in humans. What makes it especially useful for vaccines where it is desirable to maintain a high level of immunological readiness is that the organism is not killed. Rather, the immune system isolates the bacillus in a ball of cells called a granuloma. Even though BCG is held prisoner inside the granuloma, it can still produce substances and secrete them into the blood. Potentially, BCG can be engineered to secrete HIV peptides for an individual's entire life. Thus, if the strategy works, it carries the potential for imparting lifetime immunity. Like all of the other vaccine strategies, however, this one is still in the early experimental stages and is not yet ready for human use.

Therapeutics

As creative as vaccine research has been, scientists who work on AIDS are generally more optimistic about new advances in treating people who already have AIDS with drugs than in preventing infection with vaccines, at least for the immediate future. Broadly speaking, drugs aimed at combating HIV can do one of two things: they can augment the immune response against the virus to try to kill off as many viral particles and virally infected cells as possible, or they can sabotage the reproductive efforts of the virus so that no new particles are made.

Drugs that boost the immune system are similar to vaccines, and in many cases employ the same strategies, only they are administered after the person has become infected. In fact, some vaccines are being tested on people with AIDS to see whether they prolong the asymptomatic period or otherwise halt progression of the disease by boosting the immune response. In theory, all of the vaccines under development are potentially useful in treating people after they become infected or after they develop AIDS. In addition to the vaccines already discussed, which all seek to elicit antibody and killer T-cell production, scientists are also testing the possibility of giving antibodies directly to patients who are HIV seropositive and to those who already have AIDS. The problem with the antibody approach, say scientists, is viral mutability. If an antibody is given that helps the immune system kill off one viral variant, many more slip by undetected. Most scientists are coming to the conclusion that the most effective antiviral strategies combine many different activities. In other words, therapeutic approaches of the future may well combine drugs that boost the immune system with ones that inhibit viral replication.

A New Generation of Drug Treatments

The general hope is that a new generation of drugs may help extend the asymptomatic period from 10 to 30 years or more. While this is not as desirable as completely eliminating the virus from a person's system, it will allow an HIV seropositive individual to enjoy a longer disease-free life and may also limit the severity of any symptoms should they develop.

Antimicrobial drugs each work differently, but they all share the same fundamental goal—to sabotage the biochemical machinery that produces the progeny of the infecting microogranism. Since each machine is different, each therapeutic monkey wrench must be tailor-made to the machine it is trying to inhibit. A big problem in drug development, however, is that the host cellular machinery is often very similar to the machinery used by the replicating virus or bacteria. In fact, viruses are heavily dependent on the host cell's machinery to carry out their reproduction. So drug developers walk a tightrope trying to find agents that harm specifically microbial machines but that leave cellular machines intact. To the degree that drugs fail to be specific for their microbial targets, they cause side effects. Sometimes a drug that is very potent against a pathogen is also severely toxic to the host and is therefore not usable. With the hope of finding a balance between efficacy and toxicity, scientists try to focus on activities that the bacteria or virus carries out that the host cell does not. Scientists often dissect the steps involved in infection and reproduction and ask whether they can block each one.

For HIV the first interaction that scientists would hope to interrupt is the binding between the gp120 molecule on the virus and the CD4 receptor on the host cell. Some researchers have proposed using antibodies that recognize and bind gp120, in effect forming a cap that prevents access to CD4. The use of antibodies, however, begs the variability issue. To circumvent that, researchers have investigated the feasibility of using CD4 as a cap. They know that despite variations in gp120, it always recognizes CD4 and binds to it. Their hope was that soluble CD4 molecules would effectively tie up all of the viral gp120 molecules, and none would be available to bind to CD4 on cells. Unfortunately, this very clever strategy has not worked as anticipated. Through their studies the inventors of this strategy have discovered that to be effective, much higher quantities of CD4 must be administered than is practical.

The next opportunity to interrupt the viral life cycle would be in the uncoating stage, but no drugs have been developed to prevent this yet. It is in the next stage, where the virus converts its genes into DNA from RNA, that drug developers have had the greatest success in controlling HIV. Drugs that interrupt viral DNA synthesis act to inhibit the enzyme called reverse transcriptase, which effects the RNA to DNA conversion. AZT is the oldest of the reverse transcriptase inhibitors and is most commonly given to people when they have fewer than 500 helper T-cells per microliter of blood. This drug and its close relatives dideoxycytosine (ddC) and dideoxyinosine (ddI) work through a trick they play on the reverse transcriptase enzyme.

These drugs closely resemble the actual nucleic acid subunits that are strung together to make long chains of DNA, so the reverse transcriptase tries to incorporate them into the growing DNA molecule. However, a quirk in the chemistry of these nucleic acid analogs allows them to be added onto the DNA chain, but they themselves cannot form bonds with any incoming nucleic acid, meaning that they terminate DNA chain growth. The truncated pieces of viral DNA so produced are virtually useless and cannot direct the synthesis of new viral particles.

AZT seems to slow the progression of AIDS once symptoms do appear and may reduce the number of opportunistic infections that afflict a person with AIDS. Early research suggested that AZT might delay the onset of disease in infected but healthy individuals, but more recent studies have hinted that it may offer no advantage to asymptomatic people.

Although AZT is currently one of the best treatments available, it has many drawbacks. First, it is somewhat toxic. It also suppresses immune activity and causes some tissue damage. In addition, the virus seems to become resistant to AZT at an alarmingly quick rate—sometimes after only 1 year. For that reason physicians are now investigating the use of AZT in combination with or alternating with ddI and ddC, since all three drugs act in essentially the same way but have different sets of toxicities. These three drugs are the only ones approved so far by the Food and Drug Administration for use in treating AIDS.

The ability of reverse transcriptase to mutate in such a way as to become resistant to AZT has inspired a new strategy for drug intervention. In laboratory studies, researchers exposed the virus to three different drugs targeted against reverse transcriptase, with the hope that the enzyme would become resistant to all three at the same time. These alterations would require the enzyme to change so much that it could no longer perform its DNA-synthesizing function. In theory, the virus would successfully resist the drugs but would become ineffectual in the process. This approach has produced disappointing laboratory results, and no one yet knows whether it will be effective or safe for humans, but it is an interesting example of how knowledge of the viral life cycle can lead to new and creative strategies for cutting that cycle short. In this case that means cleverly turning a viral defensive strategy against the virus itself.

Another good target for therapeutic interventions is the integrase that incorporates viral DNA into the host chromosomes, but nothing has yet been developed that can do that. There is now a great deal of interest in agents that can inhibit gene activation by the Tat protein. While nothing truly promising has yet emerged, the hope is that when a Tat inhibitor is developed it will have the potential to block as much as 80 percent of HIV gene expression and possibly that much viral replication. If genes are expressed in the nucleus, however, a good target for drugs would be the Rev protein, which may help escort messenger RNA out of the nucleus and into the cytoplasm to be translated into protein. No Rev blockers have yet been developed.

Once messenger RNA is in the cytoplasm its access to the protein-synthesizing machinery might be obstructed by the use of so-called antisense DNA. In this technology, strands of DNA can be manufactured to specifically recognize and bind to messenger RNA sequences that encode viral proteins. Once bound, the antisense DNA molecules make the messenger RNA molecules unreadable by the protein-synthesizing machinery. The potential use of antisense technology against HIV and other viruses has generated much excitement in the scientific community. Antisense is a highly specific strategy that holds the additional promise of circumventing viral latency. In fact, antisense technology is only useful when cells pass from latent infection to actively producing viral particles.

In theory, all cells can take up the antisense molecules. As long as a cell remains latently infected, the antisense molecules would serve no purpose, but as soon as a latently infected cell starts expressing viral genes, the antisense molecules would effectively prohibit the manufacture of proteins from those genes. Without viral proteins, new viral particles cannot be made. So far, antisense technology has proved successful in the laboratory. Soon this technology will be used in human trials, where it is hoped, it will reduce the level of viral reproduction as well.

Assuming a virus reaches the stage where it does manufacture proteins, the assembly and release of newly formed viral particles would constitute the last opportunity for drug intervention. Here researchers are testing the potency of a class of chemicals called interferons, which are normally secreted by virally infected cells. Although no one is sure of the mechanism, scientists have found that in the laboratory the specific compound called alpha-interferon inhibits HIV from appropriately assembling and budding out of the infected cell. The drug has been associated with some negative side effects, so its ultimate appropriateness for human use remains to be determined in clinical trials.

In addition to drugs that interfere specifically with the viral life cycle, many new antibiotics and antimicrobial agents have been developed to fight the specific opportunistic infections that people with AIDS fall prey to (see Table 3.4). Despite the difficulty of finding a vaccine to prevent initial infection, physicians and researchers are optimistic that in the future AIDS patients will be healthier and more comfortable than in years past. In the years to come, AIDS may be a chronic disease, and people infected with HIV may be able to live with the infection for the rest of their lives. The new constellation of antiviral drugs and therapeutics directed against opportunistic infections may well mean that the relationship between infection and disease will change. A person who is seropositive for HIV may never develop AIDS or may remain disease free for many years.

TABLE 3.4 *Approved AIDS Medicines as of January 1, 1992*

Drug name	Manufacturer	Indication
Bactrim™+ trimethoprim and sulfamethoxazole	Hoffmann-La Roche (Nutley, N.J.)	PCP treatment
Cytovene® ganciclovir (IV)	Syntex (Palo Alto, Calif.)	CMV retinitis
Daraprim®+ pyrimethamine	Burroughs Wellcome (Research Triangle Park, N.C.)	Toxoplasmosis treatment
Diflucan® fluconazole	Pfizer (New York, N.Y.)	Cryptococcal meningitis, candidiasis
Foscavir® foscarnet sodium	Astra Pharmaceutical (Westborough, Mass.)	CMV retinitis
Intron® A+ interferon alfa-2b (recombinant)	Schering-Plough (Madison, N.J.)	Kaposi's sarcoma
NebuPent® aerosol pentamidine isethionate	Fujisawa Pharmaceutical (Deerfield, Ill.)	PCP prophylaxis
Pentam® 300 IM&IV pentamidine isethionate	Fujisawa Pharmaceutical	PCP treatment
PROCRIT®+ epoetin alfa	Ortho Biotech (Raritan, N.J.)	Anemia in Retrovir ®-treated HIV-infected patients
Retrovir® zidovudine (AZT)	Burroughs Wellcome	HIV-positive asymptomatic and symptomatic (ARC, AIDS), pediatric and adult
Roferon®-A+interferon alfa-2a, recombinant/ Roche	Hoffmann-La Roche	Kaposi's sarcoma
Septra®+ trimethoprim and sulfamethoxazole	Burroughs Wellcome	PCP treatment
VIDEX® didanosine (ddI)	Bristol-Myers Squibb (New York, N.Y.)	Treatment of adult and pediatric patients (over 6 months of age) with advanced HIV infection, who are intolerant or who have demonstrated significant clinical or immunologic deterioration during Retrovir® therapy
Zovirax® acyclovir	Burroughs Wellcome	Herpes zoster/simplex

SOURCE: Pharmaceutical Manufacturers Association, Washington, D.C.

Yet even though HIV-infected people may not develop symptoms of AIDS, one feature of infection will remain: They will still be able to transmit the virus to others. Here lies the final irony of the AIDS epidemic.

While the world looks to science to end the spread of AIDS, the best solution is also the simplest and requires absolutely no technology. Scientists stress that no one ever needs to become infected again, if only people would take care not to engage in behaviors that place their tissues in contact with the body fluids of an infected person. However, if abstaining from these activities is not possible, the simple use of condoms during sexual activities would greatly reduce the number of new infections. Programs that encourage the use of clean needles by intravenous drug users also would drastically reduce the number of new infections. In short, the solution to the world AIDS crisis is to change human behavior. Therein lies the greatest hope and greatest tragedy for eradicating the new plague, AIDS.

RECOMMENDED READING

Benditt, J.M., and B. R. Jasny, eds. 1993. AIDS: The Unanswered Questions. *Science,* May 28, 1993. Volume 260, pp. 1253-1293.

Mann, J., D. J. M. Tarantola, and T. W. Netter. 1992. *AIDS in the World.* Cambridge, Massachusetts: Harvard University Press.

Shilts, R. 1987. *And the Band Played On.* New York: St. Martin's Press.

DOUBLING UP
How the Genetic Code Replicates Itself

by Anne Simon Moffat

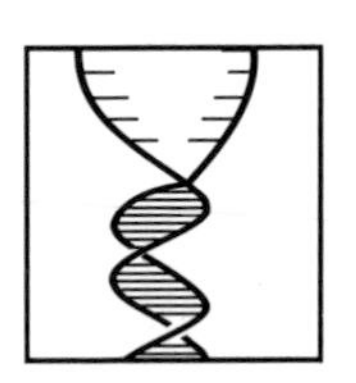

This year marks the forty-first anniversary of one of the most important scientific discoveries ever made: the finding that the universal genetic material is made of deoxyribonucleic acid, commonly called DNA. Although it had been known for hundreds of years that plants and animals had some sort of genetic material that carried information from generation to generation, what it was and how it worked remained unknown until 1953, when the American James Watson and the Englishman Francis Crick made their announcement from Cambridge, England. Their description of DNA as a twisted, ladder-like structure with rungs of complementary pairs of simple chemical bases offered a basic architecture for the genetic code. And because this physical structure involved a template where information from one DNA strand could be carried to another, it also hinted at how information could be passed from one generation to another with great fidelity. What happens is that the bases on each of the parent strands become paired with their complements on the newly formed daughter strands. Indeed, in a classic example of British—and scientific—understatement, Watson and Crick wrote in their landmark *Nature* paper that, "It has not escaped our notice that the specific pairings (of bases in DNA) we have postulated immediately suggests a possible copying mechanism for the genetic material."

This discovery was the culmination of decades of work by scores of researchers. Yet it marked the beginning, not the end, of a new era of biological research. It triggered a desire to know more about the bricks, mortar, and embellishments that filled out the basic architecture of DNA described by Watson and Crick and the precise building mechanisms used to construct it. Specific issues that were once beyond the scope of experiments—such as when does a cell know when to duplicate its genetic material and how does it do it?—could now be tackled. Moreover, broad problems—such as the way DNA duplication fits into the total cell cycle, the ongoing process of cell growth and division—could be addressed (see Box beginning on p. 100).

Over the past 40 years, the incentives for pursuing such research have been many and diverse. On one level there is the sheer intellectual excitement of getting to know the ways of life on an intimate basis. On another, such studies offer the opportunity to use new information about the structure and behavior of DNA for useful purposes: for example, improved understanding of DNA replication provides a means for controlling the unrestrained proliferation of DNA found in cancer cells (see Box beginning on p. 106).

And, over the years, experimental approaches to the study of genetic material have been equally diverse. Some researchers look at the simple genomes of viruses, which are often considered to be in a biological netherworld since they can't live on their own, while others look at the slightly larger, circular genomes of bacteria. The genetic material of yeasts, which are the simplest organisms to have DNA in organized chromosomes (and are used to make bread and beer), are the focus of still other labs. Finally, some researchers accept the challenge of looking at cells taken from advanced organisms, such as rats, frogs, plants, and man.

As with the study of most complex phenomena, the early results from studies of DNA replication and gene expression in these various organisms were murky and incomplete. It was not always clear that the piecemeal results could be synthesized into a coherent picture of what

The Big Picture: The Cell Cycle

DNA replication, although important and time consuming (many cells spend one-half their existence copying their genome), is only one stage in the cell cycle, the complex process of cell growth and division that leads to the birth of new cells.* Cells also spend considerable time simply growing in size or gathering up enough nutrients to build and maintain their subcellular apparatus, such as the energy-producing mitochondria and, in plants, the light-gathering chloroplasts. In fact, the myriad activities that must be started, completed, and integrated with others during the cell cycle would rival those found in a car assembly plant. And like the action that takes place on an assembly line, the cell cycle has an elaborate scheme of feedback controls, where the activities of one stage determine what follows.

For a better sense of the role of DNA replication in the larger scheme of cellular matters, here is an overview of the key stages of the cell cycle.

The G1 Phase

The cell cycle is defined as starting at the G1 (for first gap) phase, which lies between the previous nuclear division and the start of DNA synthesis. Once cells reach a critical size in G1, a decision is made whether to commit

*It is possible for cells to exit from the cell cycle and still be very much alive. Such quiescent cells do not necessarily increase their mass or divide, but they still carry out some very important functions, such as the transmission of nerve impulses. Nerve cells, for example, are notable because they do not normally divide. This, of course, can be a problem in the treatment of neural injuries since the inability to divide makes nerve repair difficult.

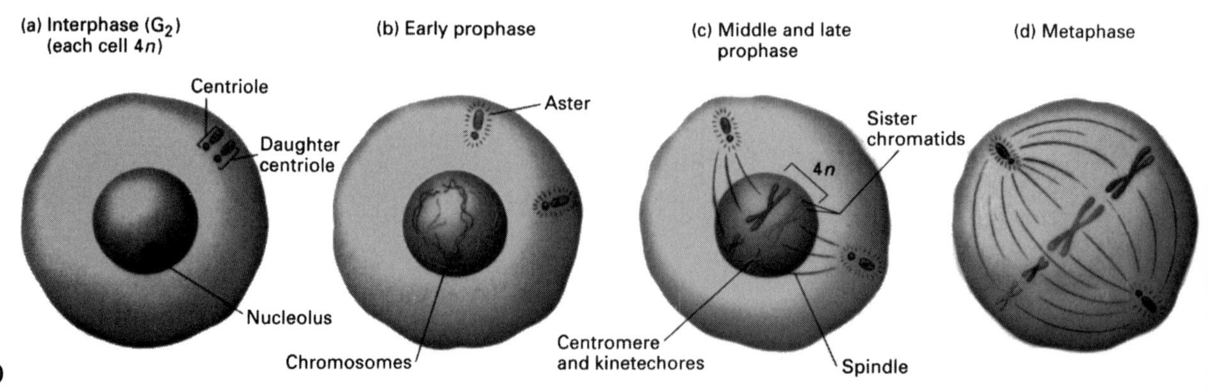

the cell to further involvement in the subsequent S or synthetic phase, leading to cell division. Some researchers refer to this part of the G1 phase as "Start." Cells that exit from the cell cycle typically do so here, and enter the so-called G0 (or resting) phase.

The S Phase (DNA synthesis)

DNA replication takes place during the synthetic or S phase. Although there is great activity in the genome during this stage of the life cycle, the chromosomes cannot be seen as distinct structures under the light microscope.

The complete cell cycle in a mammalian cell needs about 16 hours, and the S phase can take almost half that. The time devoted to DNA synthesis is even more striking in *E. coli*. These bacteria can divide every 30 minutes or less, and all but about 1 minute is often needed to complete one round of DNA synthesis.

The G2 Phase

The second gap phase is placed between DNA duplication and actual nuclear division. During this stage, cells can be very busy making proteins, lipids and carbohydrates, and other substances essential to the support of life. Also, during this period, any DNA that was damaged during the S phase is repaired.

M or Mitotic Phase

During the brief but very busy period called mitosis (Figure 4.1), new daughter chromosomes separate and the cell divides. Its various stages are divided into prophase, metaphase, anaphase, telophase, and interphase.

continued

FIGURE 4.1 *All phases of mitosis. (From* Molecular Cell Biology *by J. Darnell, H. Lodish, and D. Baltimore. Copyright © 1990 by Scientific American Books, Inc. Reprinted with permission of W.H. Freeman and Company.)*

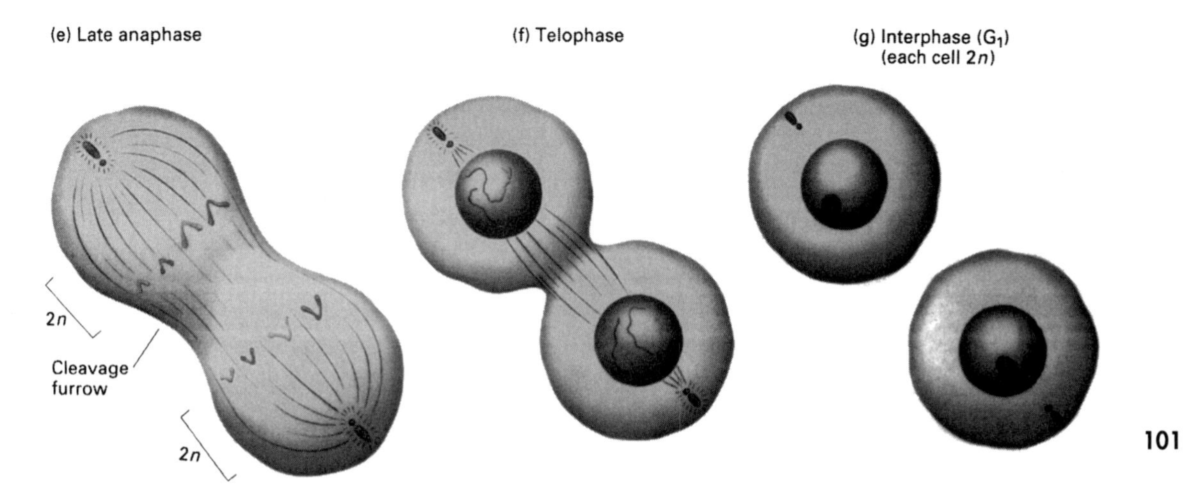

The Big Picture—*continued*

- *Prophase*—Here, the genetic material becomes condensed and the chromosomes are visible by light microscopy. The duplicated chromosomes can be seen as identical coiled filaments, each composed of the two daughter DNA molecules made during the earlier S phase. At the end of prophase, the nuclear membrane disappears, and the entire contents of the cell begin to reorganize in preparation for division. For example, each visible chromosome is made of two single-stranded chromatids held together at constricted regions called centromeres. These chromatids become shorter and more visible as the cycle passes into metaphase. At late prophase, small cylindrical particles, called centrioles, come into play. They move to opposite ends of the cell and become a sort of subcellular anchor, to which the chromosomes are eventually drawn.
- *Metaphase*—The coiled sister chromatids, which by now are loosely connected with the centrioles by microtubules, gather at the center of the cell, to the so-called equatorial plate. The sister chromatids have not yet separated.
- *Anaphase*—Here, the sister chromatids split into separate chromosomes and migrate towards one of the poles anchored by the centrioles. Thus, one copy of each chromosome is guided to each future daughter cell. The cell elongates and a cleavage furrow appears.
- *Telophase*—In this final active stage of division, the chromosomes start to uncoil and become less visible, as does the apparatus specifically associated with cell division, such as the centromeres and microtubules. However, other components of the cell, including the nuclear membrane and the nucleus, begin to reappear. Also, the cytoplasm divides between the two new cells.
- *Interphase*—This phase consists of the G1, S, and G2 phases of the cell cycle, in between each M phase.

Pulling It All Together

Although much is known about what happens in the individual stages of the cell cycle, one of the biggest challenges facing cell biology is to learn more about how the signals that guide cells from one stage of the cycle into another control events such as DNA replication. "Feedback controls" are known to exist, since in most cells inhibiting a downstream event can arrest programming of the cell cycle. For example, stopping DNA duplication keeps cells from entering mitosis. Although early studies of the cell cycle focused on external extracellular signals, such as hormones and other growth factors, in recent years many clues have emerged about regulatory signals that come from within the cell. Many of the key regulatory proteins have been positively identified.

For example, from the study of yeast mutants that are blocked through the normal progression of the cell cycle and of frog oocytes and their development into early embryos, it has become clear that one protein, called cyclin, is critical to the smooth progress of the cell cycle. Such molecules build up during interphase and are lost during mitosis. In fact, the degradation of cyclin appears essential for exiting mitosis. Cyclin was discovered as part of a large protein complex, called the maturation-promoting factor (MPF), which also includes a critical kinase enzyme, called the cdc 2 (for cell division cycle) kinase. This enzyme is positively regulated by cyclin to fuel the mitotic process. There appear to be complex regulatory schemes acting between this kinase and cyclin, and, indeed, cyclin-cdc 2 complexes also control the commitment to cell division at Start in the G1 phase of the cell cycle.

Another likely player in the regulation of the cell cycle is a protein, ubiquitin, whose name suggests its everpresent nature. It appears to have some role in the degradation of cyclin. One theory suggests that MPF stimulates an activity that makes cyclin a better substrate for ubiquitination and consequent destruction. Interestingly, there is one other known example of a protein having regulated degradation by ubiquitin, the light-activated phytochrome in plants.

It is no surprise that there are many layers of control, involving many different compounds, on the cell cycle. The challenge remains, however, to describe all such substances and how they work together to trigger cells into a smooth transition from one phase of the cell cycle to another.

goes on in living cells as they grow and multiply. But, eventually, these various experimental approaches converged on some key findings. It was recognized that DNA is the universal genetic code: evolution had fashioned a system for transferring biological information from one generation to another that works equally well for the simple genomes of viruses and bacteria as for the more complex genomes of worms, monkeys, and humans. But, equally important, it became clear that while the double helical architecture of DNA is relatively simple and straightforward, the mechanisms for building it and expressing it are not. DNA needs a large cast of supporting characters to help it get its job done. In recent years some of the most significant advances in DNA studies have been the discovery of these adjunct players—enzymes, other proteins, and various cofactors—and their roles in DNA synthesis and expression.

STARTING AT THE VERY BEGINNING

One approach to the study of DNA and its supporting players is to look at the very first events associated with DNA's replication. About 30 years ago, using a prime analytical tool of the day, the microscope, some researchers noticed large bits of chromosome that appeared to balloon up at predictable and specific times in a cell's life cycle. These so-called replication bubbles were identified in the chromosomes of animal, plant, and bacterial viruses as well as those of more advanced organisms, such as yeast and some mammals. Other researchers fed microbial and animal cells labeled isotopes at the start of DNA synthesis, with normal isotopes offered later in the same round of DNA synthesis. In both animal and yeast experiments it was found that some sections of the genetic material were preferentially synthesized before the bulk of the genome. In a variation on this research, other experimenters tagged DNA with a fluorescent compound that is similar to one of the bases of the genetic code. By looking through a microscope, with suitable light filters, small amounts of fluorescence could be seen in certain regions of the nucleus. This work offered further evidence that not all DNA copies itself at the same time and that certain bits of the DNA were preferentially copied early on in the replicative process.

At about the same time that these researchers were using microscopic techniques to examine DNA replication visually, other labs were developing biochemical strategies for examining this event. For example, cell fusion experiments done in the early 1970s showed that diffusible factors were needed for DNA synthesis. When cells in the "resting phase," the so-called G1 phase immediately prior to the stage

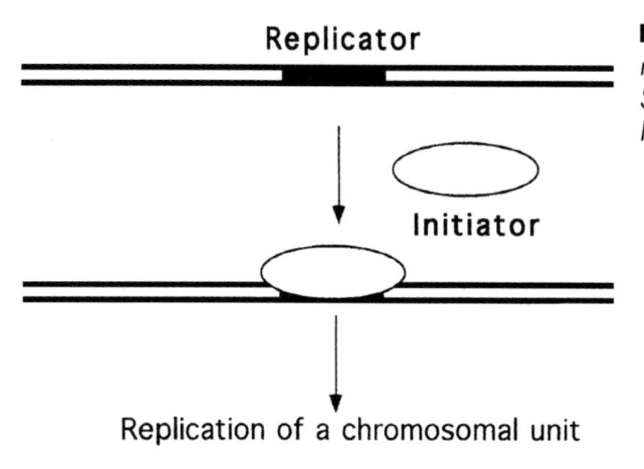

FIGURE 4.2 *The "replicon" model. (Courtesy of Bruce Stillman, Cold Spring Harbor Laboratory.)*

when DNA synthesis takes place, were fused with cells in the DNA synthetic phase, the so-called S phase, the S-phase cells did something to the G1 cells to trigger them into the S phase. Some substance present in the S phase appeared to coax the onset of DNA replication in the G1 nucleus. Moreover, this unknown trigger was shown to be measurable in the following experiment. When three or more cells were fused, the higher the ratio of S-phase to G1-phase cells in the fusion, the faster the G1 cells entered the S phase. With very high S:G1 ratios, entry into the S phase by the G1 cells was almost instantaneous.

All these experiments supported the so-called Replicon model of DNA duplication (see Figure 4.2), proposed in 1963 by Francois Jacob and Sydney Brenner, which suggested that DNA replication is controlled by a biochemical regulator, known as the initiator, which triggered the start of DNA duplication by acting on a specific collection of DNA bases, called the replicator, or replication origin. But this work did not offer definitive proof of the existence of the postulated replicator and its helpmate, the initiator. Almost a decade would pass before such proof was in hand.

To tackle this problem in mammalian cells, researchers chose to work with a tumor-inducing monkey virus known as SV40, which has a very small genome, only 1.5 microns (0.0000015 of a meter) long, and contains only about 5000 base pairs of DNA. The reasoning was that if a possible replicator origin was present, it would be easier to find it on a virus's small genome than on the genome of, for example, a human, which has about 3 billion base pairs and is about 1 meter long. The search initially involved studies of various replication intermediates taken from cells infected with SV40 to identify where in the small chromosome DNA replication started. Once this origin of DNA replication was found, researchers hunted for a specific protein, the initiator, that would bind DNA at the replication origin. The search for this protein, among millions of proteins in the cellular soup, was a nontrivial task. After many years of fractionations and purifications, where cellular

How Some Cancer Drugs Might Work

Cancer, in the broadest sense, is uncontrolled cell growth. DNA replicates and cells proliferate wildly, depriving more organized growth of needed resources. Tumorous growths can appear as amorphous, undifferentiated, and ungovernable masses. Yet despite giving the appearance of complete disarray, cancer can be the end result of just a few molecular events gone wrong. Too much, or too little, of one of the many proteins that control cell division and DNA replication can disrupt normal cell growth and division, leading to cancer.

Until fairly recently, most anticancer drugs were developed using empirical techniques, that is, by trial and error. Within the last year or two, this has begun to change. Now that some details are known about the various proteins needed to guide DNA replication, cell growth, and division in normal cells, efforts can be made to design drugs that correct some of the biochemical imbalances in cancer cells. By adding or deleting key proteins, such anticancer drugs may inject a level of management and structure that is missing in cancer cells. Numerous new biotechnology companies, many large, established pharmaceutical houses, and a number of independent research organizations worldwide are developing anticancer drugs based on this strategy. To be sure, not all cancers will be amenable to this sort of treatment, and it is unlikely that such therapies will be clinically available before the turn of the century. Still, says Bruce Stillman, the design of drugs that target key proteins and add a measure of control into a system that has run awry "represents a front line clinical approach to the treatment of cancer."

One class of nuclear proteins that has received much attention from some cancer specialists are the topoisomerases, enzymes that are required for DNA replication and expression by affecting the overall topology of strands of DNA (see pp. 120–121). One research stratagem suggests that by poisoning the activity of these essential enzymes through the use of various antitopoisomerase drugs, the uncontrolled DNA replication and cell division characteristic of cancer cells could be curtailed. In fact, Paul J. Smith, of the Medical Research Council's Clinical Oncology and Radiotherapeutics Unit in Cambridge, England, has suggested that, "Tumor cells carry the seeds of their own destruction in the form of these enzymes [and] it is up to the chemotherapist to take advantage of this unique opportunity."

In the last 5 to 6 years two discoveries were made that suggested that drugs that functioned as antitopoisomerases might work as anticancer drugs. First, newly replicated DNA was found to be a preferential target for antitopoisomerase II drugs. Second, several classes of known anticancer drugs were found to work by poisoning the topoisomerase enzymes. These drugs trap the topoisomerases on DNA, thus impairing the normal replication process. Topoisomerase II "poisons" include various drugs, such as the acridines (e.g., Amsacrine); anthracyclines (e.g., Adriamycin); anthraquinones (e.g., Mitoxantrone); and the epipodophyllotoxins (e.g., VP-16). But because only a few of these agents, such as the epipodophylloxins, trap topoisomerases in a nonproductive and stable form, the list of candidate anticancer drugs is narrowed. So, as a first step to further development of antitopoisomerase anticancer drugs, researchers are focusing on how these drugs tie up the enzymes, the character of the enzyme/drug complex, and why it is toxic to actively growing cancer cells.

One theory suggests that topoisomerase poisons trap the enzyme in a useless state, resulting in the loss of an enzyme critical to normal DNA replication. Another posits that the complex of the topoisomerase plus poison itself damages the system. But from the chemotherapist's view, the when and where of topoisomerase inactivation is at least as important as how it is done. Some evidence suggests that a collision of a replication fork with a trapped topoisomerase is the damaging event. There is also the possibility that the ability of an impaired topoisomerase to upset normal replication depends on where its sits on the chromosome. Certain sequences on the chromosome may be more vulnerable to the action of a topoisomerase poison.

Another complication that might affect chemotherapeutic strategies is the finding that inhibiting RNA or protein synthesis protects cells from the cell-killing action of the topoisomerase poison, Amsacrine. So now there is a search for factors that could inhibit, or enhance, the impact of topoisomerase poisons. "The hope [is] that such factors can be manipulated to the chemotherapist's advantage," says Smith.

Still another possibility for enhancing the effect of topoisomerase poisons as anticancer drugs is to coax cells to increase their production of topoisomerases and, hence, their vulnerability to this enzyme's poisons. This is not as farfetched as it sounds. For example, in laboratory studies with human breast cancer cells, it was found that the manipulation of estrogen could stimulate the expression of topoisomerases at a particular point in time and, consequently, the effectiveness of well-timed doses of the topoisomerase poisons, VP-16 and Amsacrine. The end result is the increased killing of cancer cells. This therapeutic strategy will soon begin clinical trials.

Smith says that in the past, "Not knowing the full nature of the cytotoxic interaction between an antitumor agent and its target cell has always undermined our ability to design more effective drugs, to target treatment modalities and to gain some therapeutic advantage by manipulating the biology of neoplastic cells in situ. This situation has now changed for several classes of potent antitumor agents which poison DNA topoisomerases." And there are signs that nuclear proteins other than the topoisomerases will soon be targets for therapeutic agents designed to fight cancer.

proteins were repeatedly divided and tested for activity as an initiator, Robert Tjian, then at the Cold Spring Harbor Laboratory (and now at the University of California, Berkeley) announced in 1978 discovery of the SV40T antigen as the initiator that specifically bound to the virus origin of DNA replication. This protein was first found with antisera from animals that had tumors produced by SV40: it was therefore called tumor antigen, or "T" antigen. When the virus gene encoding T antigen is introduced into cells that are not dividing, expression of T antigen induces the infected cells into the S phase, causing replication of the host cell's DNA and allowing virus DNA replication. In some cell types, T antigen will also transform the cells so that they now cause tumors in animals. Thus, T antigen is an initiator protein that binds to the virus origin of DNA replication and is a potent onco-protein, a cancer-causing agent.

How this T antigen initiator bound to the genome was found by examining both natural mutants and artificially mutated specimens of SV40. These and other experiments revealed that the T antigen was a relatively large protein of 708 amino acids and that 114 of these bound the origin (replicator) sequences on the DNA. It was also shown that only 65 base pairs from the whole virus genome were the target upon which the T antigen acted to trigger DNA replication. Flanking this essential core were other DNA sequences in the virus genome that stimulated DNA replication.

A key issue, of course, was what the initiator protein actually did to trigger DNA duplication. Work by several groups in the 1980s revealed that this T antigen, in the presence of other compounds that could supply energy, would unwind eight base pairs of DNA and, once unwound, the single strands of DNA could serve as templates for copying, thus starting the sequence of events that would ultimately lead to complete duplication of the DNA.

It was also realized that T antigen alone is not able to replicate the virus genome but requires proteins from the host cell. It was expected that these proteins would be the very same ones that are responsible for replicating the cell chromosomes in each S phase of the cell cycle. A key breakthrough toward demonstrating this was the development, by Joachim Li and Thomas Kelly, of the Johns Hopkins University School of Medicine, of cell-free protein extracts from monkey or human cells that would support the replication of SV40 DNA in the test tube, when T antigen was added (Figure 4.3). Subsequent work in Kelly's laboratory, as well as work by Jerard Hurtwitz, of the Sloan Kettering Cancer Center, and Bruce Stillman, of the Cold Spring Harbor Laboratory,

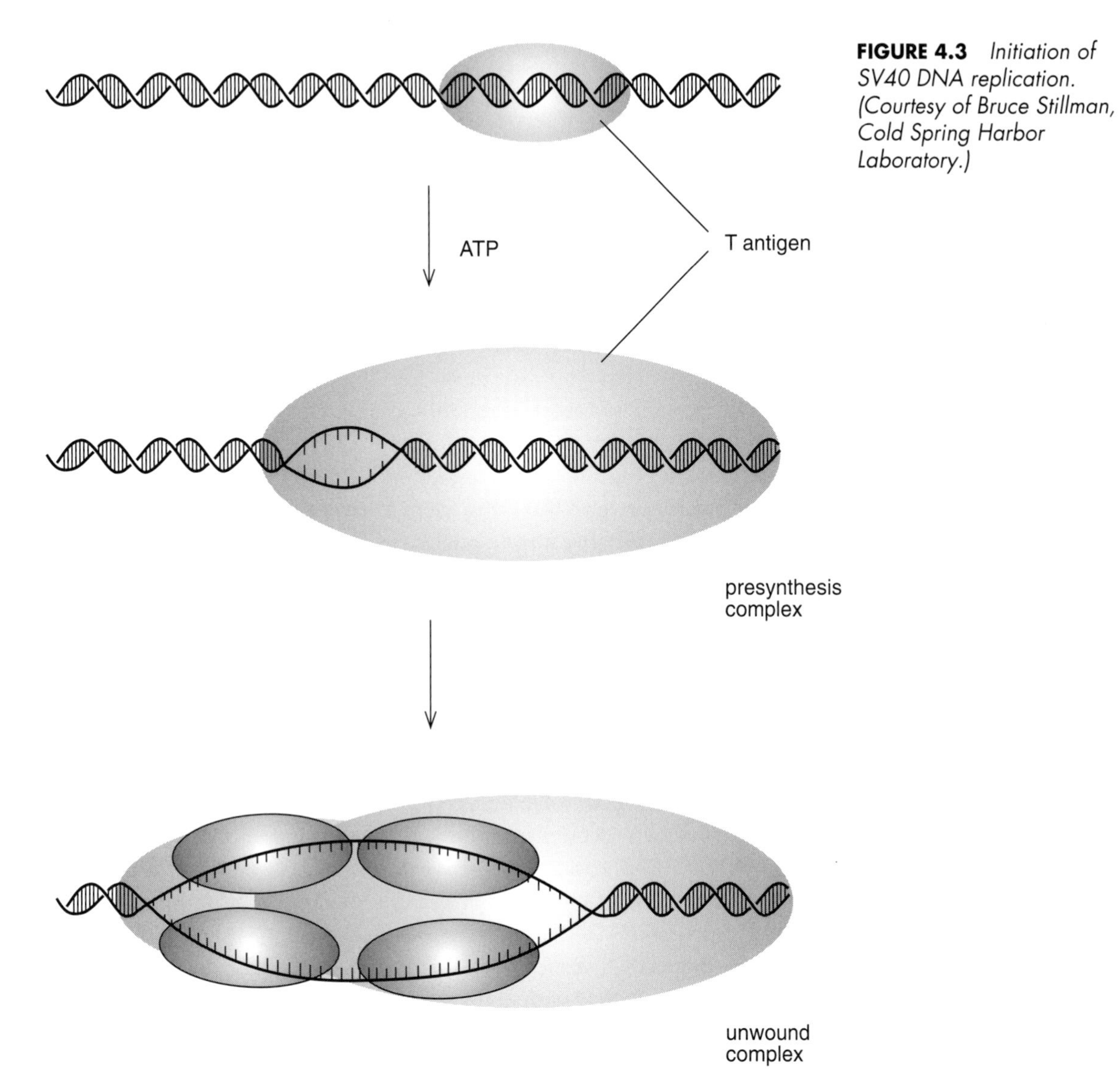

FIGURE 4.3 *Initiation of SV40 DNA replication. (Courtesy of Bruce Stillman, Cold Spring Harbor Laboratory.)*

identified these key cell proteins and revealed that they were required for replication of virus as well as for cellular DNA.

TACKLING THE NEXT STAGE OF COMPLEXITY

A second key issue was whether other organisms, especially those more complex than viruses, have replication origins and used proteins like T antigen to get the process going. Was the system used by viruses, which, after all, are not genuine living organisms, an interesting anomaly? Or was this system relevant to all biology? To help answer that question, many researchers turned to the study of bacteria, the next most advanced group of organisms. In particular, they looked at *Escherichia coli* (*E. coli*), a common intestinal bacterium with a genome of about 4 million base pairs, about 1 millimeter in length. Here, too, researchers at first sidestepped the fact of not knowing exactly where DNA synthesis began, and they focused on a protein that seemed critical for the start of DNA replication, a protein that is the product of a gene called dna A. It is a scarce protein but appeared critical for cell viability, since *E. coli* rarely survived mutations to the gene. Knowing that this protein was somehow essential to DNA duplication but not knowing its details, researchers cloned the gene for dna A in recombinant DNA vectors that allow high-level expression in bacteria by taking over their metabolic machinery and directing them to churn out proteins. This gave scientists reasonably ample quantities of this usually scarce protein to work with. (It should be noted that developing such production schemes can be tough. Duplicating in vivo production schemes in alternate hosts can require considerable tinkering to get the desired results.) It was found that this protein binds to a particular site on the bacterial chromosome, in between two genes known to code for two key enzymes. The initiator protein dna A binds tightly to a small number of such nucleotide sequences, "seeding" the cooperative binding of 20 to 40 more copies of dna A, to form a large nucleoprotein complex. This complex serves as a sort of molecular pliers, pulling apart the double strand of DNA. Replication begins at this site and proceeds in both directions along the strand. This origin of replication site in bacteria is about 245 nucleotides in length. Moreover, research has shown that within this short DNA stretch, two-thirds of such nucleotides occur at the same position in at least five bacteria studied, including the distant relative *Vibrio harveyi*, a marine bacterium. This information suggests that these genomic sites for initiating DNA synthesis have been conserved during evolution. Such conservation among species hints that these nucleotides are important for biological function and may well interact with initiator proteins that are also conserved among the species. A key challenge, then, was to demonstrate that this general strategy for starting duplication of DNA—having a protein identify a specific origin sequence, among millions of others on the double helix, and then having it uncoil the DNA—also worked for the more advanced organisms known as eukaryotes, organisms that have a nucleus and have the genome divided among multiple chromosomes.

SNARING KEY EVIDENCE

But proof was not in hand until 1992, when Bruce Stillman of the Cold Spring Harbor Laboratory in Cold Spring, New York, and a speaker at the 1993 "Frontiers of Science" Symposium, together with his colleague Stephen Bell, captured an initiator protein from yeast, the simplest of the eukaryotes, which, like humans, have true chromosomes enclosed in an organized nucleus. It seemed a reasonable hypothesis that yeast, plants, monkeys, and man, all eukaryotes, would share a general scheme for duplicating DNA that involved replication origins on their chromosomes and initiator proteins, like the viral T antigen or *E. coli*'s dna A protein. But there were also signs that, because of the size of their genomes, eukaryotic systems had complexities that are unknown in viruses and bacteria. For this reason scientists could not safely conclude that the replication scheme that worked so well in lower prokaryotic organisms would work the same way at the next level up, in the eukaryotes. For example, in human cells the complete cell cycle can take about 24 hours, with about one-third of that time devoted to duplicating the cell's DNA. With about 3 billion bases to replicate in 8 hours, such cells need more than one replication origin to get the job done in time. "The way nature solves this problem," says Stillman, "is to plump a lot of these initiator sequences along the chromosome [see Figure 4.4]. Then

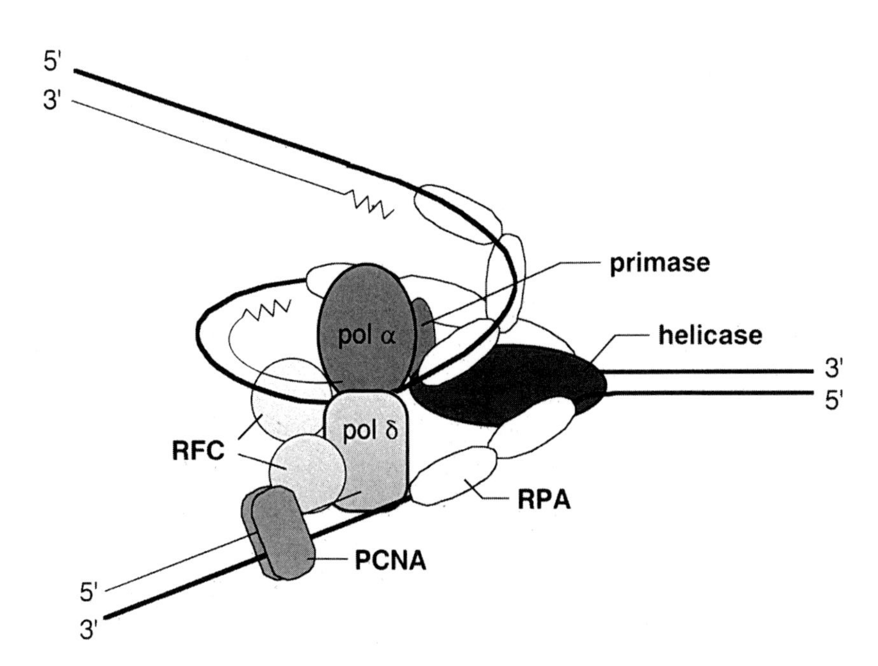

FIGURE 4.4 *The multiprotein complex of the replisome keeps DNA synthesis progressing smoothly. In eukaryotes it includes two polymerases, alpha and delta; primase; helicase (T antigen in the case of SV40 virus DNA replication); replication protein A (a single-stranded binding protein); replication factor c; and the proliferating cell nuclear antigen. (Courtesy of Bruce Stillman, Cold Spring Harbor Laboratory.)*

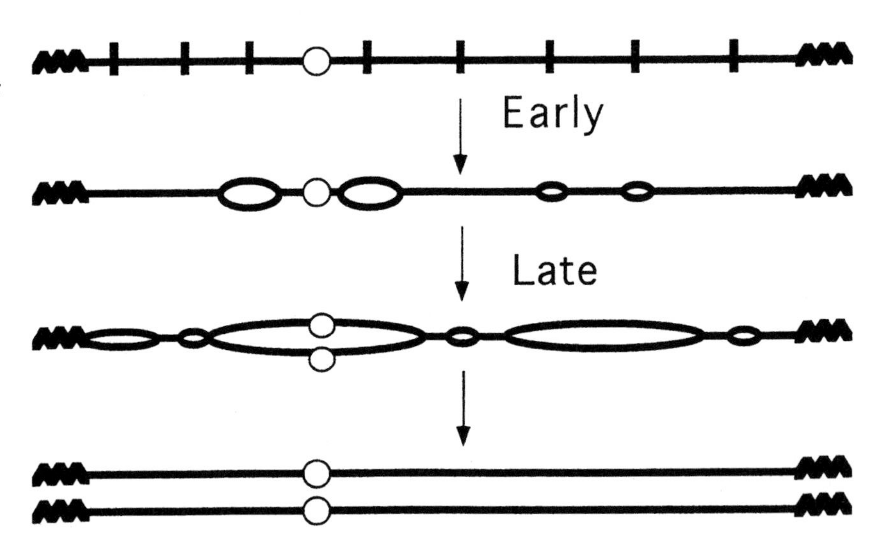

FIGURE 4.5 *Multiple replication origins in chromosomes. (Courtesy of Bruce Stillman, Cold Spring Harbor Laboratory.)*

you initiate from all of them and you'll replicate the genome." Like a parallel-processing supercomputer, and very unlike a simple virus, eukaryotic cells have hundreds, thousands, or even tens of thousands of replication processes going on simultaneously (see Figure 4.5). Indeed, it is now known that in the particular experimental yeast favored by researchers, called *Saccharomyces cerevisiae*, at least 400 replication origins exist in its 16 chromosomes. Moreover, the fruit fly *Drosophila melanogaster* can have 6000 replication segments at one time, just in its largest chromosome.

Thus, while it seemed that there are some strong parallels between DNA replication in higher and lower organisms, there also seemed to be some key differences. The challenge was to find out just what they are.

Early clues as to the nature of DNA replication in yeast came in the late 1970s when general studies of this organism revealed that yeast chromosomes need several distinct structures to duplicate, including special DNA sequences known as autonomously replicating sequences (ARs). In 1979 Ron Davis and his colleagues at Stanford University Medical School and John Carbon and his colleagues at the University of California, Santa Barbara, showed that select DNA sequences could confer a high rate of stability when reintroduced into yeast cells. A subset of these ARs was subsequently shown to act as origins of replication in chromosomes. Deletion analysis, which involves mutating yeast chromosomes by selectively removing some bases in the DNA, had shown that at least 100 bases are needed for these ARs sequences to work efficiently. By fine tuning their studies of the ARs, researchers also discovered that a minimal 11-base-pair sequence is essential for origin activity. Known as the "A" element, which appears common to all origins from *Saccharomyces*, the 11-base-pair essential element is always accompanied by "B" elements that stimulate origin activity. Last year, after an intense search by many labs for this protein, Stillman and Bell, also of the Cold Spring Harbor Laboratory, announced discovery of a multiprotein complex in yeast that could act as an initiator by binding the essential 11-base-pair element of the ARs.

The search required arduous testing of many proteins present in yeast cell extracts. What Stillman calls "molecular shrapnel" was shot at the yeast DNA, splitting it into random pieces. However, if the initiator protein is bound over the origin, it protects the DNA from being cut. With this information, a "footprint" of the initiator protein binding to a particular set of DNA sequences was pieced together, and, using special biochemical techniques, the protein was extracted and purified from the complex. This protein complex, it was found, bound to all of the yeast's replication origins. Moreover, although this initiation protein complex binds strongly to natural, wild-type ARs sequences, it does not bind to inactive DNA elements whose ARs have been tinkered with, and destroyed, by mutations.

Like the initiator protein T antigen, but unlike most other DNA binding proteins of viruses, this newly discovered protein complex needs the addition of energy from the compound adenosine triphosphate (ATP) to work properly. In yeast the initiator protein appears to bind the DNA by wrapping the DNA around itself "and is reminiscent of the pattern observed when the *E. coli* initiation protein dna A binds to its cognate origin," says Stillman. But the development of new reagents, such as antibodies against particular protein subunits and cloning of the genes encoding the multiple protein subunits, "will help us understand its many roles in initiating DNA replication," Stillman adds.

IMPLICATIONS FOR THE FUTURE

Discovery of a yeast initiator protein complex is significant for many reasons, as has been recognized by the editors of several leading scientific publications, who rushed commentaries on this finding into print. First, the presence of at least six protein subunits suggests that the triggering of DNA replication may be controlled in a complicated way. In fact, Stillman has noted that there is precedent for the use of multiple proteins to recognize an origin of DNA replication. In 1991 it was discovered by Michael Botchan and his colleagues at the University of

California, Berkeley, that the bovine papilloma virus initiator proteins, E1 and E2, recognize their origin of replication cooperatively.

Second, Stillman suggests there may be important parallels between the mechanisms for initiating duplication in DNA and the mechanisms for controlling expression of the replicated DNA. Both may use an array of multipartite switches, in which a critical number are turned on before the system is set in operation. In particular, Stillman "proposes that an origin of DNA replication in yeast, element 'A,' provides an initiator function which, like the TATA box of a eukaryotic promoter, is essentially inactive unless stimulated by the other 'B' elements, called activator elements."

Discovery of the origin recognition complex will also allow other experiments that may reveal why the initiation of DNA replication at each origin is limited to one round only per cell cycle. It has been suggested that the initiator protein function is destroyed following each initiation event or that a newly replicated DNA strand is made insensitive to the presence of the initiator. With an initiator protein complex in hand, these ideas can be tested. In fact, Joachim Li of the University of California, San Francisco, and Bruce Alberts, formerly at that institution and now president of the National Academy of Sciences, have noted that, "By following the fate of the *Saccharomyces cerevisiae* initiator protein during the replication reaction and throughout the cell cycle, we can expect to learn how the cell prepares itself for a new round of initiation and how it prevents those preparations from occurring prematurely."

Another possibility is to use information about the yeast origin recognition protein to fill gaps in our understanding of the cell cycle (see Figure 4.6). For example, much attention has been given to the study of the various biochemical events of the G1 phase immediately prior to the synthesis of DNA. It is known that certain genes that are active during this phase produce enzymes known as kinases, which are important in the energy transfer reactions known as phosphorylations. Using the initiation protein complex as a starting point, it may be possible to work backwards to define a relationship between this complex and the important kinases produced during the preceding G1 phase. Perhaps some of the products produced in the G1 phase help with the synthesis or activity of one of the proteins that make up the origin recognition complex.

Still another strategy for learning more about the many proteins needed for eukaryotic DNA duplication is to work backwards from the yeast origin recognition complex to the genes that code for it. Joel A. Huberman, of the Roswell Park Cancer Institute, Buffalo, New York, has

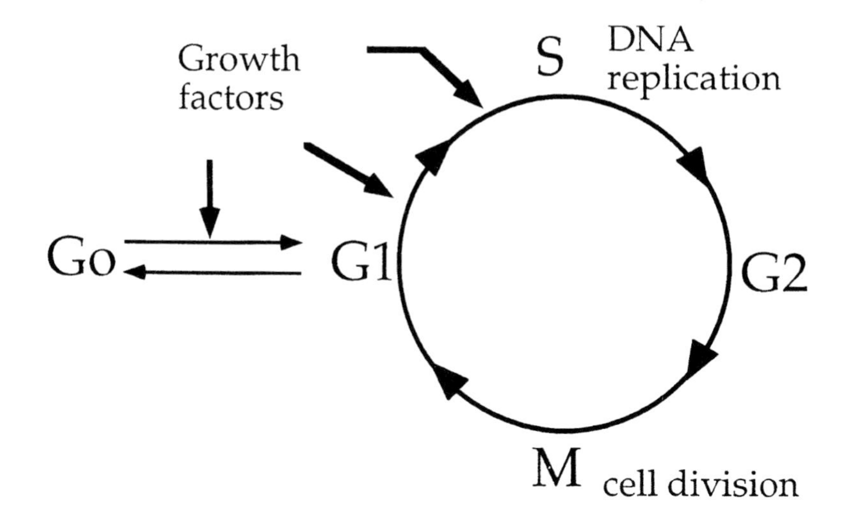

FIGURE 4.6 *Cell cycle and regulation of DNA replication. (Courtesy of Bruce Stillman, Cold Spring Harbor Laboratory.)*

- DNA replication only occurs in S phase
- Regulation at the G1 to S phase transition
- Replication only once per cell cycle
- Temporal control throughout S phase
- Checkpoint and damage controls

noted that, because Bell and Stillman have already succeeded in getting pure preparations of the origin recognition complex (ORC), "We won't have long to wait until the individual polypeptides are purified and their genes cloned." Then, Huberman adds, "It will . . . be possible to compare the predicted amino acid sequences of the ORC polypeptides to those of known proteins; any similarities that turn up will suggest roles for the individual polypeptides in the initiation process." And, add Li and Alberts, "The discovery of homologues to the ORC, and analysis of their DNA binding sites, could provide a shortcut to defining replication origins in [other] higher eukaryotes."

Finally, identification of the yeast ORC may permit a scheme for replicating eukaryotic DNA in the test tube, a procedure that would greatly ease its study. "Learning how to unleash them [the ORC] in vitro could therefore reveal how this protein is regulated in vivo," say Li and Alberts.

In short, discovery of the multiprotein initiation factor in yeast provides a long sought after key for unlocking the secrets of the genomes of advanced organisms. For the past 40 years necessity often demanded that most research on the molecular details of DNA replication and expression be done on relatively simple, prokaryotic organisms. Now, a new era of biological studies could begin, one that puts more emphasis on the workings of the genome in advanced organisms, including humans.

EPILOGUE

Just as the discovery of the T antigen, dna A protein, and yeast origin recognition complex have yielded important clues about how DNA replication starts, the discovery of other proteins has revealed other details about the later stages of DNA replication. In most cases such information has been gained by analyses of mutant microbes that fail to replicate properly and are defective in a single protein. But the search can be very difficult since some proteins are only partially active in isolation and may have detectable activity only if found with large and, often elusive, protein complexes. Still, despite such difficulties, knowledge of the large number of proteins needed to complete DNA replication in *E. coli* is quite detailed, and there has been reasonable, but slow, progress in defining those proteins that have a role in the replication of eukaryotic DNA.

Here is a rundown of the key accessory proteins—sometimes called the protein machine—that help DNA to copy itself. Although the polypeptide components of the machinery may differ in detail from organism to organism, enough is known about them to conclude that some key polypeptides fill certain roles in DNA replication in all organisms. The proteins described here go into action after the initiator proteins described earlier.

Proteins That Keep Replication in Order

Once DNA is unraveled and replication begins, the active site of replication advances in both directions along the DNA strand until it meets an advancing replicating segment from a neighboring origin or the entire chromosome is copied. The spot where the newly synthesized

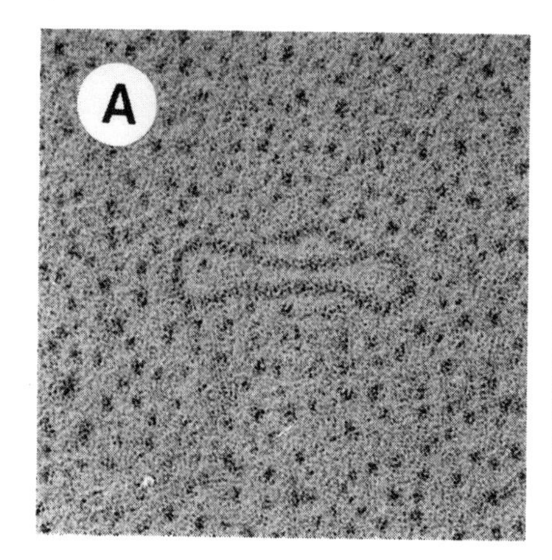

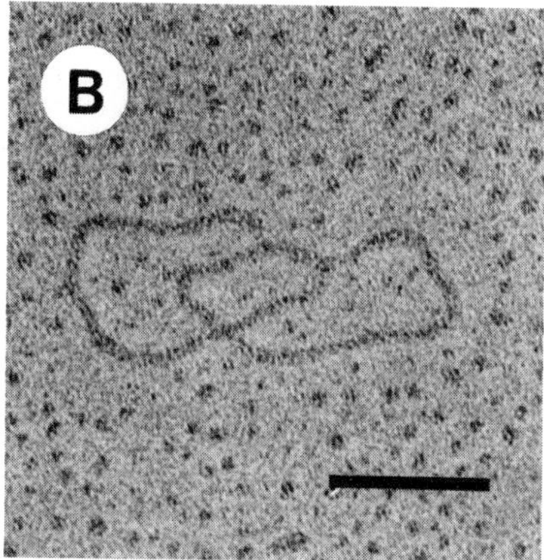

FIGURE 4.7 *Electron microscopy of in vitro replication intermediates. DNA was mounted for electron microscopy by the method of Davis et al. The extent of replication of molecule A is approximately 20 percent and that of molecule B is approximately 90 percent. Bar = 0.2 µm. (Reprinted by permission from Li and Kelly, 1984.)*

strands meet the original unreplicated region is called a replication fork (see Figure 4.7). At each growing fork, one new DNA strand, called the leading strand, grows continuously in the same direction of the fork; the second strand, the so-called lagging strand, grows in the opposite direction (see Figure 4.8). At the critical juncture of the growing fork, a large collection of accessory proteins is needed to keep the complex process in order.

Helicases and Single-Stranded Binding Proteins

Among the first proteins to go into action are the helicases, which unwind the DNA and prepare the template strand for copying. These enzymes are found at every replication fork. A related protein, known as the single-stranded DNA binding protein or the helix destabilizing protein, also comes into action early on in replication. It binds to a site where the double helix has unwound but is not yet copied. It is thought to work by preventing the reformation of double helix, single-stranded tangles, or odd structures, such as hairpin loops, all of which would upset replication. The single-stranded DNA binding protein speedily dissociates as a different enzyme, DNA polymerase, approaches.

The Polymerases

The DNA polymerases were first recognized for their ability to add deoxynucleotides into DNA. All DNA polymerases discovered so far can only elongate preexisting chains: they can't start them. These enzymes were first purified from *E. coli*, where at least three different

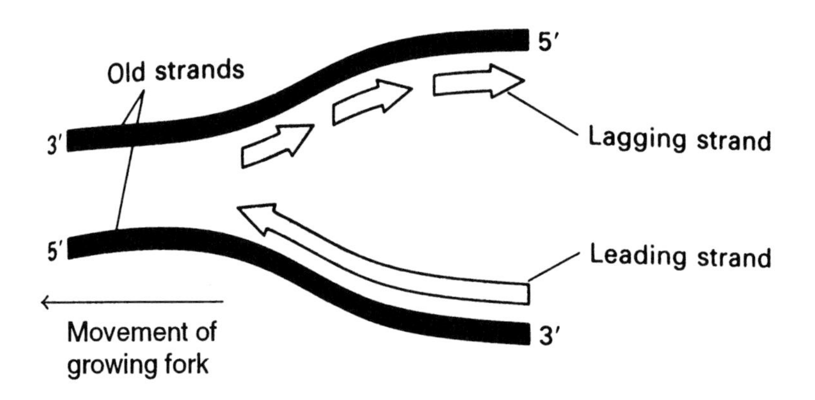

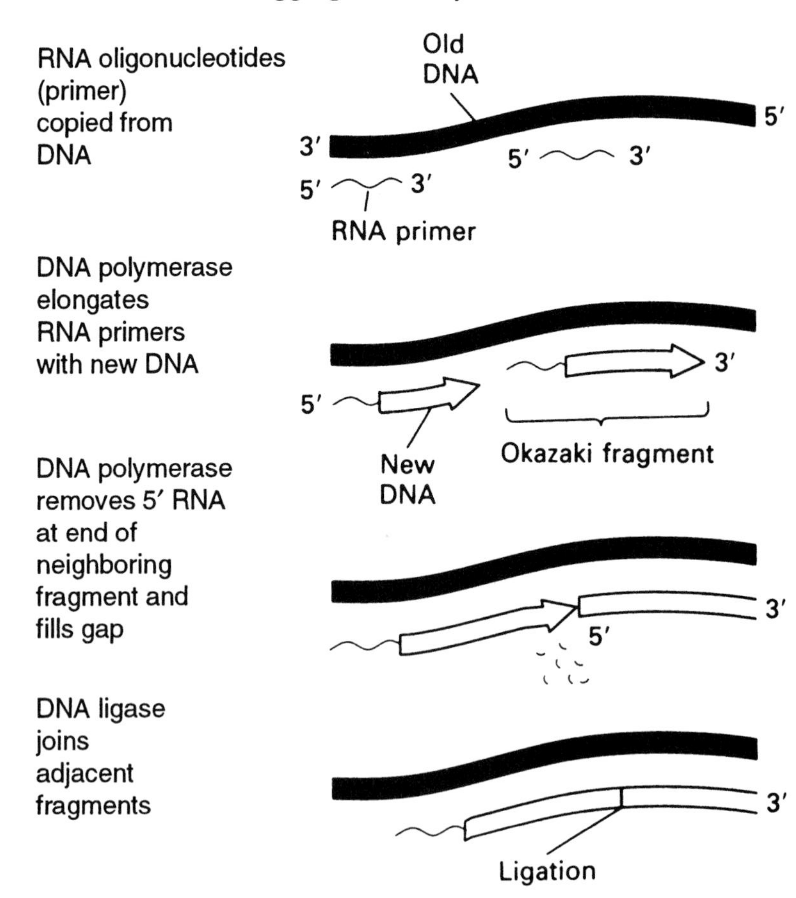

FIGURE 4.8 *The overall structure of a growing fork (top) and steps in the synthesis of the lagging strand. (From* Molecular Cell Biology *by J. Darnell, H. Lodish, and D. Baltimore. Copyright © 1990 by Scientific American Books, Inc. Reprinted with permission of W.H. Freeman and Company.)*

polymerases have been described, and others have since been extracted from more advanced organisms, too. In fact, many years after DNA polymerases were purified from *E. coli*, four were identified from mammalian cells. Some of their roles are known, although they remain to be as well described as their cousins in *E. coli*.

In *E. coli* one complex polymerase, known as the DNA polymerase III holoenzyme, is responsible for the elongation of DNA chains at the growing fork. It is made up of two multiunit polymers, and it speeds along the parent chain, residue by residue, assisting with the addition of nucleotides, one by one. One segment of the complex, known as the alpha subunit, seems to catalyze the addition of individual nucleotides, and it does so most efficiently, at a rate inside the cell of about 1000 nucleotides per second. This is one of the fastest polymerization reactions known, in vivo. One factor that assists DNA polymerase III is its tenacity. It is believed that the giant asymmetric polymerase III clamps on to the growing DNA chain throughout the polymerization, instead of dissociating and binding randomly to other growing chains.

Exonucleases, Polymerases, and Ligases That Piece Together the Leading and Lagging Strands

As mentioned earlier, two different types of DNA strands are made at the replication fork: the leading strand builds in the direction of the replication fork, and the other, the lagging strand, builds in the opposite direction. These two strands are made by different means, requiring different enzymes.

The primer for synthesis of the leading strand is thought to be a loose end from a newly made strand at the origin of DNA replication, which then grows continuously. In contrast, synthesis of the lagging strand is carried out in a discontinuous manner. Each DNA fragment on the lagging strand, known as an Okazaki fragment, is begun with a ribonucleic acid (RNA) primer. Many such fragments are attached to primers extending along the lagging strand template and, eventually, the RNA primers are cut off by enzymes known as exonucleases. Then what comes into play is a particular type of DNA polymerase, known as DNA polymerase I, that fills in the gaps left by the removal of the RNA primers by adding deoxynucleotides. Finally, the various pieces of the lagging strand are cleanly sealed with the aid of another enzyme, DNA ligase.

The Replisome Protein Complex Keeps All in Order

Keeping synthesis of the leading and lagging strands in phase is nontrivial, and it is thought that a special protein complex, called the replisome, is needed to orchestrate fast and clean synthesis. It is made of a large collection of accessory proteins: Two DNA polymerase III holoenzymes, one for the leading strand and another for the lagging strand; a primosome complex, which includes the enzyme primase (the enzyme that makes the short RNA primer on the Okazaki fragments); dna B helicase, a second helicase (sometimes called a rep protein), the single-stranded DNA binding protein; and others. The concerted action of these various accessory proteins keeps synthesis of the leading and lagging strands clean flowing.

The Topoisomerases Adjust Swivel to DNA

The protein apparatus of the replisome appears to prevent tangles at the outset of synthesis, but another set of proteins, the topoisomerases, probably help resolve tangles that are created later in synthesis and exist in newly replicated DNA: they seem essential to the successful conclusion of replication and the separation of finished chromosomal copies.

It is well established that DNA has an unusual twisted, double helical structure. The tightness of its coiled form is described in terms of the linking number, which describes the number of times one strand of a helix winds around another; twisting, which refers to the periodicity of the winding; and writhe, the reciprocal of twist. Typically, the linking number equals twist plus writhe. Underwinding yields negative supercoils, and overwinding yields positive supercoils. Once it was realized that DNA replication and its subsequent packaging into chromosomes depended on supercoiling, a search began for those enzymes that could affect DNA topology. They are the topoisomerases.

The first of these to be discovered was the omega protein in *E. coli*, which can remove negative supercoils. This type I topoisomerase does this by breaking a single strand of DNA, passing the intact strand through the gap, and then resealing the gap, thus changing the extent of strand interwinding. If this process is carried out repeatedly, supercoiled DNA can be relaxed. The so-called *E. coli* topo I is less active on positively supercoiled than on negatively supercoiled molecules. However, those topo I enzymes found in eukaryotes are more versatile: They relax both positive and negative supercoils. Bacteria and yeast with mutant topo I enzymes suffer a slower rate of growth, but the mutant is not lethal. These enzymes also seem involved in the transcription or "reading" of DNA.

Another group of topoisomerases, found in all cells, is called Type II or, in bacteria, gyrase. They can change the linking number by cutting and then resealing a split double helix. There appear to be two forms of the Topo II, alpha and beta, which differ in a number of biochemical properties. The alpha variety seems to bind better to DNA segments that are rich in the bases adenine and thymine; and the beta variety prefers DNA sequences that are rich in the bases guanine and cytosine. Most important, the topoisomerases appear essential for segregating the replicated daughter chromosomes, and they may also be needed to resolve complicated, interlocked, or knotted DNA strands.

Using Protein Editors to Correct Errors

An unusual feature of replicating DNA that requires the help of yet more accessory proteins is its ability to correct errors. DNA is the only cellular macromolecule that can be self-repaired, probably because the cost of damaged DNA outweighs the costs of repair to the cell. If accumulated errors, both those made during the act of replication and those brought on by environmental factors, such as thermal stress and chemical mutagens, are left uncorrected, growing cells might accumulate so much genetic misinformation that they could no longer function. Faulty proteins would be produced at an unacceptable rate, barring normal cellular activities. Moreover, DNA in sex cells might be so damaged as to prevent the production of viable offspring. So a system is needed to identify errors in DNA and to make corrections.

What happens in *E. coli*, for example, is that one segment of the DNA polymerase III holoenzyme, the ε subunit, proofreads newly synthesized DNA by recognizing the distortions produced by an incorrectly paired base. This is probably done by some physical means where the enzyme detects a malformation in the three-dimensional structure of the DNA. For example, if ultraviolet radiation, which can disrupt thymine bases by causing the dimerization of their pyrimidine structures, skews a nascent DNA strand, this can be "felt" by a probing enzyme. The damaged bit is then cut out by a special enzyme, known as an endonuclease, before extension of the strand is complete. Repairs are then made by the now familiar enzymes, DNA polymerase I and DNA ligase. Such inspections and repairs are made with exceptional efficiency, assuring that DNA is replicated with great fidelity. For example, a wrong base is added by the holoenzyme to the growing chain about once every 10,000 steps, and the correction activity itself errs about once every 1000 steps. This gives an overall error rate of 10^{-7}, one of the lowest rates for any enzyme.

DNA repair in more advanced organisms is less well understood. But studies of humans with xeroderma pigmentosum, a malady characterized by easily pigmented skin and vulnerability to skin cancer, has suggested that they lack the ability to repair DNA that has been damaged by ultraviolet radiation. Some other diseases, such as the leukemias, Franconi's anemia and Bloom's syndrome, are associated with heightened sensitivity to agents that can damage DNA.

Additional progress in learning about DNA repair in eukaryotes has come from studies of mutant yeast and Chinese hamster cells. For example, yeast mutants sensitive to damage by ultraviolet light have been used to locate several radiation-sensitive genes, including one that directs production of a helicase, an enzyme that unwinds DNA and is essential to the normal replication. Moreover, ultraviolet-sensitive Chinese hamster cells have been manipulated to reveal a human gene that is involved in DNA repair.

Both these lines of study may reveal more molecular details about DNA repair in advanced organisms, including humans. But they could also offer an explanation of how uncorrected damage to DNA leads to the uncontrolled proliferation of DNA and cell growth characteristic of cancer. Such research offers a clear link between the erudite aspects of DNA synthesis described earlier and the more general goal of using basic science to reduce human suffering.

BIBLIOGRAPHY

Bell, S. P., and B. Stillman. 1992. ATP-dependent recognition of eukaryotic origins of DNA replication by a multiprotein complex. Nature 357(May 14):128–134.

Diffley, J. F. X., and B. Stillman. Dec. 1990. The initiation of chromosomal DNA replication in eukaryotes. Trends in Genetics 6(12):427–432.

Darnell, J., H. Lodish, and D. Baltimore. 1990. Molecular Cell Biology, Chapter 12. Scientific American Books. pp. 449–487.

Fangman, W. L., and B. J. Brewer. 1992. A question of time: Replication origins of eukaryotic chromosomes. Cell 71(Oct. 30):363–366.

Glotzer, M., A. W. Murray, and M. W. Kirschner. 1991. Cyclin is degraded by the ubiquitin pathway. Nature 349(Jan. 10):132–138.

Huberman, J. A. 1992. Quest's end and question's beginning. Current Biology 2(7):351–352.

Li, J. J., and B. M. Alberts. 1992. Eukaryotic initiation rites. Nature 357(May 14):114–155.

Li, J. J., and T. J. Kelly. 1984. Proceedings of the National Academy of Sciences USA 81:6976.

Marahrens, Y., and B. Stillman. 1992. A yeast chromosomal origin of DNA defined by multiple functional elements. Science 255(Feb. 14):817–823.
Murray, A. W. 1992. Creative blocks: Cell-cycle checkpoints and feedback controls. Nature 359(Oct. 15):599–604.
Smith, P. J. 1990. DNA Topoisomerase dysfunction: A new goal for antitumor chemotherapy. BioEssays 12(4).
Tsurimoto, T., T. Melendy, and B. Stillman. 1990. Nature 346(Aug. 9):534–539.

RECOMMENDED READING

Murray, A. W., and M. W. Kirschner. 1991. What controls the cell cycle. Scientific American (March):56–63.
Stillman, B. 1989. Initiation of eukaryotic DNA replication in vitro. In Annual Reviews of Cell Biology 5:197–245.

MAGELLAN'S VENUS
A World Revealed

by Andrew Chaikin

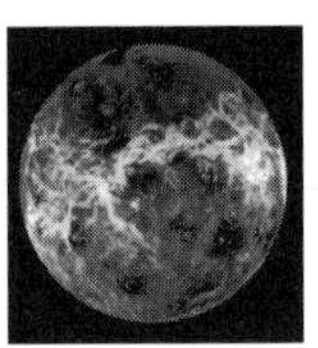

August 1990 was a time of shrinking horizons for planetary scientists. For a generation, unmanned space probes had revealed the most spectacular places in the solar system. They had orbited Mars and photographed volcanoes nearly three times as high as Mt. Everest, a canyon that would stretch the length of the United States, and dry river valleys that spoke of an ancient watery epoch on this desert world. They had flown past Io, the tiny moon of Jupiter, and found it to be so volcanically active that it is literally turning itself inside out. And, most recently, in 1989, Voyager 2 had made a thrilling reconnaissance of Neptune and its moon Triton, an ice world at the edge of the solar system. Indeed, every major world in the sun's family (save Pluto) and a bevy of moons had been orbited, flown past, or landed on. A great era of discovery had ended. In contrast to the budgetary heyday of the 1960s that had spawned these voyages, austerity was now the rule. With only a handful of missions approved for the coming decades—often fraught with delays and budget cuts—planetary scientists seemed to have little to look forward to. The arrival of a robotic explorer called Magellan in orbit around Venus in August 1990 seemed almost like an afterthought.

Except, that is, to the scientists who had spent the better part of two decades working to make this happen. To them, Magellan offered a unique chance to study the one planet long considered by many scientists to be most like our own. Equipped with a radar-imaging system to pierce the cloudy veil, Magellan was designed to map the planet's surface in unprecedented detail. High on the list of the Magellan science team's questions was the most intriguing one of all: Is Venus still geologically active?

To be sure, much was known of Venus prior to Magellan's arrival. Soviet landers had proved that it is a place of broiling heat and crushing atmospheric pressure. The planet's surface features had been glimpsed in the preceding two decades, first using Earth-based radar and then from orbiting Soviet and American probes. But Magellan's pictures showed features as small as a few hundred meters across, far more detail than any previous views of Venus—in fact, better resolution than that available for vast areas of Earth, namely the ocean basins. From the first images the spacecraft radioed to Earth, the Magellan team of scientists—many of whom were veterans of several planetary missions—were stunned. The sheer variety of features was breathtaking. Everywhere, it seemed, there were volcanoes, ranging in size from tiny peaks at the limits of resolution to giant shield volcanoes hundreds of kilometers across. The smaller variety, several kilometers in diameter, numbered in the tens or even hundreds of thousands. Most of the planet's surface was covered with smooth, flat plains, also attributed to outpourings of molten lava. At first glance, it seemed that much of Venus' history was written in volcanism.

Elsewhere, huge belts of fractures sliced through the equatorial plains, extending nearly all the way around the planet. Looking closely, the scientists saw much finer fractures distributed over almost the entire planet. Elsewhere, they found complex jumbles of hills and valleys, now known as tesserae, that seemed to be in the process of destruction by tectonic forces. And there were some features found nowhere else in the solar system: giant welts called coronae, whose margins are jumbles of fractured rock but whose centers seem to have been formed by the buildup of lava. The smallest coronae are less than 100 kilometers across; the largest, christened Artemis, is 2500 kilometers in diameter, roughly the size of Greenland. Even more mysterious was an incredibly long and narrow channel, nicknamed the River Styx by Magellan scientists. Roughly 2 kilometers wide and some 6800 kilometers long, more than the distance from Havana, Cuba, to Anchorage, Alaska, the River Styx is a puzzle: What carved it? With no water available, geologists speculated on a variety of other fluids, including liquid sulfur; today molten lava of an unusual composition is considered most likely.

The biggest surprise, however, was the crispness of the landscapes in nearly all the Magellan images. Almost every feature seemed to have been preserved in pristine condition from the time of its formation. The surface showed virtually no evidence of erosion, a fact attributed to the absence of water in Venus' atmosphere (and, to a lesser extent, to the calmness of the surface winds). In some places, different sets of features were superimposed on top of each other, like a multiple-exposed photograph. It was immediately evident that the surface of Venus had a profoundly rich and detailed story to tell. By the middle of 1991, Magellan had finished its first 243-day mapping cycle and had begun its second, and the scientists were overwhelmed. "The trouble with this mission," said one Magellan team member, "is that we get so much data it's mind-boggling." Said another, "It's as if we'd never seen Earth, and we lived on the far side of the moon, and all of a sudden we've got a 100-meter-resolution picture. . . . It's a whole new planet out there." Today, more than 3 years after Magellan began its mission, decoding that story has challenged some of their most basic preconceptions about Venus and about the nature of planetary evolution.

A VEILED WORLD

Like most scientific pursuits, astronomy demands more of those who practice it than curiosity, and from the beginning Venus has challenged astronomers' abilities to be detectives. A near-twin of Earth in size (Venus' diameter is 12,102 kilometers; Earth's diameter is 12,756), mass (81.5 percent of earth's mass), and bulk composition, Venus is our nearest planetary neighbor;[1] for many decades it was also the most mysterious. Veiled by opaque clouds composed mostly of carbon dioxide, Venus displays only a featureless brilliant white disk in the telescope. Even to the unaided eye, however, it is among the most beautiful of celestial objects; gracing predawn or early evening skies, it can shine brightly enough to cast shadows.

Before the space age, some astronomers speculated that beneath the clouds Venus might be a steamy tropical world much like prehistoric Earth, but nothing could have been further from the truth. In 1963 Mariner 2, history's first planetary probe, showed that temperatures on Venus were too high to support liquid water. A decade later, Mariner 10 surveyed the Venusian upper atmosphere and clocked winds up to 300 kilometers per hour (Pasachoff, 1990, p. 179). Remarkably, while Venus rotates once every 243 days, its atmosphere whips around the planet in the other direction, once every 4 days.

To find out what lay beneath those clouds, astronomers had only one tool: radar. By the early 1970s, they were using radio telescopes like the 300-meter dish at Arecibo Observatory in Puerto Rico to bounce radar signals off the planet's surface. Good observations were possible only during inferior conjunction, the period when Venus is closest to Earth. Because Venus always shows the same face to Earth at this time—a result of a near commensurability of Venus' rotation rate and Earth's orbital period—Earth-based radar surveys were limited to about 25 percent of the planet. Images made from the radar data showed just enough to tantalize: circular features that seemed to be impact craters, bright spots thought to be volcanoes, and banded terrain that might be the result of crustal folding. Clearly, Venus' surface had been shaped by a variety of geologic processes, but no one would be able to say more without getting a better look.

Meanwhile, in 1975 the Soviet Union scored an astonishing success when two unmanned probes, Veneras 9 and 10, parachuted through a dense atmosphere laden with droplets of sulfuric acid and touched down on the torrid surface. The Veneras found surface temperatures of 450° C and pressures of 90 bars (i.e., 90 times the pressure at the surface of Earth). And in the brief minutes before they succumbed to these hellish conditions, the probes sent back black and white images of a rock-strewn surface. At Venera 9's landing site the rocks were surprisingly sharp edged, testifying to an almost total absence of erosion, while Venera 10's images showed rocks with a pancake-like appearance. In 1982 the Soviets bettered this spectacular achievement with Veneras 13 and 14, each of which survived on the surface for more than 2 hours. In addition to sending back higher-quality images, including some in color, each lander actually drilled a sample of the surface material and brought it inside a sealed chamber for chemical analysis. The results indicated a composition consistent with basalt, the iron- and magnesium-rich volcanic rock most common on the surfaces of the earth and moon.

As is so often the case in science, a single piece of information—that there is probably basalt on Venus—had great significance. Geologists believe that Earth coalesced from a swarm of smaller bodies called planetesimals some 4.6 billion years ago. Intense heat from impacting debris, and from the decay of radioactive isotopes contained in the planetesimals, melted the interior, allowing it to separate into layers of different composition. Heavier elements such as iron and nickel sank to the middle to form the molten core, while lighter species, such as the calcium-rich mineral feldspar, floated to the top and cooled into a primordial crust. In the middle, iron- and magnesium-rich minerals like pyroxene and olivine formed the hot and plastic layer called the mantle. This process, called differentiation, is the first major step in a planet's evolution from a primitive body. Laboratory experiments have shown that basalt is derived by partly melting dense mantle rocks. On Earth, basalt covers 70 percent of the surface: it forms the crust underneath the oceans, emplaced by the continuous outpouring of molten rock at the mid-ocean ridges. We know from the Apollo lunar samples that basalt is also pervasive on the moon's smooth dark lowland regions called the *maria*, proving that these areas formed volcanically. The bright highland areas, on the other hand, are rich in calcium feldspar and are thought to be the remnants of the primordial crust. Clearly, the moon also underwent differentiation. And from the Venera analyses, scientists could say confidently that Venus too had reached that important evolutionary milestone.

A MATTER OF HEAT

But that came as no surprise. Indeed, scientists would have been astonished had Venus proven to be a primitive body, simply because of its size. Ultimately, a planet's geologic history is largely shaped by its efforts to get rid of the excess heat in its interior. One important factor in determining a planet's initial heat supply, and how fast that heat can be shed, is size. How fast can a planet generate heat? That depends on its volume: bigger planets start out with more heat left over from the process of accretion, as well as a greater supply of heat-producing radioactive elements. But the rate at which a planet loses heat is proportional to its surface area. Smaller planets, whose ratio of surface area to volume is higher, deplete their heat supplies sooner than large ones.

A survey of the rocky worlds that populate the inner solar system, the so-called terrestrial planets,[2] illustrates a direct relationship between size and degree of geologic activity. Our moon, 3476 kilometers across, started out with less heat than Earth did. It also cooled faster, mostly by means of conduction through the rigid outer shell and to a lesser extent by the volcanic outpourings that formed the *maria*. Most scientists agree that by around 3.2 billion years ago the moon's internal fires had cooled enough to end volcanic activity. From that point on, the moon was considered geologically moribund. Its interior may have remained warm for some time, but there was no longer any expression of interior activity on the surface. Unlike Earth, where volcanic activity, erosion, and deformation of the crust have erased all but traces of its earliest history, the moon today is a museum. Its surface still displays the scars of its infancy: countless impact craters formed by bombardment from meteoroids and asteroids, the leftover debris from the solar system's formation. The largest craters on the moon, termed impact basins, are hundreds of kilometers across. The smallest, pinpricks blasted by micrometeorites, are visible on the surfaces of lunar rocks.

Mercury, the next-largest terrestrial planet (4878 kilometers in diameter) has a cratered, moon-like surface that suggests it also did not evolve much beyond chemical differentiation and *mare* volcanism. But Mars, still larger (6786 kilometers across), is clearly more evolved than Mercury and the moon. While it too has heavily cratered areas resembling the lunar highlands, Mars's spectacular volcanoes, canyons, and dry river valleys attest to a more extensive and varied geologic history. But is Mars still active today? That question might seem impossible to answer in the absence of rock samples to provide absolute age measurements (a handful of meteorites are thought to have been blasted off the Martian surface but probably very early in the planet's history). However, a technique called crater counting allows geologists to estimate the ages of planetary surfaces from photographs alone, in particular by measuring the frequency of craters of different sizes. The principle is simple: like a target in a shooting gallery, a planetary surface will display more craters the longer it is exposed to bombardment from cosmic debris. For example, the lunar highlands, which date back to the formation of the solar system 4.6 billion years ago, are much more heavily cratered than the *maria*, which formed roughly a billion or more years later. With some assumptions about the flux of impacting bodies over time, geologists have estimated that on Mars the oldest cratered terrains are quite a bit younger than the moon's, having formed roughly 3.8 billion years ago. The freshest surfaces, including some volcanic plains, may have formed in very recent geologic time. It is possible that Mars's geologic activity has continued to the present but at very low levels. The simple pattern that emerges is that the larger terrestrial planets held on to their heat longer and thus had longer geologic lifetimes.

What of Earth? Roughly twice the size of Mars, Earth is still vigorously active, and unlike the other terrestrial planets has evolved well beyond chemical differentiation and simple volcanism. Earth manages its heat budget by means of an "engine" that we call plate tectonics. The surface of Earth is divided into a set of about 10 moving, interlocking plates, some thousands of kilometers across, whose interactions with each other and the interior account for most of the major crustal, or tectonic, processes. This engine is powered by convection within the hot plastic mantle. Because Venus is so similar to Earth in size, scientists speculated that it too might still be geologically active. In fact, Soviet landers had measured the concentrations of the radioactive isotopes of uranium, thorium, and potassium in Venus' rocks and found that they contain the same amounts as terrestrial rocks, suggesting that Venus' interior is as hot as Earth's. Could Venus have developed plate tectonics? That question would go unanswered until scientists could obtain a global view of its surface.

ORBITERS LIFT THE VEIL

In 1978 NASA's Pioneer Venus Orbiter (PVO) was dispatched to monitor Venus' atmosphere and to map the planet's surface elevations using a radar altimeter. PVO proved to be a true stalwart; it transmitted data to Earth until the fall of 1992, when it burned up in the atmosphere. A decade before that, however, the altimeter had mapped more than 90 percent of the planet with a resolution of 100 kilometers or better, enough to show gross topographic differences. PVO's portrait showed that 60 percent of the surface is pool-table flat, in no place deviating from the mean planetary radius of 6051 kilometers by more than 500 meters. Rising up from these rolling plains are a handful of highland regions, the largest of which, Aphrodite Terra,[3] is about the size of Africa. Another highland area, named Ishtar Terra, is roughly the size of Australia and is one of the most geologically spectacular regions on the planet. While the western part of Ishtar is a vast plateau some 2000 kilometers across, its eastern portion contains a range of mountains that tower up to 12 kilometers above the planet's mean radius—greater (34 percent more) than the height of Mount Everest above sea level. Ishtar bears a tantalizing resemblance to the Tibetan plateau, where due to plate tectonics India is colliding with Asia, pushing up the Himalayas in the process.

At first glance, Ishtar and the other highland regions seem to suggest a similarity to Earth's continents. But a closer look at the data reveals a fundamental difference. The *hypsometry* of Earth, that is, the statistical distribution of its surface elevations, is strongly bimodal: the continents cluster around 0.5 kilometers above sea level; the ocean basins at about 4.5 kilometers below it. But Venus' hypsometric profile is unimodal. Truly high regions are rare; only 5 percent of the surface lies at elevations greater than 2 kilometers above the mean radius. Lowlands are even rarer; less than 1 percent of the surface lies more than a kilometer below it. The total relief on Venus, about 13 kilometers, is similar to that on Earth, but there are few sharp transitions. At least on a scale of tens of kilometers, most features on Venus grade smoothly into one another.

This has important implications for Venus' surface composition. On Earth, composition is closely linked to elevation, since continental crust is rich in granite and oceanic crust is basaltic; since granite is less dense than basalt, the continents tend to be more buoyant than the ocean basins. Purely on the strength of Venus' hypsometry, geologists have proposed that its surface composition is probably fairly uniform and, as the Venera landers showed, basaltic.

Furthermore, absent on PVO's map were clear signs of features suggesting plate tectonics. There were no analogs of mid-ocean ridges or subduction zones. Attempting to explain an absence of plate tectonics, scientists noted the planet's high surface temperature, which heats surface rocks roughly halfway to their melting points. As a result, researchers hypothesized, the Venusian lithosphere would be substantially more plastic than Earth's and would lack the rigidity necessary to form a series of global interlocking plates. It seemed likely that if Venus were still active, it would have to have developed a form of geology substantially different from that of Earth. PVO's data, however, lacked the resolution to resolve the uncertainty.

In 1983 the Soviet Union returned to Venus, this time with a pair of orbiters equipped with imaging radar. Veneras 15 and 16 mapped about 25 percent of the northern hemisphere of the planet showing features as small as a few kilometers, comparable to the best Earth-based radar views. Their images gave scientists a first glimpse of Venus' impressive array of volcanic and tectonic landforms. It remained for Magellan to finish the unveiling.

CONTROVERSY IN CRATERS

If there were any doubts that Venus had experienced a rich geologic history, Magellan removed them. As expected, Magellan found no signs of active plate tectonics on Venus, and no unambiguous signs that such activity had ever occurred in the past. It did not verify the hypothesis of Brown University's James Head and Larry Crumpler that Aphrodite might turn out to be an analog of a terrestrial mid-ocean ridge, where new crust forms as two plates are pulled apart. It was also clear that deciphering the extraordinarily detailed history written on the Venusian surface was going to be a long and difficult job. The first and simplest step was to attempt to establish the age of the surface by means of crater counting. The Venera images had already suggested a paucity of craters on Venus, and Magellan gave confirmation. In part, this is due to the fact that Venus' atmosphere has a density only 40 times less than the meteoroids that enter it. Projectiles 100 meters or less in diameter are destroyed before they can strike the surface. In some cases, meteoroids appear to have exploded only a few kilometers above the ground, producing round blotches of apparently fractured terrain that show up as dark spots in Magellan's radar images. But in general, craters smaller than about 35 kilometers, which make up the vast majority of the craters on the moon and other terrestrial planets, are far more scarce on Venus. Craters less than about 2 kilometers across are absent altogether. This reduces considerably the size of the statistical sample available to crater-counting geologists and makes any estimate of Venus' surface age more uncertain. Nevertheless, some 912 craters have been identified, and according to Gerald Schaber of the U.S. Geological Survey in Flagstaff, Arizona, the story they tell is extraordinary. For one thing, they appear to be spatially random, precisely the distribution one would expect if the entire planet were plunked down in the cosmic shooting gallery and allowed to accumulate craters. But for how long? How old is the surface of Venus? According to Schaber, the number and size of the craters are consistent with an average age of about 500 million years—give or take a few hundred million. It could not have been a more provocative result, for it suggested that a few hundred million years ago the slate of Venusian history was wiped clean, then begun anew.

What could have obliterated all traces of geologic evolution for the previous 4.5 billion years? Schaber, along with the University of Arizona's Robert Strom and several colleagues proposed that Venus suffered an episode of volcanic activity so pervasive and so intense that nearly the entire planet was covered in fresh lava. In recent years, geologists have theorized that massive volcanic outbursts resurfaced parts of Earth, including a basalt "mega-eruption" 250 million years ago that covered much of what is now Siberia. But Schaber says, "We're talking about volcanism on Venus on a scale that makes even those pale in comparison." Furthermore, Schaber found that the vast majority of Venusian craters—some 61 percent—appear pristine. Of the remaining craters, only 4 percent have been invaded by newer lava flows (Figure

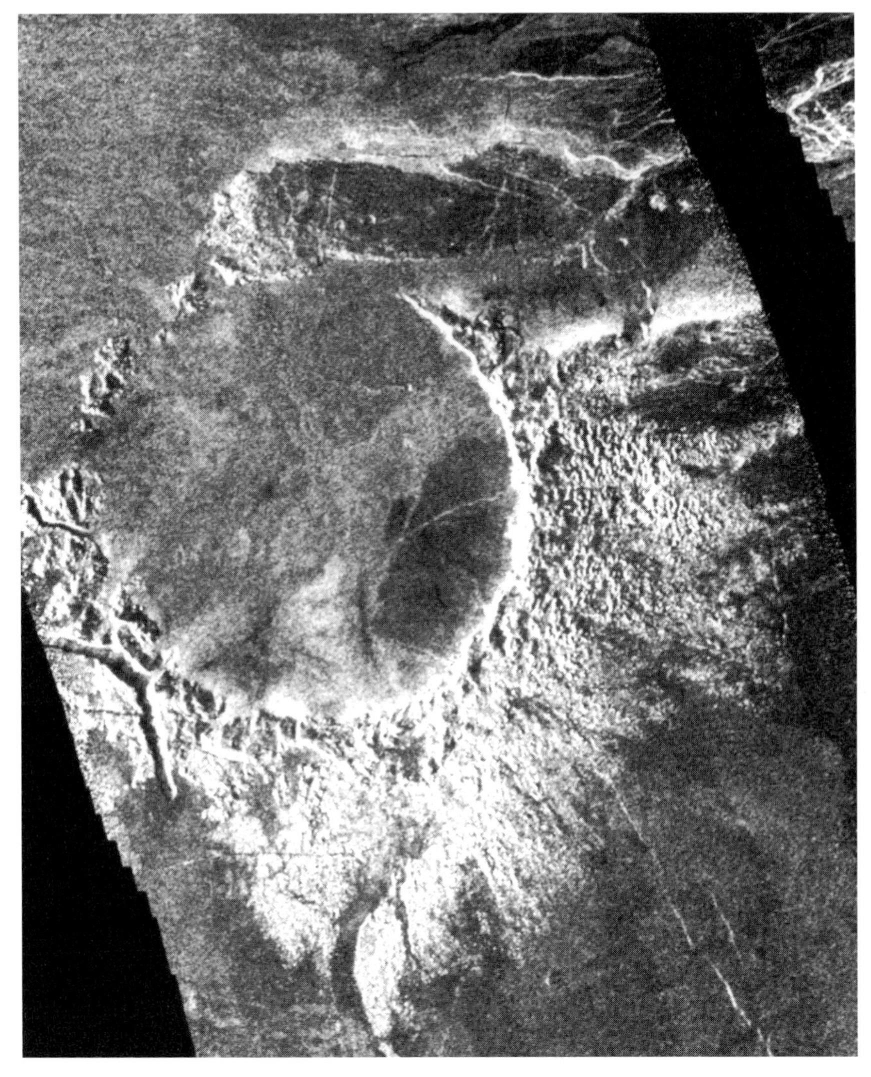

FIGURE 5.1 *Craters like this one, flooded by one of Venus' pervasive lava flows, are relatively rare in Magellan images. Bright features in this and other Magellan images are better reflectors of radar energy, usually due to steeper slopes or higher surface roughness. (Courtesy of the Jet Propulsion Laboratory, NASA.)*

5.1). That suggests that the catastrophe ended abruptly; otherwise, more partly buried craters would be seen. Since the resurfacing event, Schaber proposed volcanic and tectonic activity could continue at reduced levels, especially along the fracture belts that connect the broad low rises in the equatorial highlands.

The main attraction of Schaber's so-called global resurfacing model was that it neatly explained the observed crater distribution. And yet the idea that the planet completely repaved itself and then became relatively quiescent was immediately controversial. At the very least, the idea seemed to explode the notion that Venus and Earth had traveled similar evolutionary paths. At worst, it raised troubling questions. What could have caused the enormous volcanic outburst Schaber was proposing? Why did the subsequent level of activity drop so sharply afterward? For that matter, was the 500-million-year age implied by the cratering statistics valid to begin with?

"It's important not to get trapped in statistics," said Roger Phillips, now at Washington University in St. Louis. In 1991 Phillips, who had collaborated with Schaber on initial analyses of the Magellan results, challenged his colleague's hypothesis. The distribution of craters on Venus may be *random*, he stressed, but that doesn't mean it represents a surface of uniform age. In fact, he pointed out, the planet has areas where craters are noticeably more common than average, and others where they are relatively sparse. Now, that kind of "clumpiness" occurs in any random sample. But was it possible that Venus' craters weren't randomly placed after all, that these clumps and gaps were telling us something about the planet's evolution?

As a geophysicist, Phillips doesn't normally concern himself in great detail with the pictures taken by a planetary probe. While geologists study what has happened on the surface of a planet, geophysicists like Phillips are more concerned with what has gone on inside it. But Magellan has required geophysicists like Phillips to cross over into geology, to try to make sense of Magellan's strange findings.

By the fall of 1991 Phillips and some of his colleagues had spent time studying the Magellan images, in particular looking for craters that had been cracked by faults or partially invaded by lava flows. To be sure, such modified craters are in the minority on Venus; like other landforms, most craters look as fresh as the day they were formed. But to Phillips's surprise, work by his collaborator Richard Raubertas at the University of Rochester showed that the modified craters were not randomly sprinkled around the planet but concentrated in areas where "normal" craters were few in number. The connection was telling. "To me, that's the first clue that those areas of low crater density are not that way due to chance, but in fact, something is eating craters on Venus." Phillips proposed that if craters on Venus were being *destroyed* randomly over time—here, buried by a lava flow; there, chewed to bits by a fault zone—the result would still be a random distribution of craters but on surfaces with many different ages. The only constraint, Phillips said, was that each of these small resurfacing events must act in such a way as to produce a crater distribution that appears random. (In 1992, as a "mathematical convenience," Phillips considered a model in which there was an equilibrium between resurfacing and crater production; he later rejected this model.)

But planetary scientists are hardly in agreement; each hypothesis has its supporters and skeptics among the Magellan scientists. Phillips' computer simulations of Venus' evolution, designed to show that a spatially random crater distribution could be preserved despite small scattered resurfacing events, met with mixed reviews. As Schaber says, "The cratering record is the easiest [aspect of Venus' history] to interpret, but, boy, it sure has caused a lot of controversy."

In part, the controversy stems from Venus' relatively small number of craters and the ambiguity that results from counting them. But there is another factor, namely the differing perspectives of the geologists, like Schaber, and geophysicists like Phillips. Two geophysicists who are attempting to unravel the mystery are Suzanne Smrekar of Caltech's Jet Propulsion Laboratory and Robert Grimm of Arizona State University. Officially, they are not members of the Magellan team, and being in their early thirties, they are considerably younger than most of those researchers; they belong to a new generation of planetary scientists. "It's important to realize," Grimm says, "that what we see on the surface isn't the whole story."

As a geophysicist, Grimm finds the global resurfacing model difficult to accept. "Here's Venus going like gangbusters, and suddenly it comes grinding, screeching, choking to a halt," he says, explaining that the current level of geologic activity is only a tiny fraction of what would have been required during the resurfacing event. "That's geophysically curious to me."

Grimm stresses that, while Schaber's global resurfacing hypothesis is the simplest *geologically* speaking, it challenges geophysicists' notions of how planets behave. From Grimm's point of view, the simplest hypothesis is that Venus—which had seemed so much like a twin sister before the spacecraft reconnaissances—really is like Earth on the inside, with a mantle that is vigorously convecting and causing geologic activity in the overlying lithosphere. But is this borne out by what we know about Venus? Though it is hidden from our view, geophysicists have developed some clever techniques to try to find out.

PROBING THE INTERIOR

Geophysicists like Smrekar and Grimm have relied primarily on seismic data to probe Earth's interior. Variations in the velocity of seismic waves picked up at widely spaced receiving stations reveal the presence of layers of differing density, which are the result of differing temperature, chemical composition, or crystal structure. For example, part of the upper mantle is less viscous than the mantle rocks deeper down. This zone, named the asthenosphere, extends from just below the lithosphere (about 125 kilometers down, on average) to about 250 kilometers. This layer is thought to behave fairly plastically, possibly because it may also contain pockets of partially molten rock. The asthenosphere plays an important role in plate tectonics: it cushions the overlying lithospheric plates as they ride along. In effect, it decouples them somewhat from the convective action of the mantle, thus moderating the intensity of surface tectonic activity.

On Venus, geophysicists have had only one probe of the interior: a map of the planet's gravity field compiled during the Pioneer Venus mission. By recording minute Doppler shifts in the radio signals from the orbiting spacecraft, scientists were able to monitor subtle changes in the craft's motion and from this to construct a low-resolution gravity map. As on Earth, tiny variations in the strength of the field are detected, but the pattern of these variations is markedly different. On Venus, areas where the field strength is slightly greater than average, called positive gravity anomalies, occur wherever there is a topographic high such as Aphrodite or Ishtar. Negative anomalies (places where the field strength is slightly diminished) are detected within lowland basins like Lavinia Planitia. This is in sharp contrast to the situation on Earth, where gravity anomalies bear little relation to gross (continent-scale) topography and are assumed to be caused by density variations deep within the mantle. We know, however, that the gross topography of Earth's surface reflects variations in thickness within the upper reaches (i.e., the upper several tens of kilometers) of the crust, rather than the state of the deep mantle.

A more systematic way of comparing the gravity anomalies of Earth and Venus is provided by a measurement called the apparent depth of compensation (ADC). Consider an iceberg floating in the ocean. The small visible portion is supported by a much larger submerged icy mass that is in turn held up by isostatic pressure from the surrounding, higher-density fluid. If the iceberg is stable—if it is not sinking or rising—it is said to be *isostatically compensated*. Similarly, a topographic high such as a mountain must also be supported in some fashion, lest it sink under its own weight. The principle of isostatic compensation of mountains was discovered in 1865 by Sir George Airy, England's astronomer royal, who used it to explain a discrepancy between two sets of surveying measurements made in India.

Some terrestrial mountains, such as the Himalayas, are supported by subsurface masses, or roots, made of crustal rocks whose density is lower than the surrounding mantle rocks; the Himalayas, like the iceberg, are isostatically compensated. Others, like the Hawaiian swell—the broad topographic rise surrounding Hawaii's volcanic islands—are supported instead by an upwelling of hot buoyant material from the mantle. The Hawaiian swell is said to be *dynamically compensated.* In both cases the positive gravity anomaly of the surface topography is offset to some degree by the negative anomaly of the underlying hot lower-density material.

The apparent depth of compensation is computed in three steps. First, the profile of a topographic feature is inverted to form a mirror image. Next, the vertical scale of this mirror image is exaggerated until it simulates the shape of a crustal root required to support the observed topography. This exaggeration is necessary because the root is not a very efficient source of buoyancy. For example, the density contrast between ice and sea water is 10 times smaller than the density of ice, which is why only 10 percent of an iceberg's total mass is above the water. For crustal roots and surrounding mantle rock, the density contrast is four to five times smaller than the density of the load at the surface. Finally, the depth of this simulated root is increased until the resulting gravity anomaly matches the observed value. The deeper the root, the less effect it has on the gravity field measured at or above the surface. The result, the apparent depth of compensation, is not a literal value but an index of the depth of the maximum density contrast, whether due to differences in composition (as for a crustal root) or temperature (as for Hawaii). If the compensating material were compressed to a point, the ADC tells how deep it would have to lie to produce the observed gravity signature.

The surprising result from PVO's gravity data is that the apparent depths of compensation for many features on Venus range from 100 to 300 kilometers, much greater than on Earth. This suggests that these areas are dynamically compensated simply because the high temperatures found at great depths would cause a crustal root to weaken and flow away over geologic time. The simplest explanation is that the ADC values reflect the thickness of the lithosphere. And therein are the makings of a dilemma. Cornell geophysicist Donald Turcotte points out that other evidence implies that Venus' lithosphere is much thinner. Turcotte starts with the premise, supported by the Soviet data on radioactive isotopes, that Venus' heat budget is similar to Earth's. On average, Earth's interior temperature increases with depth at a rate of 25° C per kilometer. If Venus' thermal gradient is the same, then as Turcotte points out, we can solve for the thickness of the lithosphere since we know the temperatures at both ends of the gradient. The surface is at 450° C; the base of the lithosphere would correspond to approximately 1250° C, the temperature at which basalt becomes so plastic that it can be swept up in the convecting action of the mantle. Again assuming a thermal gradient equal to Earth's, the 800-degree difference between the two corresponds to a lithosphere only 32 kilometers thick. "There's no quibble about these numbers," says Turcotte; "these numbers just have to be." A 32-kilometer lithosphere would be unusually thin compared to the terrestrial average of 100 kilometers. A thicker lithosphere, Turcotte says, would explain how Venus' highlands are supported. Furthermore, he notes, there are the high ADCs to contend with. They make sense, he says, if they represent the bottom of the lithosphere. For that reason, Turcotte believes that in most places the lithosphere of Venus is several hundred kilometers thick; however, he says, it's only temporary. While Turcotte believes Venus today is far less geologically "alive" than Earth, he does not believe the planet is dead or even dying; rather, it is in a state of quiescence or, as Sue Smrekar terms it, "hibernation." Turcotte believes a hibernating Venus is the explanation for the global resurfacing model.

SCENARIO FOR AN ANCIENT CATASTROPHE

To start, Turcotte focuses on the question that is central to any planet's evolution, namely how Venus has gotten rid of its excess heat. First, he says, consider how well our own planet manages that task. Earth's plate tectonic engine is so finely tuned that it accomplishes 85 percent of the heat extraction needed to keep the planet's interior stable over geologic time (i.e., many millions of years or more). That means having just the right amounts of subduction zones and mid-ocean ridges, just the right plate velocities. In other words, the rates of production and destruction are ideal—and no one can say how or why that is the case.

Now imagine that Venus' heat engine is not nearly so finely adjusted. Instead of running smoothly for a billion years or more, it goes in fits and starts. There are periods of intense geologic activity, perhaps similar in form to plate tectonics but perhaps not. During these times, heat is released at a rapid rate, allowing the interior to cool. With cooling, geologic activity wanes in intensity, and the lithosphere—up to now, perhaps as thin as 15 to 25 kilometers over much of the planet—begins to thicken. After about 100 million years, the lithosphere has grown so thick that it blocks all volcanism. The planet enters a state of geologic quiescence. With its thick unbroken lithosphere, Venus is now more like Mars than Earth.

Because volcanism has stopped, heat builds up in the interior. Plumes of hot mantle material rise to meet the lithosphere, thinning and buoying it, causing topographic highs. In these places some geologic activity may resume. Meanwhile, the rest of the lithosphere continues to thicken, releasing its own heat to the surface by means of conduction. Heat transfer is so efficient, Turcotte says, that the lithosphere can grow to a thickness of perhaps 300 to 400 kilometers without being melted by the underlying mantle material.

But the lithosphere does not continue to grow indefinitely. At some point, the stress due to the lithosphere's great weight makes it susceptible to fracturing, especially in places where plumes of hot material rise up from the mantle. Fractures open up, and pieces of the lithosphere bend, then break, then sink into the underlying mantle. The fractures propagate, until most of the lithosphere has fractured and been subducted into the mantle. Without knowing whether this ever develops into plate tectonics, we can use the term "lithospheric recycling," of which plate tectonics is just one variety.

Meanwhile, volcanism resumes, flooding the planet with fresh lava. After another 100 million years, when the interior has recooled and the lithosphere rethickened, there is another period of quiescence, and the cycle continues. On average, Turcotte envisions a new resurfacing event about every 500 million years, which means that Venus should be due to wake up fairly soon.

If this sounds decidedly unearthly, that is exactly the point. "The thing one has to realize about Earth," Turcotte says, "is the importance of the continents." He explains that the continents, made of relatively weak, granitic rocks, have done their part to keep plate tectonics going by breaking apart several times over geologic history, allowing new ocean basins to open up. The difference in Venus' tectonic style may be due largely to the fact that it has no weak granitic crust. In any case, the remarkable thing, Turcotte says, is not that Venus is different from Earth but that our own world's geologic engine is able to run as smoothly as it does.

Without knowing what kind of geologic activity Venus has during its active periods, Turcotte is unable to specify precisely how it stops. But looking at today's Venus, he says, we see clues to how it starts. Some researchers have proposed that the strange features called coronae are actually sites of incipient subduction. The fractures that ring these

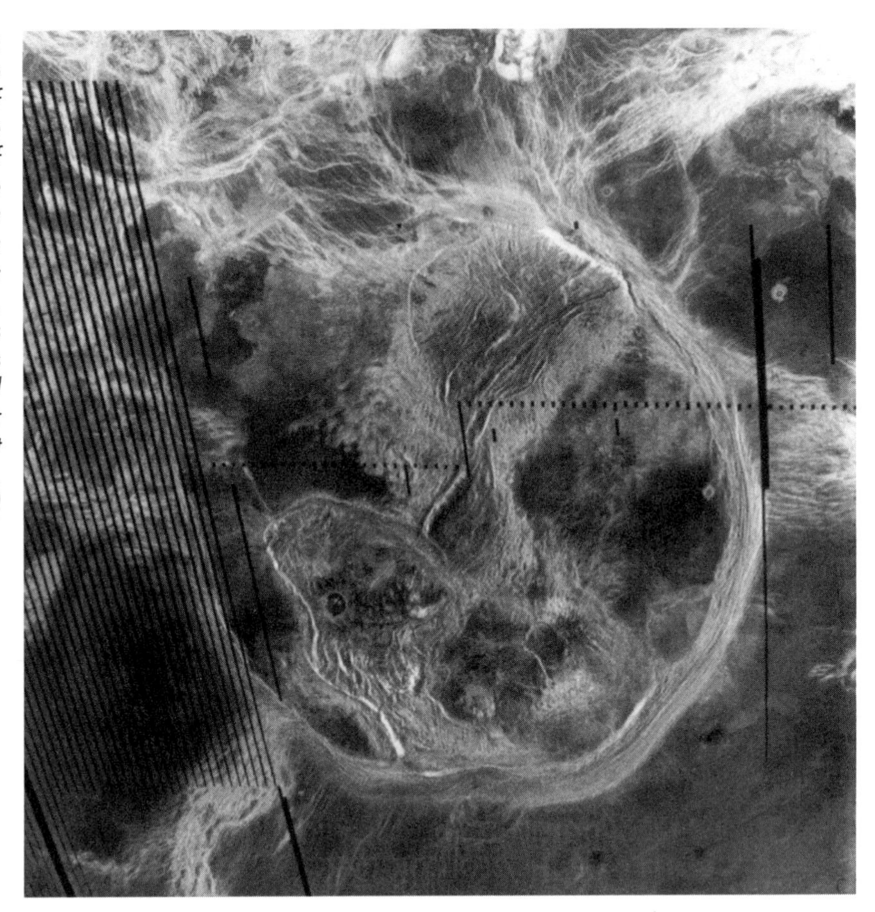

FIGURE 5.2 *Like welts on Venus' skin, coronae are an enigmatic class of features thought to have formed from upwellings of molten rock beneath the surface. Most display lava flows encircled by a fractured ring of tectonic ridges. Artemis corona, the largest such feature, is some 2100 kilometers wide, and is surrounded instead by a wide circular trench. (Courtesy of the Jet Propulsion Laboratory, NASA.)*

features, say the scientists, are places where the lithosphere is bending and sinking into the mantle. The coronae in Magellan's images, Turcotte suggests, represent places where lithospheric recycling tried to get started but couldn't. As with so many ideas about Venus, this idea is controversial (see Figure 5.2).

VENUS IS ALIVE

One of the recurring themes of Magellan's Venus is that the same data may draw widely differing interpretations from different scientists. In the case of the ADC values indicated by the gravity data, Bob Grimm and Sue Smrekar take the opposite view from Donald Turcotte. They do not believe Venus has a thick lithosphere. At the 100- to 200-kilometer depths indicated by the ADC values, Grimm notes, temperatures would almost certainly be too high for rocks to remain solid over geologic time. But if the lithosphere is relatively thin, what is holding up Venus' highlands? Isostatic compensation by means of a low-density crustal root seems to be out of the question, for the same reason: the thickness of crustal rocks required would also be unable to withstand such high temperatures. That leaves only one alternative: Venus' highlands are dynamically supported by mantle convection. Is that feasible?

"No geophysicist doubts that the interior is convecting," Grimm says, noting with an unintentional pun, "It would have to be stone cold inside for it not to convect." The question, he adds, is whether this causes geologic activity. The answer seems to be yes. In 1991 Smrekar and Roger Phillips studied two-thirds of Venus' landforms using PVO's gravity data. They found that half of the 10 Venusian highland regions they studied had ADCs in excess of 150 kilometers. These regions also resemble areas on Earth called hot spots, places where hot mantle material is welling up and deforming the lithosphere directly above it. Hot-spot volcanism, for example, is thought to have formed the Hawaiian islands; over time the motion of the Pacific plate over the hot mantle plume resulted in a series of volcanoes arrayed along a line. At least five of the Venusian uplands studied by Smrekar and Phillips occupy broad, dome-like uplifts and seem to be places where volcanism and crustal rifting have occurred. Because of these characteristics, so much like terrestrial hot spots, the two geophysicists concluded that the Venusian features are probably hot spots as well. Are they active today? Smrekar says, "Even a fairly sluggish planet might occasionally be able to squeeze out one plume. For example, Mars produced one whopping plume that created the Tharis Mons volcano, which is not thought to be active today." But on Venus, she says, the presence of five probable hot spots strongly suggests that Venus' mantle is anything but sluggish.

More recently, Grimm and Phillips made a case study of the upland region called Eistla Regio. Its western and central regions, which comprise an area about the size of Brazil, are a pair of broad, dome-shaped highs separated by rift valleys. Each of the domes is topped by two volcanoes. Grimm's desire was to determine what kinds of forces might be acting on these features as a result of temperature and density variations in the underlying mantle. Here again the Pioneer Venus gravity survey provided the key. In computer simulations Grimm used the gravity data for Eistla to calculate the forces of tension and compression that would be experienced by the surface if Eistla were being supported by mantle convection. His results predicted that in the trough the crust should be pulling apart, while the surrounding plains should experience compression. In fact, this is reflected by the geologic features we see in Magellan images: a rift zone some 100 kilometers wide cuts through the middle of Eistla, while on the surrounding plains the low ridges attest to compressive forces. Grimm believes a plume of molten rock is ascending directly under Eistla Regio. Presumably, it was the source of the lavas that built the two large volcanic shields. As the plume material crests and spreads laterally, it causes the overlying lithosphere to be torn apart. At the margins, where the plume material is descending, the surface above is compressed. The plume's buoyancy also accounts for the broad uplift that forms Eistla. Furthermore, Grimm believes, the existence of a significant gravity anomaly at Eistla means that uplift is still going on today. He suspects that a variety of similar scenarios are at work in Ishtar, Aphrodite, and the other Venusian highlands.

The idea that Venus' topographic features may arise from convective action within the mantle has broad implications. In a general sense, Grimm says, this tells us that Venus' lithosphere is mechanically more strongly coupled to the mantle than on Earth (which in turn, he says, would cause high ADC values). He and Smrekar believe the reason for this is that Venus lacks a low-viscosity asthenosphere. This makes sense if we consider how the Earth's asthenosphere is formed: When oceanic crust subducts into the mantle, it carries seawater down with it, which drastically reduces the melting temperature of the surrounding mantle rocks. As a result, the rocks of the uppermost mantle—the asthenosphere—exist almost at their melting temperature and probably contain pockets of molten rock. In other words, the Earth has an asthenosphere because it has oceans. But Venus' water is thought to have boiled away during an episode of so-called runaway greenhouse heating, perhaps a billion years after the planet formed, changing the planet into a bone-dry furnace. Just why Venus has no asthenosphere is controversial, but if it is true, Grimm and Smrekar say, it explains the high ADC values, because the mantle would be more strongly coupled to the surface. It also has major implications for the planet's geologic style, as we will see.

Meanwhile, for more evidence that Venus is geologically active today, Smrekar looks to the enigmatic coronae. These features, she says, are thought to begin life as volcanoes that eventually flatten into a more gentle topographic high. "If the coronae stopped forming 500 million years ago," Smrekar says, "there should not be any left in the early, volcano-like stages." But, she notes, Magellan has found coronae in all stages of development. In other words, Venus is currently volcanically active.

But on that point there is unanimous agreement among the Magellan

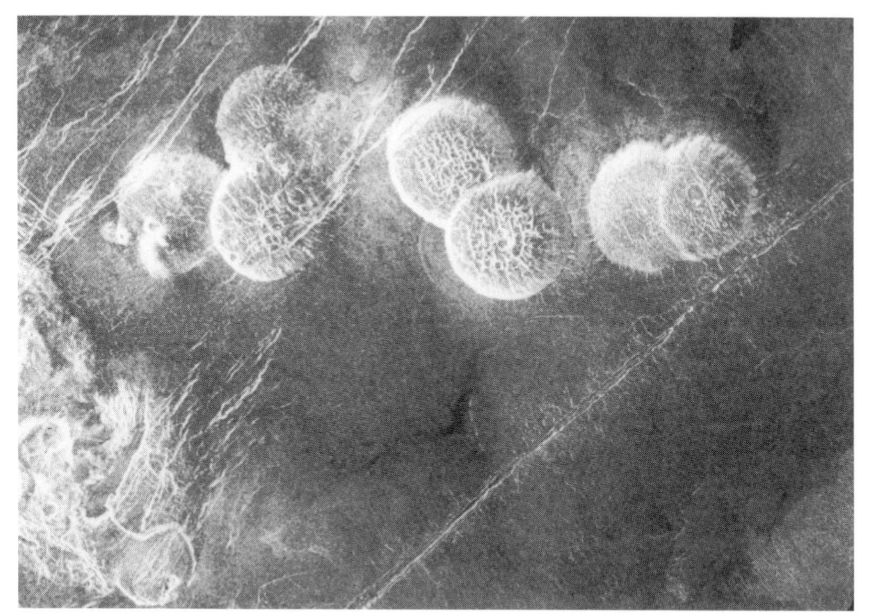

FIGURE 5.3 *Pancake-like volcanic domes, each roughly 25 kilometers across, pepper the surface on the eastern edge of Alpha Regio. Thousands of volcanic features, ranging from gigantic shield volcanoes to tiny cinder cones, have been spotted in Magellan's images. (Courtesy of the Jet Propulsion Laboratory, NASA.)*

scientists if for no other reason than that 80 percent of the surface contains volcanic landforms. As James Head explains, "It would be incredibly presumptuous of us, with all this evidence of volcanic activity, to say, 'Oh, the damn thing shut off, just as [Magellan] went into orbit.' I mean, we're talking about 4½ billion years of evolution dominated by volcanism" (see Figure 5.3).

And yet, according to Head, Venus' current yearly volcanic output is far lower than many scientists expected. To account for the scarcity of craters invaded by lava flows (a mere 4 percent, by Schaber's count), Head estimates that between 0.1 and 0.5 cubic kilometers of lava is brought to the surface each year, as compared with the 20 cubic kilometers per year produced on Earth, mostly on the ocean floors. This suggests that volcanism has not played a major role in erasing Venus' craters in recent geologic history. If Roger Phillips is right—if something is eating craters on Venus—scientists must look for another explanation.

THE TECTONIC ALTERNATIVE

Michael Malin, a San Diego-based geologist who formed his own space science firm, is a veteran of planetary missions going back to the Viking Mars landings in 1976. He too is puzzled by the idea of a planet-

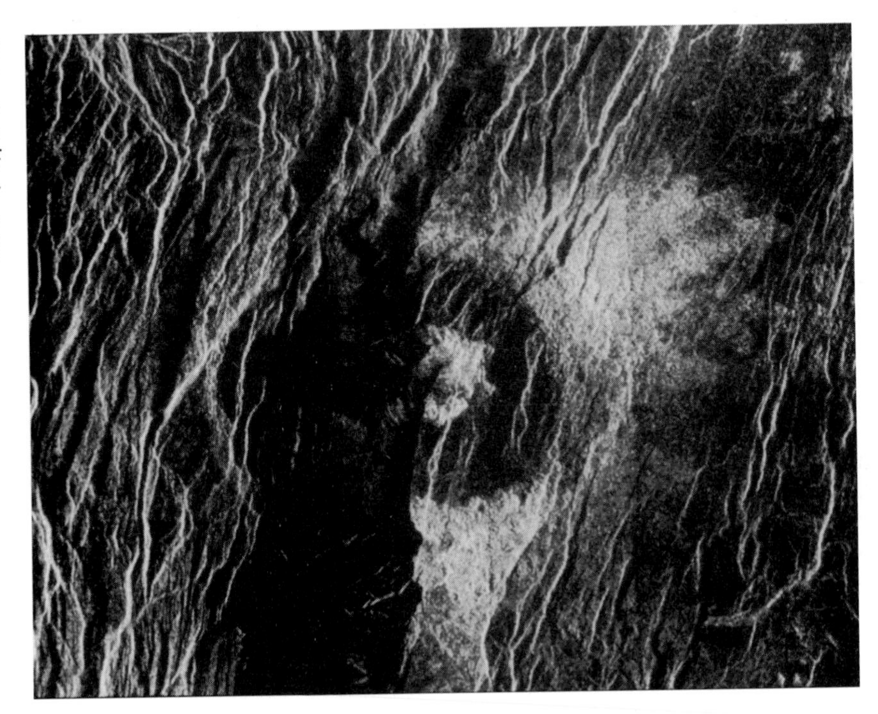

FIGURE 5.4 *Disappearing act: In Beta Regio a crater caught in the act of being destroyed by fractures attests to Venus's history of tectonic activity. (Courtesy of the Jet Propulsion Laboratory, NASA.)*

wide catastrophe in recent geologic time, an event that would make Venus unique among the terrestrial planets. Malin asks, "How do you resurface a planet?" With volcanism apparently not a viable mechanism, he suspects tectonic activity. "The only way to answer this," says Malin, "is to look at the craters themselves."

Even then, it is not an easy question to answer. Something as simple as the number of craters cut by faults or fractures is a matter of considerable debate; Schaber's team has estimated 34 percent, while Phillips and his colleagues count half as many (see Figure 5.4). In any case, tectonically altered craters far outnumber the 4 percent of craters that appear to have been flooded by lava. Ultimately, Malin and collaborator Robert Grimm would like to conduct a detailed study to see how many craters are found within tectonically disturbed regions. For now, the pair have conducted a "quick and dirty" analysis using a parameter called RMS (root mean square) slope, which is a component of Magellan's radar data. Simply put, the RMS slope is a measure of the roughness of the surface. The higher the mean slope value, the rougher the surface. And you would expect blocky surfaces in such tectonically altered regions as rift valleys, coronae, and ridge belts, Malin says. Thus, a map showing values of RMS slope ought to serve as a rough tectonic map of Venus.

According to Malin's RMS slope map, Venus is dominated by tectonic activity; some 60 percent of the surface is folded, fractured, faulted, or otherwise tectonically disturbed. Perhaps, he says, Venus is much more tectonically active than Earth. Or maybe it just seems that way: perhaps the difference is that on Earth so many other forces conspire to alter the landscape—vegetation, erosion by water and wind, burial by sediments—that tectonic features seem lost in the shuffle.

Having made this finding, Malin compared the map to the crater locations and found that tectonically deformed areas are somewhat deficient: they contain only 40 percent of known craters. The remaining 60 percent of the craters are found on the 40 percent of the planet that hasn't been tectonically altered. A close look at the tortured regions known as tesserae shows that they have fewer craters than the surrounding plains; the enormous fracture belts that girdle the planet's equator have even fewer craters. To Malin, the implication is clear: craters on Venus meet their doom not by burial under boiling lava but by the slow, steady disintegration of the very ground on which they lie. But not too slow: the fact that geologists see very few severely faulted craters means that the process acts relatively quickly, perhaps within a few million years. Some yet-unknown tectonic process, Malin says, has been erasing Venus' craters. While he considers this an important result, Malin stresses that it is preliminary; more definitive answers will have to wait for detailed mapping. In any case, he says, one of the most important implications is that craters on Venus are *not* randomly distributed, if one considers the type of geologic terrain they appear on.

For now the puzzle of exactly how tectonic activity has destroyed craters is part of the much larger question about the origin of Venus' tectonic features. Until scientists can explain how these features formed and developed, the nature and intensity of Venus' tectonic activity will be unknown. But there is evidence that Venus remains tectonically active today. The Maxwell Montes, the planet's highest mountain range, tower some 12 kilometers above the surrounding plains. If Maxwell Montes seem like Venus' answer to the Himalayas, the comparison is not unwarranted. How these mountains formed is highly controversial; one of many suggestions is that they formed from a block of crust that was thrust upward by compressive forces within the lithosphere. To some scientists the wonder is that they are there at all. Sue Smrekar believes Maxwell Montes offer the most compelling evidence of all that Venus is geologically alive. She notes that the extreme slopes of these mountains, 30 degrees in some places, are astonishing, considering the fact that the surface temperature is halfway to the melting point of basalt. Such inclines would also experience tremendous pressure from the millions of tons of rock contained in these mountains. Weakened by the blast-furnace heat at the surface, Smrekar says, a slope that steep "wants to flow away like crazy" over a span of tens of million years. Had Maxwell Montes stopped forming 500 million years ago, Smrekar says, those slopes would not exist today. "The only possible way to retain these awesome mountain fronts (they are steeper than any mountain range on Earth) for half a billion years is to make Venus incredibly cold—about as cold as we think Mars is." Obviously that is not the case, but there is still some uncertainty about the true strength of Venus' rocks because of the fact that they contain no water. But Smrekar believes that even on a very dry Venus "slopes of 30 degrees are still not going to hang around for 500 million years." She estimates the age of Maxwell Montes at 10 million years or less.

But as if to emphasize the ambiguity with which Venus confronts planetary scientists, the crater Cleopatra, 100 kilometers across, sits on the high slopes of Maxwell Montes (see Figure 5.5). The crater shows no signs of deformation, indicating that it formed after the mountains did. The problem is simple: Craters that big just don't come along very often, on Venus or Earth. When they do, they are significant events; a 300-kilometer-wide crater in the Yucatan is believed to be the scar from the

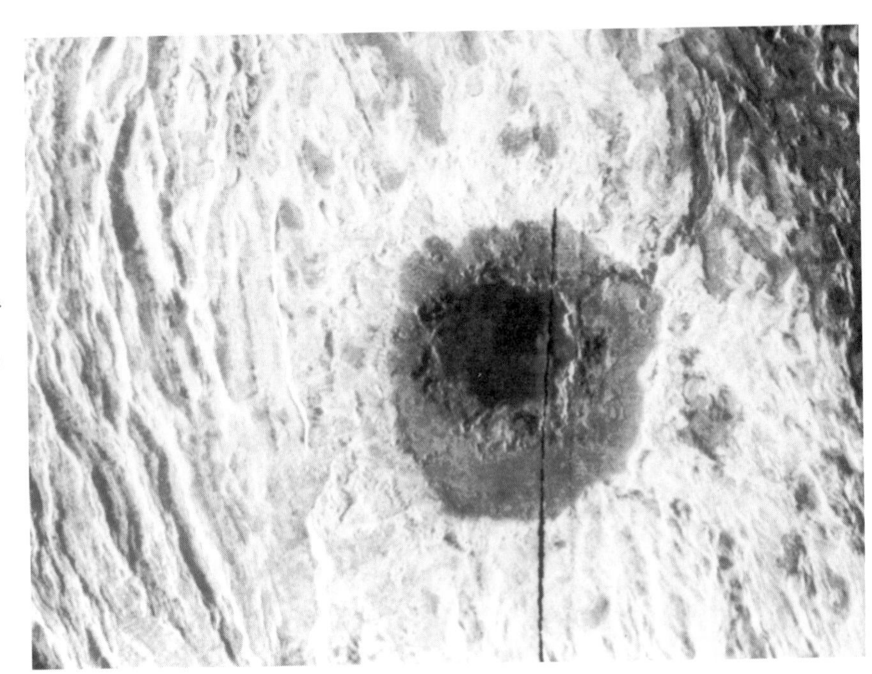

FIGURE 5.5 *Cleopatra crater, 105 kilometers across, lies on the high slopes of Venus' Maxwell Montes. Its very presence is a dilemma for scientists who claim these mountains formed in recent geological time. (Courtesy of the Jet Propulsion Laboratory, NASA.)*

impact that ended the reign of the dinosaurs 65 million years ago. Even if Maxwell's present slopes are 50 million years old, as estimated by MIT geologist Noriyuki Namiki, Cleopatra's presence is problematic. Namiki estimates that there is only an 8 percent chance that a 100-kilometer crater would form on an area the size of Maxwell within a span of 50 million years. Namiki's former adviser, geophysicist Sean Solomon, now at the Carnegie Institute of Washington, D.C., says, "[Cleopatra] is an embarrassment to the hypothesis that Maxwell is a very young feature." On the other hand, if the mountains are ancient, scientists may cast their votes with Turcotte, who says that on Venus the only way to maintain such extreme relief for aeons is to have a very thick (and therefore strong) lithosphere. Smrekar counters by noting that it would be much harder to fracture and compress a thick lithosphere to produce such enormous mountains.

Another intriguing tectonic feature of Magellan's Venus is the existence of landslides inside the planet's large canyons. Mike Malin believes these landslides were triggered by Venusquakes; some of the canyon slopes are so steep, he says, that landslides there appear inevitable. Before Magellan arrived at Venus, Malin predicted that if Venus were as seismically active as Earth at least one major landslide ought to occur during the projected mission. However, Magellan's third mapping period was cut short, and the "smoking landslide" has not been found.

MEANWHILE, IN THE UPPER MANTLE

For Smrekar the question remains: Why does tectonic activity seem to have overshadowed volcanism in recent Venusian history? Smrekar is developing a hypothesis to explain the formation of Venus' hot spots that considers not only the physics involved but also the chemistry. First, consider the fact that rocks deep within mantle stay solid despite their high temperatures; this is because of the tremendous pressure at great depths. When a hot plume of mantle rock rises toward the surface, Smrekar says, it encounters steadily decreasing pressure; this allows it to succumb to the high temperatures. It does not melt completely, however. The liquid produced is basaltic in composition, rich in iron and magnesium. If it can rise to the surface, it will erupt as lava. The rock that remains behind, called residuum, is deficient in iron and magnesium and thus is less dense than the surrounding mantle rocks. On Earth much of this low-density residuum is carried back into the mantle. But on Venus geophysicists have proposed a different scenario, first suggested by Roger Phillips and Bob Grimm, and fully developed by Brown

University's Marc Parmentier and Paul Hess. In this scenario the absence of an asthenosphere may allow the low-density residuum to collect immediately beneath the lithosphere. Over time this buoyant material would spread out into a layer of fairly even thickness.

Smrekar is studying in particular what would happen if a hot spot formed beneath this residuum layer. Because of this layer, a hot plume of mantle material would not be able to reach the lithosphere, thus reducing the amount of volcanic activity. However, the presence of the residuum layer beneath a relatively thin lithosphere would not inhibit tectonic activity. And the high ADC values would be explained by the fact that buoyant mantle plumes would be stalled at depth. Thus, Smrekar believes, this hypothesis can explain not only the crater statistics but also the gravity signatures. Over the life of an individual hot spot, she notes, its ADC value would gradually decline as the hot mantle plume comes closer to the surface and finally dissipates. Smrekar and three colleagues are studying Venus' hot spots in the hope of determining their stage of evolution.

RETHINKING THE CRATERING RECORD

In the summer of 1992 Roger Phillips and geologists Ray Arvidson and Noam Izenberg went back to the work of studying the Magellan images and, once again, found evidence that the "clumpiness" of the crater distribution is not simply random chance. Areas that are bright in the radar images have fewer craters than average. The craters that are present have been cut by fractures or are partly covered by lava flows. The prime example is the belt of tectonic and volcanic features that extends through the Aphrodite highlands and eastward toward the highland region called Beta. It is no surprise that craters should be missing here; as Jerry Schaber and others had pointed out earlier, this appears to be a region where the surface has been rifted apart and volcanism has taken place. These regions, Phillips believes, may be some of the youngest surfaces on Venus.

Conversely, as Phillips and his colleagues found, regions that appear radar dark have more craters than average. The most prominent dark areas lie in Venus' northern hemisphere, such as those north of Aphrodite. A closer look showed that the craters in these areas were surrounded by halos of radar-dark ground, each of which extends for many tens of kilometers. The halos are probably dark because they are places where the ground is covered with fine-grained dust, which scatters radar waves poorly. Specialists who study impact craters mention two possible origins for the radar-dark halos. In one theory the shock wave generated by the plummeting meteoroid pulverized the region around the impact site. In the other, fine dust produced during the impact was carried into the atmosphere, then settled on to the ground.

When Phillips studied the dark areas north of Aphrodite, he found that they were dark because they were covered with craters whose dark halos nearly coalesce. Some dark regions have twice as many craters as average, implying ages of 1 billion years; Phillips believes they are the planet's oldest regions. Furthermore, they show no sign of having been altered by subsequent geologic activity that would have disturbed the dark halos. In general, Phillips says, terrains on Venus are not the same age but fall into three basic groups:

- heavily cratered, and thus ancient, radar-dark regions; based on Phillips's rough estimate, these occupy 20 to 30 percent of the planet;
- young (crater-poor) radar-bright areas, including the Aphrodite upland region; these also cover perhaps 20 percent of the globe; and
- the rest of the planet, perhaps 50 percent, could contain a variety of different ages, Phillips says; more detailed studies are needed to be more certain. For now, however, Phillips believes this assortment of terrains, with sharply different crater ages, shows that Venus' history cannot be explained in terms of one major resurfacing event. Phillips's finding seems to challenge the premise that is the foundation of the global resurfacing model, that Venus' craters are simply a result of half a billion years in the cosmic shooting gallery.

THE GEOLOGIC LINK: DISTRIBUTED DEFORMATION

It remains to forge a link between the observations that Venus' terrains differ in age and that tectonic activity appears to be the primary agent for removing craters and the geophysical premise that Venus' interior is like Earth's. That link is offered by Grimm in the form of a hypothesis he calls distributed deformation. He believes that what is going on in Eistla Regio and, by implication, the other highland areas is telling us how Venus' geology works. On Grimm's deformation map of Eistla, stresses seem to be evenly distributed; the transitions between regions of tension and compression appear to be smooth. But this is just what we would expect if, as the ADCs suggest, Venus' lithosphere is strongly coupled to the interior. Imagine for a moment that Venus

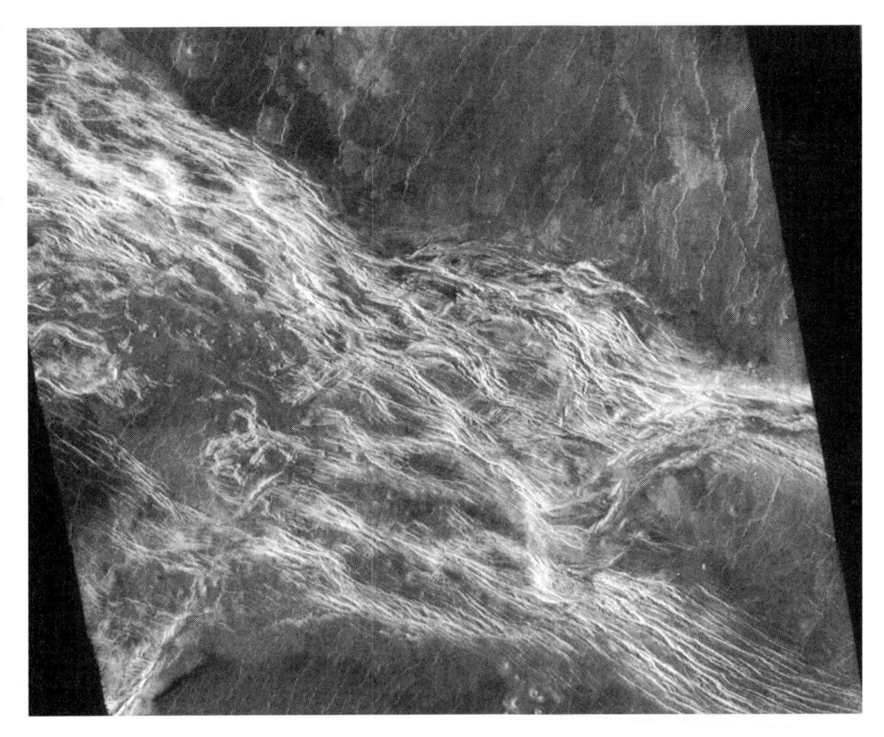

FIGURE 5.6 *Belts of ridges and rifts cross Lavinia Planitia. Do they represent the boundaries of Venusian "miniplates"? Width of image is 430 kilometers. (Courtesy of the Jet Propulsion Laboratory, NASA.)*

started out with large moving plates like Earth's. With no asthenosphere to act as a buffer, the plates would bear the brunt of the vigorously convecting mantle. Like scum on a pot of boiling water, a large plate would break apart into smaller pieces. At that point, Grimm says, plate tectonics would no longer exist.

This, Grimm believes, is just what we see on Venus today. If a global network of tectonic features ever existed on Venus, there is no trace of it in Magellan's images. Instead, Grimm says, tectonic features are widely distributed across the planet. In some areas we see tectonic belts encircling "islands" of undeformed terrain (see Figure 5.6). A picture emerges of a Venus whose lithosphere is divided into a series of blocks, each on the order of a few hundred kilometers across. (Notably, this is roughly equivalent to the average spacing between craters on Venus.) There may be tectonic activity or volcanism at the margins of these blocks, but the blocks themselves are too tightly packed to undergo significant lateral motion. Nevertheless, Grimm says, such a configuration would not limit the extent of Venus' geologic activity. Over a span of tens or hundreds of millions of years the pattern of mantle convection would change and with it the location of geologic activity. Grimm and

Phillips believe that if you could view a time-lapse movie of Venus' evolution you would see different parts of the planet wink on and off, erupting with lava flows or suffering a spasm of mountain building or being rifted apart by hidden stresses in the interior. At any given moment, much of the planet would be quiescent; over a span of a billion years, however, the whole planet would be affected.

Not surprisingly, the hypothesis has already generated controversy; Jerry Schaber questions the likelihood of a Venus "conveniently divided into a series of blocks, each of which is just the right size" to avoid disrupting a spatially random crater distribution. Grimm admits that there is a long way to go before the distributed deformation model is on solid ground. Roger Phillips points out that geophysical models are so sensitive to slight changes in the assumed parameters that almost any answer is obtainable just by twiddling the knobs, so to speak. Having a model that explains a hypothesis, he says, doesn't prove the hypothesis, whether it is global resurfacing or a competing theory.

"I think the story is very complex," Phillip says. "I certainly do not claim to know all the answers at this point. I probably never will. But I don't think a model as simple as [global] resurfacing can be right. There's enough evidence to rule out that model, or models close to it, as far as I'm concerned."

That said, geologists may be edging toward consensus. Turcotte says, "The uniform age is obviously just a first approximation, and one has to look at variations on that. . . . But there do not seem to be large [masses] of very old rock on Venus. And I think that is significant."

TOO SOON FOR CERTAINTY

No matter which hypothesis you mention, Venus maddeningly seems to offer both support and contradiction. For example, Donald Turcotte says his "hibernating Venus" hypothesis leaves plenty of room for mantle convection to support upland regions like Eistla Regio, just as Grimm suggests. The reason for this, he says, is that Venus' lithosphere varies greatly in thickness, just as it does on Earth. For example, the lithosphere under the eastern United States reaches a thickness of 175 kilometers, while in the western portion of the continent the lithosphere is less than 30 kilometers thick and consists only of crustal rocks; consequently, the western United States is a site of volcanic and tectonic activity. On Venus, Turcotte says, the thickest lithosphere could be 300 or even 400 kilometers thick; the thinnest, at the center of some active highland regions, could be only 12 kilometers thick.

But for Smrekar the very existence of Maxwell Montes is enough to demonstrate that Venus is tectonically, as well as volcanically, alive. That, she says, is the reason the lithosphere of Venus cannot be several hundred kilometers thick, as Don Turcotte maintains. "I can't prove to you that there's not a thick lithosphere," Smrekar says. "But if there is, it's extremely difficult to have ongoing volcanism and tectonism. That's really the heart of the argument."

Every one of the scientists mentioned here has stressed that it is too soon to draw firm conclusions from Magellan's wealth of data. Detailed studies, like the geologic mapping that Michael Malin plans, should help pin down the details of Venus' evolution. And scientists are eager to see NASA, the National Aeronautics and Space Administration, approve a special seismic lander to touch down on the Venusian surface to probe the planet's interior. For now, they are hopeful that Magellan will provide a critical key to the puzzle in the form of new, high-resolution data on the gravity field. These data, they say, would help answer detailed questions about the structure of the lithosphere and the nature of the interior. At this writing, they were still hopeful that the spacecraft—many of its systems malfunctioning, threatened with termination by NASA—would achieve that longed-for goal. But it is safe to say that in the years to come scientists will continue to reap the fruits of Magellan's rich scientific harvest. And the cloud-hidden world that shines so brightly in our skies will grow less and less mysterious.

NOTES

1. Venus orbits at a mean distance of about 108 million kilometers, approaching to within 41 million kilometers of Earth.
2. Most geologists consider the moon a planet by virtue of its relatively large size, compared to other planetary satellites, and its significant geologic history.
3. By convention of the International Astronomical Union, many features on Venus are named for goddesses and famous women.

BIBLIOGRAPHY

Beatty, J. K., and A. Chaikin (eds.). 1990. The New Solar System. 3rd Edition. Sky Publishing Corp., Cambridge, Mass.

Cooper, H. S. F., Jr. 1993. The Evening Star. Farrar, Strauss, and Giroux, NewYork.

Grinspoon, D. H. 1993. Venus unveiled: Has the Magellan spacecraft detected plate tectonics on Earth's twin ? The Sciences (July/August):20–26.

Magellan at Venus. Special issues of the Journal of Geophysical Research. Part 1, vol. 97 #E8, 13063–13689 (August 25, 1992). Part 2, vol. 97 #E10, 15921–16382 (October 25, 1992). Research papers written by Magellan scientists.

Morrison, D. 1993. Exploring Planetary Worlds (Scientific American Library). W. H. Freeman and Company, New York.

National Aeronautics and Space Administration (NASA). 1984. The Geology of the Terrestrial Planets. NASA SP-469. U.S. Government Printing Office, Washington, D.C.

Newcott, W. 1993. Venus revealed. National Geographic (Feb):36–59.

Pasachoff, J. M. 1990. Astronomy: From the Earth to the Universe, 4th ed. Saunders College Publishing, Philadelphia.

Press, F., and R. Siever. 1974. Earth. W. H. Freeman and Company, San Francisco.

Solomon, S. 1993. The geophysics of Venus. Physics Today (July):48–55.

CLOCKS IN THE EARTH? The Science of Earthquake Prediction

by Addison Greenwood

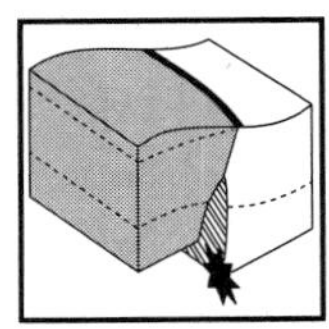

Earthquake country. In America this epithet usually evokes images of western California, where the San Andreas fault—perhaps our most notorious geological feature and probably the most intensely studied by scientists—appears for most of its length as the thinnest of lines in the dirt. The fault was indubitably the cause, however, of the earthquake and consequent raging fire in 1906 that devastated San Francisco (see Figure 6.1). After this natural disaster, at least 700 lay dead (the San Francisco city archivist recently reexamined many of the city's records and concluded that the number of deaths may well have been many hundred more) and many of the city's structures were toppled.

More than a mere memorial to past events, the San Andreas fault is a harbinger. For the present population of San Francisco, for millions of other Californians who also live along the fault zone, and for hundreds of millions more throughout the world living near other plate boundaries, the past is prologue. The strongest natural forces on our planet, earthquakes are not accidental but incidental to the seething, ceaseless dynamics of the great heat engine that is the earth's interior. The next major earthquake in California, "The Big One" natives call it cavalierly, is as inevitable a natural phenomenon as the sunrise, though unfortunately not as predictable. When it arrives, the dead could well number in the tens of thousands, a scenario that charges modern earthquake scientists, called seismologists, with a mission and a sense of purpose far beyond the pursuit of pure science.

THE RIDDLES AND RHYTHMS OF 1992: EARTHQUAKE SCIENCE IN THE MIRROR

In 1992 the American seismological community was galvanized and challenged by several dramatic events. On the morning of June 28, a meeting attended by many of America's most eminent seismologists was about to get under way at the University of California-Santa Cruz. Their purpose was to review and begin to evaluate a U.S. Geological Survey (USGS)-run project nearly a decade in the running: The Parkfield Earthquake Prediction Experiment. For reasons that will be explained shortly, some seismologists were virtually "certain" (in statistical terms they expressed a 95 percent confidence) that a major quake was due to occur near a small town in Central California before too many years. Writing in *Science* in 1985, Bill Bakun and Al Lindh of the USGS predicted that an earthquake would occur before 1993 near Parkfield. As of June 1992 the predicted quake had not arrived. But the project had focused on that particular section of the San Andreas fault a level of scientific attention and study that was unprecedented in America; many valuable insights had been collected, and the scientists were meeting to consider these and other implications of the Parkfield experience.

Then on the morning the Parkfield meeting was to convene, the earth began to shake. The quake was not the magnitude (M) 6 earthquake predicted for Parkfield, however, but one many times larger (see the Box on p. 158), an M 7.3 quake emanating from a small desert community called Landers, dozens of miles to the south. The Landers quake relegated to a distant second place (in terms of the magnitude of energy released) the state's most famous recent quake—the Loma Prieta M 7.1 quake in 1989, which, as the World Series telecast had just begun, millions experienced vividly via television. As California's largest quake in 40 years, Landers was big news to scientists not because of its devasta-

FIGURE 6.1 *San Francisco, on the morning of April 18, 1906. This famous photograph by Arnold Genthe shows Sacramento Street and the approaching fire in the distance. (Photograph courtesy of the Fine Arts Museums of San Francisco, Achenbach Foundation for Graphic Arts.)*

tion—fortunately, its effects remained largely in the desert—but because it occurred in the south. The Loma Prieta quake had given seismologists important information about the San Andreas fault around San Francisco, but Landers provided a message to Los Angelenos, whose city's fate rests upon a network of underground faults that share with Northern California only the fundamental fact that all are part of the major fault system defining the boundary between the Pacific and North American plates.

Why does the occurrence of a quake many miles from Los Angeles carry a prophetic message for the nation's second-largest city? Will "The Big One" strike there or at San Francisco, or somewhere else along the hundreds of miles of faults throughout California? Will the Parkfield prediction pan out, and when the quake does arrive, will the unprecedented experimental effort devoted to that region pay off? What is happening underground during an earthquake, and why, and, most importantly, where and when? These questions all point to a bottom line that few would dispute, for from a successful methodology of earthquake prediction could come the conservation of billions of dollars and the survival of many people who otherwise might perish in these unavoidable natural disasters. And why unavoidable? The question belongs in the same category as, "Why does the sun rise in the east?" The exigencies of a lawful universe, as elucidated by Newton and many more since, provide the scientific context for a sunrise. So too are earthquakes, on Earth at least, inevitable. And many find it ironic that though so much closer and theoretically more accessible to empirical examination than the sun, the internal machinations of the earth remain shrouded in mystery.

"The Big One" obviously concerns scientists in a vital way, since predicting it and its effects could save many lives and billions of dollars. But the mythology of "The Big One" (a boon to newspaper sales if ever there was one) has obscured from the public some of the scientific clarity developed in the last decade about earthquakes. Nonetheless, while major strides have been taken to understand the genesis of earthquakes, nature may not be so willing to deliver up her secrets. It is by no means certain that scientists will *ever* be able to accurately forecast an earthquake, says geophysicist Bill Ellsworth (USGS-Menlo Park), though he and colleague Duncan Agnew (UC-San Diego) recently compiled a state-of-the-art survey of the current methods and models used to predict earthquakes. Nonetheless, each new quake literally pulls back a veil, and, as was the case when news of Landers arrived at the Parkfield meeting, the excitement among seismologists is palpable that behind this next one may be some scientific signal of a breakthrough in prediction, or at least confirmation for their models about the mechanics of earthquakes.

THE EARTH ITSELF

Geology is the study of how the earth has changed through time. Evidence of such changes during the earth's 4.6-billion-year history are preserved and embedded in the rocks that, like an old rusting car bumping its way through city life, will contain clues of their journey through the earth's interior and down through aeons of time. Geological time is a euphemism for millions and billions of years. The journeys of the smallest of rocks and the largest of continents both ride the same juggernaut: internal currents of scorching, coursing heat that actually move masses of earth like great underground river currents, though at a pace of only inches a year.

A more accurate image than a river, suggests Ellsworth, is a boiling pot of water, where the sea of bursting bubbles on the surface indicates that convection cells have developed in the pot. For the same reasons of basic physics—convection—hotter rock in the earth's deeper mantle

Earthquake Measurement Scales

Besides where and when, a legitimate prediction must say how big—a parameter known as an earthquake's *magnitude*. Earthquakes can devastate human society, and the earliest efforts to catalog their size reflect this anthropocentric emphasis, focusing on, according to seismologist Bruce Bolt, damage to structures of human origin, the amount of disturbances to the surface of the ground, and the extent of animal reaction to the shaking. The size of nineteenth-century (and earlier) earthquakes is more than an arcane bit of data, however, even today. Recurrence models developed by seismologists to predict future earthquakes based on the pattern of past ones employ not only the date but also the magnitude of a preinstrumental earthquake.

To make use of an intensity scale, first locate all written reports and eyewitness observations whose times and dates are known sufficiently to cluster them into widely placed reactions to the same event. Next, decipher each separate report according to the scale, assign the individual data points an intensity value from I to XII, and plot them on a geographical map of the region. Connect the dots corresponding to each value, and you've constructed an isoseismal map.

With the twentieth century came a more scientific measure of magnitude. Seismographs (or seismometers) originated as elaborate and sensitive pendulum-like devices with a "pen" mounted over a continuous roll of paper, producing a jagged line drawing that reflects the most delicate shaking of the earth's surface due to seismic waves. More modern seismographs use film and digital technology, and their readings are plugged into equations to quantify the energy released in an earthquake.

The most famous of these equations was developed by American seismologist Charles F. Richter, who, writes Bolt, defined the magnitude (M_L) of a local earthquake as "the logarithm to base ten of the maximum seismic-wave amplitude (in thousandths of a millimeter) recorded on a standard seismograph at a distance of 100 kilometers from the earthquake epicenter" (Bolt, 1988, p. 112). The enormous range of energies involved requires a logarithmic measure, where a rise of one digit yields a tenfold increase in displacement. Earthquakes have been recorded as low as –2 and as high as 9 on the scale, encompassing a range of 10. Compare, for example, the 1966 Parkfield quake (M 6) to the 1989 Loma Prieta (M 7.1) event that hit the San Francisco area. Loma Prieta released 30 times as much energy as Parkfield.

Intensity value	Description*
I.	Not felt. Marginal and long-period effects of large earthquakes.
II.	Felt by persons at rest, on upper floors, or favorably placed.
III.	Felt indoors. Hanging objects swing. Vibration like passing of light trucks. Duration estimated. May not be recognized as an earthquake.
IV.	Hanging objects swing. Vibration like passing of heavy trucks; or sensation of a jolt like a heavy ball striking the walls. Standing cars rock. Windows, dishes, doors rattle. Glasses clink. Crockery clashes. In the upper range of IV, wooden walls and frame creak.
V.	Felt outdoors; direction estimated. Sleepers awakened. Liquids disturbed, some spilled. Small unstable objects displaced or upset. Doors swing, close, open. Shutters, pictures move. Pendulum clocks stop, start, change rate.
VI.	Felt by all. Many frightened and run outdoors. Persons walk unsteadily. Windows, dishes, glassware broken. Knickknacks, books, etc., off shelves. Pictures off walls. Furniture moved or overturned. Weak plaster and masonry D cracked. Small bells ring (church, school). Trees, bushes shaken visibly, or heard to rustle.
VII.	Difficult to stand. Noticed by drivers. Hanging objects quiver. Furniture broken. Damage to masonry D, including cracks. Weak chimneys broken at roof line. Fall of plaster, loose bricks, stones, tiles, cornices, also unbraced parapets and architectural ornaments. Some cracks in masonry C. Waves on ponds, water turbid with mud. Small slides and caving in along sand or gravel banks. Large bells ring. Concrete irrigation ditches damaged.
VIII.	Steering of cars affected. Damage to masonry C; partial collapse. Some damage to masonry B; none to masonry A. Fall of stucco and some masonry walls. Twisting, fall of chimneys, factory stacks, monuments, towers, elevated tanks. Frame houses moved on foundations if not bolted down; loose panel walls thrown out. Decayed piling broken off. Branches broken from trees. Changes in flow or temperature of springs and wells. Cracks in wet ground and on steep slopes.
IX.	General panic. Masonry D destroyed; masonry C heavily damaged, sometimes with complete collapse; masonry B seriously damaged. General damage to foundations. Frame structures, if not bolted, shifted off foundations. Frames racked. Serious damage to reservoirs. Underground pipes broken. Conspicuous cracks in ground. In alluviated areas, sand and mud ejected, earthquake fountains, sand craters.
X.	Most masonry and frame structures destroyed with their foundations. Some well-built wooden structures and bridges destroyed. Serious damage to dams, dikes, embankments. Large landslides. Water thrown on banks of canals, rivers, lakes, etc. Sand and mud shifted horizontally on beaches and flat land. Rails bent slightly.
XI.	Rails bent greatly. Underground pipelines completely out of service.
XII.	Damage nearly total. Large rock masses displaced. Lines of sight and level distorted. Objects thrown into the air.

*To avoid ambiguity of language, the quality of masonry, brick or otherwise, is specified by the following lettering:

Masonry A—Good workmanship, mortar, and design; reinforced, especially laterally, and bound together by using steel, concrete, etc.; designed to resist lateral forces.

Masonry B—Good workmanship and mortar; reinforced, but not designed in detail to resist lateral forces.

Masonry C—Ordinary workmanship and mortar; no extreme weaknesses like failing to tie in at corners, but neither reinforced nor designed against horizontal forces.

Masonry D—Weak materials, such as adobe; poor mortar; low standards of workmanship; weak horizontally.

FIGURE 6.2 *Modified Mercalli Intensity Scale (1956 version from Richter, 1958, pp. 137–138). (From* Earthquakes *by Bruce A. Bolt. Copyright © 1988 by W.H. Freeman and Company. Reprinted with permission.)*

region (just above the white hot liquid outer core) moves upward, displacing some of the cooler upper-mantle material (which dives back down to deeper regions), creating a kind of great circular movement. Corresponding to the surface of the pot in this model is the interface where the earth's crustal plates (generally 40 kilometers or so thick, though nearly twice that under some mountains) can actually separate from the earth just below. The chemistry and rheology of this upper-mantle region, called the asthenosphere, permit the convection of heat in the earth to be transmitted to the rigid plates above in the form of movement.

The crustal plates that form the earth's surface can be said to slide around on the viscous surface, reacting to these internal heat currents in the earth somewhat as small plastic poker chips might to the convection cells in the boiling pot of water. Instead of water bubbles bursting in the pot and releasing vapor into the air, the rising mantle material cools and forms new crust, as the plates slide around and jam into each other at their edges. The energy from this motion is temporarily stored elastically in the rock sections that press and lock together under friction, though "straining" to continue the motion powered by the heat currents. Eventually, the strain energy stored in this system overcomes the frictional resistance of the rock being crammed together, and the system "breaks," releasing the energy in the form of an earthquake.

This model emerges from the reigning paradigm of plate tectonics that has revolutionized modern geology. About a dozen major plates constitute the surface of the earth. The visible continents we inhabit are themselves merely passengers frozen in place atop these slowly shifting rocky rafts of earth. As an entire plate moves only a few centimeters in a year, it is not surprising that people hundreds or thousands of kilometers from a plate's edge, say in Dubuque, Iowa, can remain largely oblivious to the implications of the land beneath their feet not—relative to the center of the earth—being definitively anchored. But those on the very edge of a plate, including millions of Californians, experience plate tectonics viscerally, as patches of rock at the western edge of the North American plate undergo stress, accumulate elastic strain, and eventually and inevitably break loose from rock of the adjacent Pacific plate that is being propelled in a different direction. All such underground shifts are earthquakes, though most occur at levels undetectable except by instruments. Understanding exactly how this occurs, where and when it is most likely, and how much energy will manifest at the surface is the focus of Earth scientists from a number of specialties who come together in search of the keys to earthquake prediction.

Elementary school children now hear this fairly straightforward description of why earthquakes are inevitable, and yet 30 years ago geophysicists were just beginning to piece it together, marking a new era in seismology, a field born at the end of the nineteenth century with the development of recording instruments designed to measure the shaking of the earth (seismometers or seismographs), and the "pictures" they produce (see Box on p. 162). But even the seismograms don't reveal the picture of plate tectonics described above. They simply don't contain that information. "Very clever people," says Ellsworth, "working with very sparse information, cracked the plate tectonics puzzle in the early 1960s, and the dramatic advances in seismology since then rest on better data gathering through advances in tools and technology, the enhanced power to compute complex numerical solutions, and advances in the theory."

What Agnew and Ellsworth call the first "modern" theory of earthquake prediction, by G. K. Gilbert in 1883, was conceived without benefit of the plate tectonics model. Gilbert's notion, as idealized in the block-and-spring model, was developed by another American, Harry Fielding Reid, into what has come to be known as the elastic rebound theory. Reid's study of the fault region that broke during San Francisco's notorious 1906 earthquake provided the first solid evidence for the elastic rebound of the crust after an earthquake. After Reid, scientists began to look at individual earthquakes as passing through a seismic cycle that they could examine and analyze quantitatively, searching for patterns that could be used to develop a prediction of the next rupture. As a generalization, stress accumulates slowly over time on a given fault, it reaches a critical point of failure, the fault slips, and energy is released as an earthquake. Go into the lab—or, more often these days, boot up your computer for a simulation—attach a spring to a block, and exert a pull, and you can watch the basic idea on which Gilbert and Reid built the earthquake cycle hypothesis in action. You will find that the amount of force required before the spring slips and the distance the block moves are predictable.

But the crucial question must be probed: Predictable in what sense? Ellsworth says that in the late 1970s, as elaborate experiments with block and spring models were being conducted in laboratories around the world (this construction of analog earthquake models is still going strong in such places as the Institute for Theoretical Physics in Santa Barbara), an important conceptualization came from Japanese seismologists Shimizaki and Nakata. They realized that, assuming the spring is being pulled at a constant rate, if the block always moves when a given force is

Seismic Waves

Seismologist Thorne Lay from UC-Santa Cruz defines seismology as "the study of elastic waves and the various sources that excite such waves in the earth and other celestial bodies" (Lay, 1992, p. 153). Tectonic forces develop from convective recycling of the earth's interior, heat conducted to the crust that is converted to energy stored in rocks, under strain, which eventually fracture: An earthquake. Seismographs reveal that earthquakes are constantly occurring, several hundreds of thousands each year. Most rumble at levels undetectable by our unaided senses.

Seismographs also show that the energy waves emitted from these earthquakes take several distinguishable forms, each traveling at characteristic speeds. One type, the body waves, travel through the earth, with the primary or P-wave able to traverse both the liquid and solid parts of the interior. The other type of body wave is slower. Called secondary waves (S-wave), these waves shear material sideways as they travel and therefore cannot propagate through the liquid zones of the earth's depths. The other major type of wave moves across the surface of the planet and also comes in two types, the Love wave and the Rayleigh wave.

Basic models and equations were developed to recognize and analyze the travel times each of these waves took—beginning at the source where rocks fracture in the earth's crust—to arrive at various recording seismographs spread over the surface of the planet. By triangulation of the times that a specific wave arrived at several stations, a given event could be located within the earth. By comparing discrepancies between the times the various types of characteristic waves were predicted to—and then actually did—arrive at a given station, scientists gleaned a three-dimensional model of the earth's interior (Figure 6.3), and computer power has further enhanced the process, refining the approach now called seismic tomography.

For instance, shear waves will not propagate through liquid, but primary body waves will; waves of any type pass more quickly through a colder region than a hotter one; and bulk mineral material adopts patterns of crystal alignment in its atoms that can enhance or retard the flow of waves. "These are just a few examples," said Lay (1992, p. 172), "of the ways in which seismological models have provided boundary conditions or direct constraints on models of planetary composition and physical state."

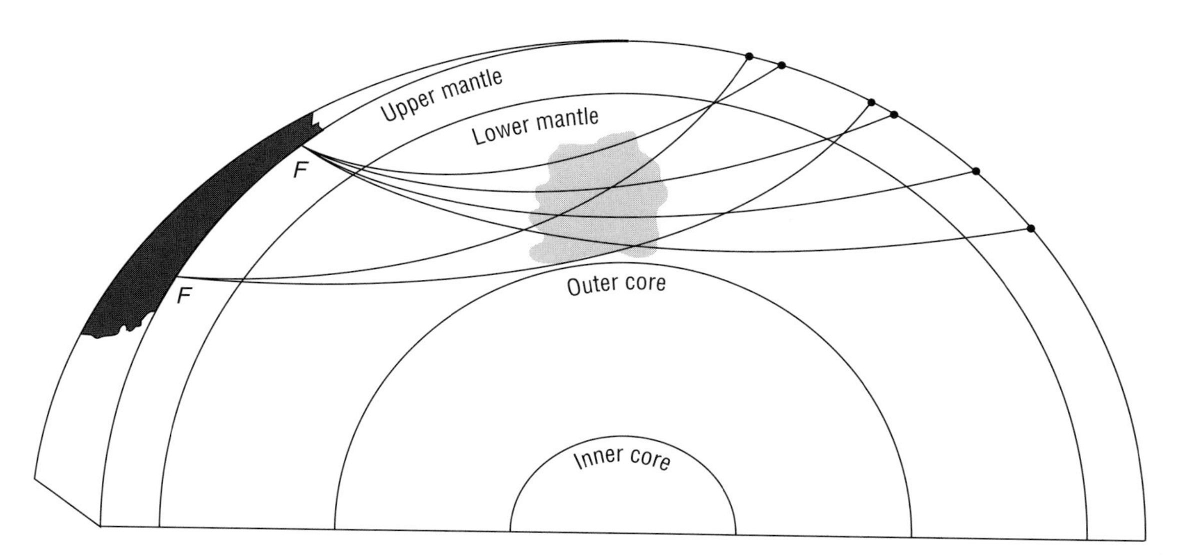

FIGURE 6.3 *Paths of earthquake waves through the earth's mantle from earthquake sources at F to seismographs at the surface. All the seismic paths pass through the shaded region and provide a tomographic scan of it. (From* Earthquakes *by Bruce A. Bolt. Copyright © 1988 by W.H. Freeman and Company. Reprinted with permission.)*

reached, one can predict *when* this will occur—what they called time-predictable behavior. Conversely, if the block always springs back to the same point, the result is slip predictable, though it can't be said when (over the course of exerting the force by pulling the spring) this will happen. In terms of prediction, says Ellsworth dryly, the slip-predictable model is very pessimistic. Conversely, as geophysicists continue to improve their ability to actually measure the tectonic stresses that build up in fault zones, the time-predictable model could one day lead to a more precise means of forecasting.

THE LANGUAGE OF EARTHQUAKE PREDICTION

What is an earthquake prediction? "Any serious prediction," say Agnew and Ellsworth, "must include not only some statement about time of occurrence but also delineate the expected magnitude range [see Box on p. 158] and location." Without these elements, a prediction could not with certainty distinguish between what is being foreseen and what might actually occur, given how much seismicity occurs near plate boundaries, so seismologist and writer Bruce Bolt of the University of California-Berkeley adds a fourth element: "A statement of the odds that an earthquake of the predicted kind would occur by chance alone and without reference to any special evidence" (Bolt, 1988, p. 160). Seismicity means the distribution of earthquakes in space and over time. "We currently record an average of 40 earthquakes per day in Southern California," said Tom Heaton (USGS-Pasadena), "and there is every reason to believe that there are many more that are too small for us to record."

The current definitive overview of seismicity in California, edited by geologist Robert Wallace, was published recently by the USGS as Professional Paper 1515. *The San Andreas Fault System, California*, describes the geological (rocks and their formation), geomorphic (surface features), geophysical (the behavior of matter and forces), geodetic (surveys of the surface), and seismological (waves emanating from earthquakes and explosions in the earth) aspects of this network of faults that extend along much of the length of California, about 1300 kilometers, and perhaps as wide as 150 to 200 kilometers. There is actually an even larger fault system of which the San Andreas is but a part, running up the edge of the Pacific plate from the Gulf of California all the way north to Alaska.

The San Andreas fault was discovered in the 1890s by Andrew Lawson of Berkeley. The northern end of the fault and the zone around it begin about 280 kilometers beyond San Francisco in the Pacific Ocean. It runs south-southeasterly just off the California coast, reaching landfall north of the city, and continues under metropolitan San Francisco, not far from the heart of the city. Continuing to the south it runs parallel to the coast about 60 kilometers inland until reaching the San Gabriel Mountains east of Los Angeles and then curls inland a bit more to the south, perhaps 150 kilometers east of San Diego. From an aerial or satellite photo, Wallace has said, the San Andreas looks like "a linear scar across the landscape," anywhere from several hundred meters to a kilometer in width, an effect of the erosion of rock that has been broken underground and then settled, over the aeons. The San Andreas fault proper is the principal plane where the local horizontal displacements in the earth's crust have occurred. In the zone surrounding this shear plane, geophysicists use such terms as fault branches, splays, strands, and segments to describe details in the fault geometry. Whatever shows at the surface is called the fault trace.

With the fault terrain thus delineated, a legitimate earthquake prediction must locate where on a given fault an event is expected. Most large earthquakes that occur at plate boundaries are categorized as shallow-focus earthquakes, occurring less than 70 kilometers deep. The word focus is usually replaced by the seismologist's term hypocenter, meaning the point where the rupture begins. Projecting the hypocenter directly above to the earth's surface locates an earthquake's epicenter, which is usually what appears on most seismic maps. The much less frequent intermediate-focus earthquakes occur from 70 to 300 kilometers deep, and the fewer still deep-focus earthquakes below that, though none are known to have originated deeper than about 680 kilometers.

The deeper earthquakes can be large but are rarely destructive to humans, said Marcia McNutt, a geophysicist at the Massachusetts Institute of Technology who is exploring how the mantle forces act to deform the earth's crustal plates.

The third and most problematic element of an earthquake prediction is time: When will the event of a predicted magnitude occur at a particular fault segment? Though the Japanese have a much larger billion-dollar experiment in place southwest of Tokyo, the most dedicated American effort ever marshaled to answer this question is currently under way at a small town in central California.

STALKING ELUSIVE PATTERNS

The Characteristic Earthquake Hypothesis

If the time-predictable model is to prevail and provide planners with information upon which useful precautions can be based, a discernible pattern should emerge from the sequence of ruptures at a given point on a particular fault. It was their attraction to this assumption that drew three young seismologists to Parkfield, California, in the late 1960s and early 1970s. Al Lindh was an unabashed child of the sixties, carpenter, garbage route impresario, and an itinerant student for years up and down the West coast, a cultural explorer who eventually got his B.S. in geophysics at UC-Santa Cruz and joined the USGS in 1973. He then began work on his doctorate at Stanford, near his job at the USGS in Menlo Park. His dissertation took him to the small town about midway between L.A. and San Francisco that was the site of a half dozen historical earthquakes M6 or so in size. More importantly, it was where the San Andreas, running south, appeared to "lock up" again, just south of what is called the creeping section. This is where, simply put, the earth on either side of the fault in this short region seems to be moving freely without accumulating any strain. Lindh showed that the fault took a five degree turn, which seemed to isolate it from the creeping section and put it back in the line of fire of traditional tectonic accumulating stress.

Lindh then graduated from working essentially alone and planted the seeds for a much grander notion: To establish at Parkfield a focused experiment that would monitor the area for as many signs of impending earthquakes as geophysicists could develop instruments to measure. "A lot of my energy went into talking to people for three or four years. It was plain old science as usual. You write papers, you give seminars, you talk to friends. Some of them believe you, some don't; but they keep hearing the same old story. After a while some of them even decide they want to get involved" (Heppenheimer, 1990, p. 175). One who did was Bill Bakun, another geophysicist whose graduate work focused on Parkfield. In 1979, together with Tom McEvilly (a professor of geophysics at UC-Berkeley), based on the Parkfield sequence, Bakun had sketched out what was to become the cornerstone of the characteristic earthquake hypothesis.

At Parkfield they perceived similarities between the 1934 and 1966 quakes that led them to a more detailed analysis of the four previous quakes on the fault in 1922, 1901, 1881, and 1857. Analysis of historical, geodetic, and seismographic data showed a remarkable similarity in the rupture pattern. They found, said Ellsworth, that it seemed to start at the same place, and seemed to produce the same amount of slip regardless of when it occurred. But *when* it occurred also interested Bakun and McEvilly, who saw a tantalizing suggestion of pattern to the time sequence as well. They finally selected a statistical model they believed could characterize the sequence of data point/dates, and it produced a result of 22 ± 5 years. If the fault were to rupture again during that period, they were fairly certain it would be a repeat performance of, at least, the six previous ruptures they had studied. If the physical characteristics were close enough to be called "the same," or characteristic, and the time interval also fit the pattern they had discerned from their statistical modeling, strong claims could be made that this—and perhaps many other—earthquakes were characteristic in *all* of their major parameters—in sum, a tightly controlled block and spring experiment conducted by nature.

Bakun and McEvilly's statistical analysis of the Parkfield events led them to develop a recurrence model with a 95 percent reliability that predicted the quake "should" occur within 5 years (plus or minus) of 1988. That this window has now closed and the Parkfield segment has not yet ruptured casts doubt on Bakun and McEvilly's model, and seismologists have joined a debate about the characteristic earthquake hypothesis. "In fact," said McNutt, "Paul Segall at Stanford has argued from his geodetic research work that the so-called repeated Parkfield earthquakes broke different sections of the fault." Ellsworth points out that his USGS colleagues developed their prediction for the time interval to the next quake with a *particular statistical model*, a choice influenced by the time series they had to work with. But with only six data points, this model could be "the wrong one," as compared to a different choice that might have been based on a richer sequence of not only those six data points but hundreds of others as well. Looking (and guessing) in the

1980s and not in the year 4980, however, they didn't have hundreds of others but only six. Thus, the Parkfield controversy is somewhat off the point, Ellsworth believes, and the failure of the next Parkfield event to arrive on Bakun and McEvilly's schedule has little, if anything, to do with the characteristic earthquake hypothesis.

Ellsworth emphasizes that the interevent interval assumes its significance only in the "surprising" context "that the Parkfield segment appears to have produced nearly the same earthquake each time it has broken. This was a surprising result, as one might expect that one earthquake so changes the stresses that the next will bear no relation to the former, and certainly won't be identical." What drives geophysicists is their deep interest "in what happens when the fault breaks in a characteristic event, whether or not the stress levels reach the same level each time, what initiates the rupture, and just how the rupture proceeds along the fault." There has been sufficient data gathered about such events, he believes, to state that the characteristic earthquake model "is quite strong for some parts of the San Andreas fault," notwithstanding dissenting opinions about Parkfield. As evidence, he relies less on the Parkfield sequence and the timing of the next rupture there than on the extant record of smaller earthquakes (M 1–5) throughout California, many of which *do* conform to the model. He believes the question is best attacked by concentrating on those earthquakes that rupture—as can be discerned in a study of their seismograms—in an almost identical pattern underground, even if they are much smaller than Parkfield or other larger magnitude quakes that draw our attention because of the threat they pose.

Though stronger evidence continues to accumulate, it is still possible that these attempts at forecasting and subsequent work may eventually reveal the absence of pattern: Perhaps regularities in earthquakes are not generic but rather indigenous to a particular type of plate boundary, even further to more specific local conditions that have yet to be defined. It may even turn out that there is no such animal as a characteristic earthquake, but rather—to mix scientific metaphors—only an order, with species as rich and diversified as those on the Galapagos. At present, however, the characteristic earthquake idea frames much of the thinking about long-term prediction.

The Parkfield drama thus provides a concentrated lens to view seismology, and several interesting issues come into focus. First, how can scientists expand the empirical data base available for their probability models? While monitoring nature's own ongoing laboratory experiments by attending closely to the details of quakes as they occur, the only other place to look is the past, to recorded history and, with the intriguing new science of paleoseismology, into prehistory as well. Second, as people living in fault zones—and the politicians who represent them—go about their business under the shadow of "The Big One" and other less mythic earthquakes, they want to know what to do. To scientists and the USGS falls the role of providing the best information possible, both to forecast quakes and to try to mitigate their effects. Third, geophysicists are developing other analytical and theoretical ideas (besides the block and spring and characteristic models), laboratory techniques, and monitoring strategies to try to better understand the genesis and mechanics of earthquakes. Finally, though the Parkfield quake didn't erupt in October 1992, another furor of sorts did, shining a spotlight on both the science and sociology of earthquake prediction.

Hitting the Trail into Past Time

As any scientist will readily concede, your conclusions are only as good as your data. And for models as heavily reliant on the particularities of data as are the seismic cycle, elastic rebound, and characteristic earthquake theories, interevent times for a particular fault shine like diamonds in a coal mine. The more precisely known the year of past events, the closer can be the statistical analysis of the spectrum of data. The greater the number of events or data points, the more likely that subpatterns will be expressed and the greater confidence scientists can take in their conclusions. In recurrence models, scientists reflect their subjective judgments about the quality of the data used to produce a particular prediction with a reliability measure (such as A through E for strongly to weakly confident).

A major hurdle faced by those trying to unravel the mystery of earthquakes is the recency of recording devices. Since John Milne began to install his worldwide network of seismographs in 1896, there is less than a century of seismograms to pore over for evidence of repeating earthquakes. The elastic rebound theory predicts that the larger the quake, the longer the time necessary for the fault to reaccumulate that amount of stress; thus, the interevent times for quakes above M 6 are measured in centuries, and not even one full cycle has run since seismology ushered in the era of instrumentation. Ellsworth, Agnew, and many others have extended their research farther back into the preinstrumental historical record to see what settlers had to say about nineteenth-century earthquakes.

By adding history to seismology, the potential time series for the

San Andreas system was about doubled but still covered only just a bit more than one cycle for any given fault. McNutt wonders if better archeology of American Indian culture in this region might not provide evidence even farther back. But for now, the written record from the American West provides scant material for historians before the early nineteenth century, and so other scientists returned to the basic science of geology, examining the rocks and strata near the earth's surface for evidence of the history of past earthquakes.

Paleoseismology

The Cenozoic (or present) geological era coincides approximately with the appearance and rise of mammals some 65 million years ago, but geologists searching the record for evidence of prehistoric earthquakes are content for now to rummage around in the Holocene epoch, approximately the last 10,000 years. One of the lead scouts on this expedition into the past is Kerry Sieh, who modestly explains away his first National Science Foundation grant (awarded before he had even begun graduate school at Stanford): "Really, I was just in the right place at the right time." Ellsworth notes that while geologists had been gathering fault slip data since G. K. Gilbert's day, "Kerry really saw what to do with these clues in the geologic record" and made an enormous investment in time and energy on what to many others at the time was little more than a scientific hunch. Now, 20 years later, a significant body of crucial prehistoric data has been developed by Sieh (supported early on by his mentor Richard Jahns, the late dean of the School of Earth Sciences at Stanford) and others following his lead into the new subfield of paleoseismology, which he defines as "the recognition and characterization of past earthquakes from evidence in the geological record."

A paleoseismic exploration begins by excavating the earth at the fault trace to create a large pit and reveal the fault broadside, thereby unveiling a mural deep enough to look back centuries in time. Ancient organic remains exposed in the mural could be dated to establish when the layer in which they were embedded was deposited. What Sieh typically saw in such cutaway views was remarkably similar to another earthquake signature seen at the surface of the earth—offsets. Imagine a creek or river running across a fault: When the fault moves suddenly and produces an earthquake, the opposite shore banks are displaced up to several meters, moving the creek bed too, and producing a zigzag pattern through time of the course of the water. Seen from the air, even to the untrained eye, the San Andreas reveals hundreds of such offsets of streams and other obvious surface features. Similar offsets in Sieh's rocky mural provided analogous intimations of plate movement, and read like chapters in the history of a particular fault, a story written by the earth and translated by geologists. Ellsworth provides a kind of synopsis of this story: "*Some* earthquakes and some faults are not totally random in their behavior, but rather have a history, which we are learning to read."

Sieh's first major discoveries were at an area known as Pallett Creek, a tiny stream flowing down from the mountains into the southern Mojave Desert, east of Palmdale, near the town of Pearblossom. When he saw from aerial photos that the mountain stream seemed to jog in an offset of about 130 meters at the fault, he believed he had found a stream older than the Holocene and that he might unveil 10 millennia or more worth of earthquake data. But the radiocarbon analysis (see Box on p. 171) of peat just below the streambed indicated it was only 2000 years old. Sieh was incredulous that the large offset could have developed in that short a time. He eventually discovered that the stream had been redirected not by a fault but by a natural rise in the land. When he finally revealed the trace and completed a painstaking examination of hundreds of square meters of stratigraphic layers of soil and peat, he began to refine a methodology using radiocarbon dating and other lines of analysis that could date indicative offsets and formations known as sandblows to within a couple of decades.

To Sieh's ever-developing eye, the mural at Pallett Creek began to reveal a sequence of a dozen earthquake events repeating here at the edge of a continent, half a world away from anyone who could provide historical verification. As he put it, "Geology was running concurrently with European history." Sieh has a keen sense of the paradox of specifying events in these relatively piddling pieces of historical time, while working in the field of geology where eras are routinely scored in hundreds of millions of years. "This fascinated me; I love working with geology that I can put in a human context. It's tough having a sense of time when you're talking billions of years" (Heppenheimer, 1990, p. 87). In Sieh's catalog the recent geological history of the region near Pallett Creek came alive, as earthquakes were indicated in the years 260, 350, 590, 735, 845, 935, 1015, 1080, 1350, 1550, 1720, 1812, and 1857. (see Figure 6.4 for a history of earthquakes on the San Andreas fault.) There had been thirteen significant earthquakes on a fault that, though it had not broken for well over a century, now lay within 55 kilometers of Los Angeles, the second-largest city in the country. Add the 145-year average interval to the most recent occurrence, and the significance of

Radiometric Dating in Paleoseismology

Much of the earth's thermal energy comes from radioactive isotopes created during its formation some 4.5 billion years ago. Potassium, thorium, and two species of uranium are the primary elements, heavily concentrated in the earth's crust, where 40 percent of the heat flow experienced at the surface comes from.

Radioactive is synonymous with unstable. As these elements lose electrons, they decay into what are called daughter products, stable elements such as argon, strontium, and lead but versions of these that can be identified as descendants of each radioisotope. Radiometric dating relies on the discovery that each of these elements takes a specific and known time for half of the atoms in any particular collection to decay, a period known as the half-life. Because the half-lives of these elements are in the millions and billions of years, they provide a time scale congruent with the history of the planet.

Carbon-14 (C^{14}), on the other hand, has a half-life of only 5730 years, and thus *radiocarbon* dating is better suited (and limited) to about 80,000 years. This radioactive isotope (created by cosmic-ray collisions with carbon dioxide molecules in the upper atmosphere) comprises a known portion of the carbon that is continually being incorporated into organic matter on Earth. C^{14} begins to decay when its host dies, and thus the same calculation (of the present volume of atoms versus the original) allows the fossil to be dated.

Kerry Sieh of Caltech, following the lead of Stanford's late Richard Jahns, was a trailblazer with his work at Pallett Creek (known as the Rosetta Stone of paleoseismology) and other California fault sites. By cutting away a trench and revealing a fault broadside, he was able to construct what is called a stratigraphic map of the history of that fault's ruptures. By concentrating on the peat buried at various strata, he was able to develop dates for 12 Pallett Creek earthquakes going back to the third century. His original study is reported in the text.

In 1988, with colleagues Minze Stuvier from the Quarternary Research Center at the University of Washington, and David Brillinger from UC-Berkeley, he reexamined the original samples (Sieh et al., 1989). Advances in the technique at Stuvier's lab reduced the error range of 50 to 100 years down to 12 to 20. Redating the most recent 10 quakes recalculated the average interval to be 132 years, which was precisely the amount of time that had elapsed since the last quake in 1857.

Sieh's work to millions of Southern Californians is clear. Such a dramatic impact happens infrequently in the steady incremental progress of daily science, and Pallett Creek has come to be known as "the Rosetta stone of paleoseismology."

Slipping and a "Slidin"

When Sieh first began these excavations in the late 1970s, plate tectonics had become a useful and accepted set of ideas, and the basic rate at which the Pacific and North American plates were sliding past each other had been established at about 5.6 centimeters per year for at least the last 3 million years. Given this underlying relative movement

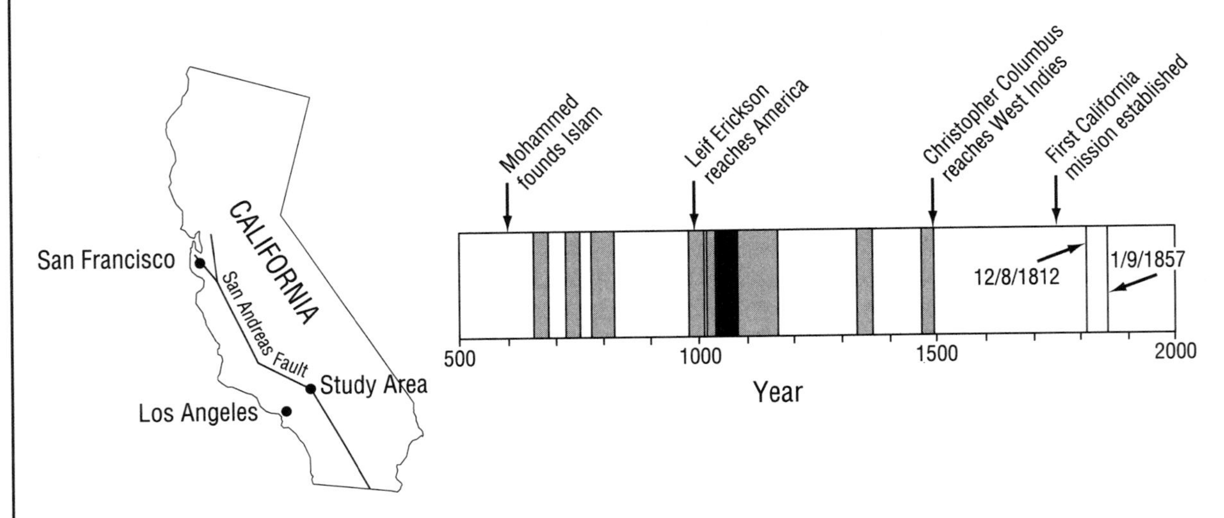

FIGURE 6.4 *History of earthquake occurrence on the San Andreas fault extended 1200 years beyond written records by geological investigations. Detailed geological studies augment the relatively short historical and instrumental records of U.S. earthquake activity and yield critical information on the timing and character of prehistoric earthquakes. At a study site on the San Andreas fault northeast of Los Angeles, earthquakes have occurred in clusters of two or three events spanning several decades separated by dormant periods of two or three centuries. Precision radiocarbon dating of buried peat horizons in a faulted sequence of marsh and stream deposits has yielded age ranges (shaded bars) for eight earthquakes prior to the 1812 and 1857 shocks, which are documented by historical accounts. The observed pattern of earthquake clustering was not apparent in earlier studies, which used less precise, conventional radiocarbon dating techniques. Knowledge of long-term patterns of earthquake occurrence is needed for probabilistic forecasting of future earthquake activity. (Reprinted from USGS, 1992, p. 26.)*

of the entire plate system, Sieh could use the elastic rebound model to provide an earthquake catalog extending deep into the prehistoric past. The calculation begins with the assumption that, if the crust were totally rigid and moving at the plate rate (which has since been revised downward to 4.8 centimeters per year), the Pacific plate would move northwestward at a steady rate, relative to the southeasterly sliding North American plate (so that in less than 40 million years San Diego would become a suburb of Seattle).

Because the fault is a frictional surface however, Ellsworth points out that sections of the fault, identified as asperities, "lock up" and the elastic crust accumulates and stores this movement in the form of strain and potential energy, breaking only when the fault's frictional resistance is overcome. Though geophysicists with increasingly powerful laboratory instruments and theoretical models have for decades been trying to specify exactly what these conditions might be, long-term forecasting still rests largely on the premise that—whatever they are, at least according to the characteristic earthquake hypothesis—they are consistent from one occurrence to the next. Thus, the plate rate provides a way to quantify the strain accumulated over time. Over vast time periods the crust must catch up with the lithospheric "raft" below (of which it is a constituent part; not to be confused with the asthenospheric boundary where the entire lithospheric plate can slide more or less freely) and thus, at any particular period, it is more or less behind. When it gets so far behind that fracture finally occurs, an earthquake is the likely result, and—at least in transform fault zones like the one where the Pacific and North American plates meet in California—measurable offsets will be carved into the earth.

Take one specific fault strand, such as Pallett Creek, and a stratigraphic map will reveal a history of such offsets for many centuries. One first measures the offset distances and (knowing the age of each offset) can then correlate them with the time it took for the plates to separate sufficiently to cause the earthquake that produced that offset. The method works best with major earthquakes, where it is presumed that all of the accumulated strain was released (following Reid's concept of elastic rebound) and the equation reverts back to zero. In fact, Reid recommended 80 years ago that more systematic measurements be taken near the San Francisco rupture, but it would be decades before this was done. Seismologists use this slip rate feature of plates as a baseline for their time-predictable models.

Another project of Sieh's illustrates the significance of this basic plate rate to forecasting earthquakes for individual faults. Using Wallace's aerial photos of a different unnamed stream that flowed out of the Tremblor Range and onto the Carrizo Plain, Sieh began by examining the offsets where the San Andreas ran directly across what he soon christened as Wallace Creek. Rather than a series of offsets, Wallace Creek at the surface showed one major zigzag of 128 meters and another larger one of 475 meters that C^{14} dated at 13,250 years. After excavating many trenches over a number of years, Sieh finally proved that the smaller offset was a measurement of the slip along the fault that had accumulated since a major mudslide about 3700 years ago. To within the limits of error, the slip rate for this and the much larger offset, after Sieh had sifted through many confounding factors, concurred at about 3.4 centimeters a year, which also matched the rate determined at Pallett Creek and elsewhere.

When the slip rates for other faults were determined, few matched the underlying annual slip rate of 4.8 centimeters per year for the Pacific plate itself, and thus other major fault zones are known to be accumulating some of the stress. This piece of data permitted slip rate calculations, based on offset distance and interevent time, to be compared throughout the San Andreas fault zone. As already discussed, one particular region of the fault in Central California, just north of Parkfield, does actually appear to be slipping steadily—a phenomenon known as creep—at the full rate for the San Andreas fault. This creeping section would seem to mitigate against anything like a superquake that could tear through all of California.

Sieh's paleoseismic studies and a study of the history of California seismicity by Ellsworth both indicated that a major event was due in Southern California, but in June 1992, said Heaton, the rupture at Landers "was a complete surprise, which cut across all known fault lines." With a seismic force of M 7.5, it was the largest California quake in 40 years and—after the 1906 San Francisco quake—the third-largest in this century. A primary question immediately arose: Did Landers make Los Angeles and other Southern California cities a safer place to live? The persistence of "The Big One" mythology was prominent in the California media in the wake of Landers, as scientists were asked repeatedly not only whether this was it, but also whether the Mojave shear zone, which contained the source of the Landers quake, was now accommodating more of the movement between the North American and Pacific plates than previously believed. The simple answer to both questions was no, but simple answers do not really convey the complexity and the extent of seismological wisdom. A working group of seismologists was convened to develop and publish information about

Southern California, post-Landers. Sieh and Agnew were two of the dozen members, which included as cochair Tom Heaton.

What the group identified as the Landers sequence illustrates the puzzle mosaic nature of earthquake science, for each major event holds the promise not only of pushing probability data farther toward certainty but also of revealing subterranean connections not readily apparent. In retrospect, it is clear that Landers culminated a sequence of events in the Mojave zone, which includes the 1975 Galway (M 5.2), 1979 Homestead Valley (M 5.6), 1986 North Palm Springs (M 6.0), and 1992 Joshua Tree (M 6.1) quakes. It can be seen as the main shock of that sequence, because of its size (M 7.5), and it is believed that stress on this fault has been significantly diminished. Moreover, not 3½ hours after Landers, a major quake occurred nearby at Big Bear, measuring M 6.5. The group decided that the Big Bear quake was an aftershock of the Landers earthquake because it was within one rupture length of the Landers main shock and had a magnitude consistent with the normal distribution of aftershock sizes for an M 7.5 main shock.

In sum, the Landers and Big Bear quakes provided an opportunity to update the overall picture of seismicity in Southern California as of June 1992 and also for the working group to reinforce the bottom line of all of the comparative statistics, of which seismologists have been aware for over a decade: "Portions of the southern San Andreas fault appear ready for failure; where data are available [such as Sieh's Pallet Creek studies], the time elapsed since the last large earthquake exceeds the long-term recurrence interval" (Ad Hoc Working Group, 1992, p. 1). Long-term forecasting revolves around the seismic cycle, and "since 1985, earthquakes have occurred at a higher rate than for the preceding four decades," the group emphasized, referring specifically to the period since 1985, which showed a rate 1.7 times greater for M 5 and larger quakes and 3.6 times greater for M 6 and larger earthquakes.

While quakes occur and energy is released on distinct fault segments, it is believed that stresses become redistributed throughout the fault zone. On the affected segment, strain will probably be diminished, but on nearby segments and faults it might well be increased. The group decided that the Landers earthquake increased the stress toward the failure limit on parts of the southern San Andreas fault, in particular on the San Bernardino Mountains segment. Contrary to some popular reports, Landers and Big Bear do not likely represent a major underground shift of stress accumulation and release to the Mojave zone. Thus, the group confirmed that over 70 percent of the effects of plate motion due to the 4.8 centimeters per year of slip between the plates is straining the southern San Andreas, an area where millions of people could be affected by a major quake.

Reading the Historical Record

Distributed stress from the Landers sequence aside, it is still crucial to look at the long-term history of large quakes in Southern California. Sieh has recently taken another route into the prehistoric record, which, though not quite as far reaching as trenching studies, seems to provide much more accurate dates: Tree rings. As with the development of stratigraphic mapping, the procedure didn't just fall out of the laboratory fully ready to be used, though botanists have known about tree rings as a measure of age for centuries. With each new spring a tree begins a new ring and thus develops a putative birth certificate. When Sieh realized that many trees shaken by prehistoric earthquakes had been damaged but not destroyed, he began to look to their tree rings to determine the precise year of the earthquakes. Whether it be damage to the tree's root system by fault rupture underfoot or to its overall health by a lightning strike, flood, or some other environmental catastrophe, exactly when a particular growth ring was affected could pinpoint an event to within several months. By looking at patterns of nearby and somewhat farther removed trees, he was able to distinguish between these various types of ancient environmental phenomena, largely based on where they stood relative to the fault and to one another. This culminated in yet another method for dating earthquakes, which could take advantage of California trees up to 500 years old.

Working with tree ring expert Gordon Jacoby from Columbia University, Sieh realized that the meager information about the December 8, 1812 earthquake came from the Spanish missions that were near the Southern California coast, and a doubt dogged him because he realized that the intensities out toward the San Andreas could have been much higher than those recorded at the mission. One night over dinner he and Jacoby were discussing the tree ring record in the context of their search for corroborating evidence of a quake on the San Andreas in the early 1700s. With his misgivings about the interpretation of the mission evidence in mind, he asked Jacoby about tree ring data after about 1810 and was told that no count had been made beyond 1812 because the rings were too close together to distinguish. "Aha" the two said in metaphoric unison, which Ellsworth thinks is "a rather good example of how science is done." Many trees, once the scientists knew where and what to look for, indicated a quake at the end of the 1812 growth season.

It could have been the earthquake known to have been centered near Santa Barbara on December 21 of that year, but in an impressive bit of deduction (given the time frame of just a couple of weeks nearly two centuries in the past) Sieh rejected that possibility, accounted for all the known data, and concluded that "there's no reason why it's not perfectly consistent with a great quake on the San Andreas" on December 8. The significance of this possibility, like the work at Wallace and Pallett creeks, is what it says about historical seismicity in Southern California and the fate of Los Angeles and other major urban centers.

With his work on the 1812 quake, Sieh had moved into a more recent area of inquiry, using nineteenth-century historical records. Working in this area of preinstrumental seismicity, scientists are limited to intricate schemes of inference and induction in order to quantify an earthquake (see Box on p. 158). Sieh and Agnew developed an isoseismal map in their exhaustive reexamination of the next major historical event in Southern California, the famous Fort Tejon earthquake of January 9, 1857, the importance of which, they point out, has long been recognized.

Their analysis suggested to them that the intensity along the fault in 1857 "must have been IX or more [corresponding on the Modified Mercalli Intensity Scale to "general panic, most masonry heavily damaged or destroyed"—see p. 159] since trees 20 kilometers west of Fort Tejon were overthrown and buildings were destroyed between Fort Tejon and Elizabeth Lake," the then-small Pueblo village that occupied the present site of downtown Los Angeles and fared much better, experiencing an intensity of VI or so ["felt by all, weak plaster cracked, dishes broken"]. Thus, they observed, "perhaps the most interesting conclusion that can be drawn from this study is that . . . were the 1857 earthquake to be repeated today, there would not be extensive damage to low-rise construction in the metropolitan Los Angeles area, [though] the evidence does suggest that there would be substantial damage to structures along the fault."

Add Sieh's interpretation of the 1812 event to this clarification of 1857, and a number of questions must be faced by the city fathers of Los Angeles. Depending on which fault ruptured in 1812, did the two major quakes in less than 45 years relieve so much strain on one fault that it may yet be many more decades before it will reaccumulate? Or were those quakes on different but nearby faults, forming a kind of "pair" sequence that could arguably be seen in San Francisco, where a major quake preceded the 1906 event by only 68 years? If so, "The Big One" could actually become "The Big Two" and strike twice during the lifetimes of many who will be fairly young for the first. Or does the

Pallett Creek sequence, based as it is on an impressive time series of events, suggest that the city is right now in the bull's-eye of the probability window? Many such questions can be better posed as a result of interpreting the historical earthquake catalog in the context of the elastic rebound theory, especially when slip measurements of the entire San Andreas system are added to the mix.

The most exhaustive catalog of California earthquakes to date—incorporating the earlier compilations and analyses of many others—was provided by Ellsworth and appears in Wallace's San Andreas overview. Ellsworth regards Sieh's work at Pallett Creek as seminal and the contributions of other paleoseismologists and historians as invaluable. Because their studies are primarily limited to earthquakes of large magnitudes, which have interevent times of many decades, even centuries, scientists have no choice but to invent complementary ways, like paleoseismology, to look back in time, the farther and more accurately the better. But Ellsworth's conviction about the importance of testing the characteristic earthquake hypothesis has led him to focus on the more numerous smaller-magnitude earthquakes that can be studied in the seismographic record. If you limit yourself to the bigger earthquakes, you're going to be data limited. Go in the other direction and look at smaller quakes, he suggests, and you will find yourself on a firm observational base with data on many earthquakes as small as M 1 from which to draw conclusions. "It's really almost an experimental machine," he marvels, "because the earth is actually turning the experiments out for you. The natural laboratory allows you to look at the cycle for small earthquakes many times within your record."

Predictability and Politics

"In the late afternoon of 17 October 1989, as the eyes of America turned toward Game 3 of the World Series at Candlestick Park, the largest earthquake in northern California since the great earthquake of 1906 struck the San Francisco Bay Area," wrote the USGS in *Science.* The television picture suddenly began to tremble and the people on camera demonstrated that unholy moment of fear [though sports commentator and San Franciscan Al Michaels was later commended for an uncommon sense of control and awareness] when an earthquake begins to rumble beneath your feet. The nearby southern segment of a fault running just east of Santa Cruz under the Santa Cruz Mountains had broken on a steeply dipping fault plane at a point 17 kilometers beneath the surface, a rupture that lasted for 7 to 10 seconds, and slipped for a couple of meters

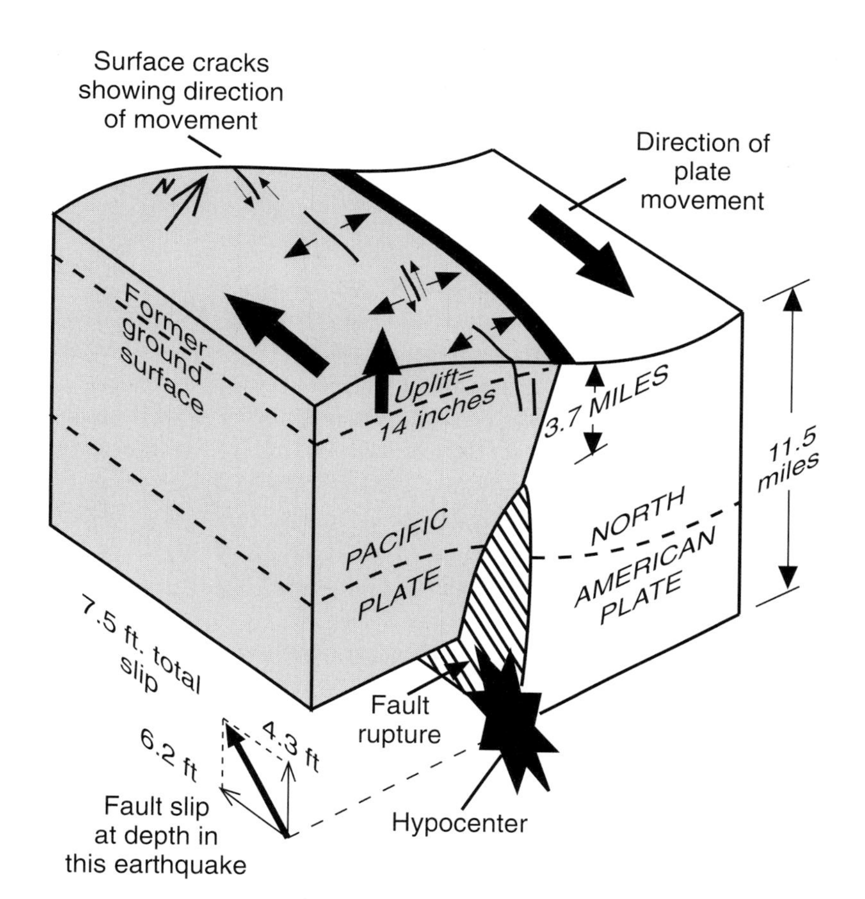

FIGURE 6.5 *Schematic diagram showing inferred motion on the San Andreas fault during the Loma Prieta earthquake. Along the southern Santa Cruz Mountains segment of the fault, the Pacific and North American plates meet along an inclined plane that dips approximately 70 degrees southwest. Plate motion is mostly accommodated by about 6.2 feet of slip along a strike of this plane and by 4.3 feet of reverse slip, in which the Pacific plate moves up the fault and overrides the North American plate. The amounts of fault slip and vertical surface deformation were determined from geodetic data. (Modified from a figure by M. J. Rymer. Reprinted from USGS, 1989, p. 6.)*

deep underground (see Figure 6.5). The fault slip itself did not reach the surface to produce a trace, but surface waves achieved a magnitude of 7.1. As reported by the USGS, the ground shaking collapsed sections of the Bay Bridge and Interstate 880 (in Oakland); began fires in San Francisco's Marina district; ultimately caused 62 deaths and 3757 injuries; destroyed 963 homes; damaged another 18,000, leaving 12,000 temporarily homeless; and ultimately will have cost some $10 billion. These statistics rate it as one of America's most serious natural disasters. As such, it raises questions about how long-term forecasting fits into the American political infrastructure, since in the context of work done by Ellsworth, Agnew, Sieh, and many others, the USGS in the *Science* article classified Loma Prieta as "an anticipated event."

Californians have felt thousands of earthquakes over the decades, and so it was natural that the politics as well as the science of prediction should center there. The latest intersection is the Working Group on California Earthquake Probabilities, organized by the National Earthquake Prediction Evaluation Council (NEPEC) under the auspices of the USGS. The group includes representatives from the USGS, academia, and private industry. Ellsworth, his colleague Lindh, and Chris Scholz (geophysicist from Columbia University and author of a definitive current text) are three of the group's dozen members. Their first two reports bracketed the Loma Prieta earthquake, and, based on the former report (Working Group, 1988), the latter (Working Group, 1990) explains why the USGS was able to "anticipate" the 1989 event. The second report establishes and discusses the future probabilities of earthquakes in the San Francisco Bay region in the wake of the strain released at Loma Prieta. In the first report, seismologists had calculated the probability that each of California's major faults would rupture in the next 30 years, expressed as a fraction. The highest probability in Northern California was assigned to the southern Santa Cruz Mountains segment, .3 (see Figure 6.6).

To the scientist familiar with statistical procedures and the Poissonian model (which was the reference case), such numbers are very compelling. To those without this perspective, however, it may seem that a .3 probability of an event occurring can provide a false sense of security based on the 7 in 10 chance the event will not occur in the next 30 years. As Ellsworth says, your chances of experiencing a quake on a given day are microscopically small, though it will be a day, if you survive it, you will never forget; what he calls a low-probability/high-impact event. To illustrate, he suggests you think about the odds of being injured driving your car, compared to your chances of being hurt in the quake, *even assuming it will arrive.* If you stay put and ride out the quake, your chances of injury are less than if you took your family for an all-day drive in the car. Thus, he concludes, "the appropriate response to such a low-probability/high-risk situation is not always obvious."

While scientists are able to vest increasing confidence in their long-term forecasts, the very nature of probability estimates will probably remain somewhat inscrutable to the general public. Those accustomed to taking their science in predigested and denatured journalistic teaspoons are not likely to develop a sudden appreciation for the subtleties of uncertainty inherent in this sort of scientific exercise, especially when their lives and possessions seem to be threatened. Thus, emphasized Ellsworth, coming to a consensus for such reports is critical "when carrying uncertain knowledge into the public arena."

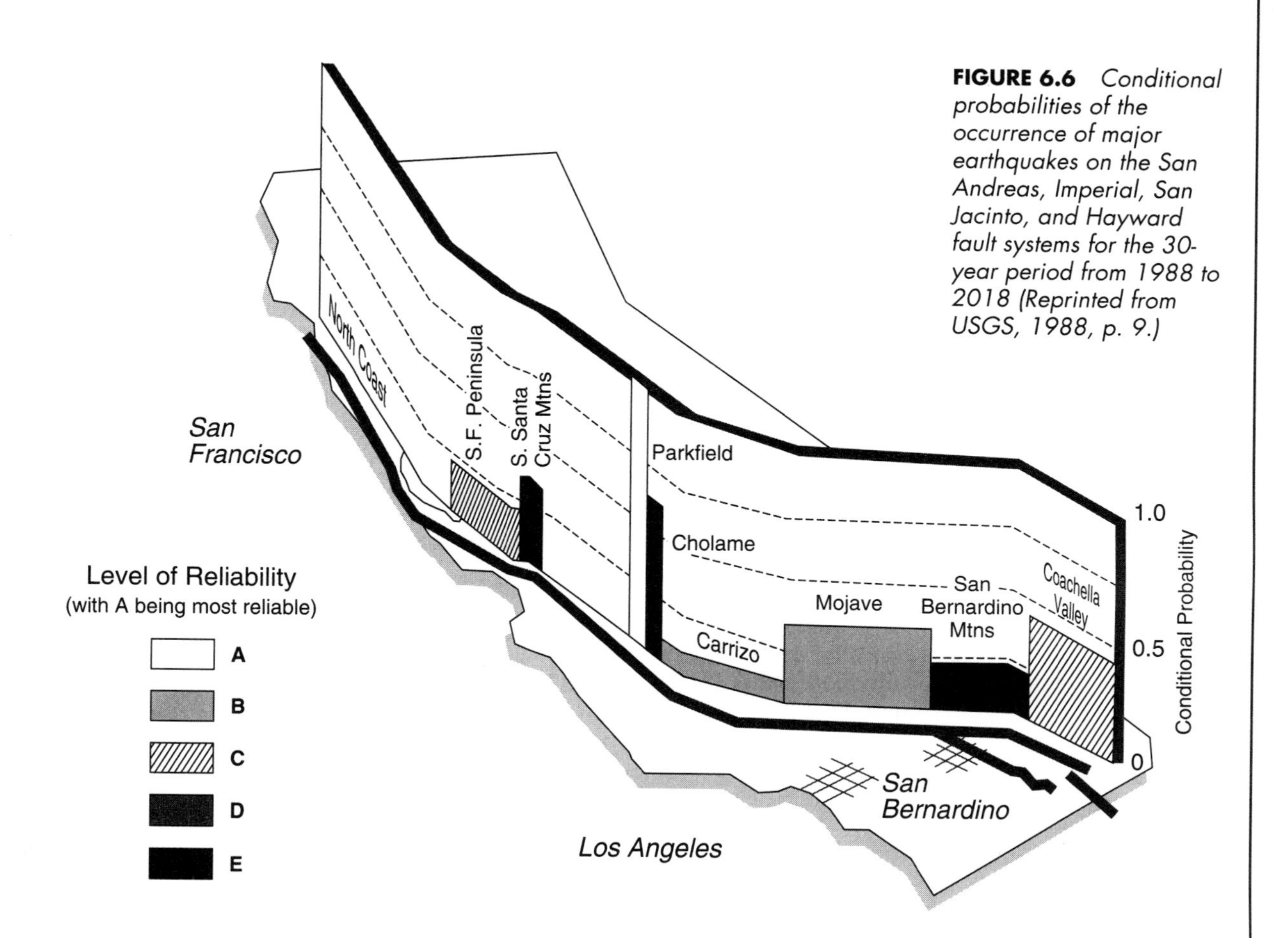

FIGURE 6.6 *Conditional probabilities of the occurrence of major earthquakes on the San Andreas, Imperial, San Jacinto, and Hayward fault systems for the 30-year period from 1988 to 2018 (Reprinted from USGS, 1988, p. 9.)*

Rather than try to unveil what may be seen as the esoteric mysteries of such statistical phenomena, recent experience suggests it is better to look broadly at the results. Using the current models, the consensus 30-year probability of a rupture for the Santa Cruz segment was .3. (Lindh in particular had focused on this segment, and expressed the probability much higher, between .47 and .83 for an M 6.5 event). Yet the earthquake ruptured in less than 2 years, many people were killed, and few precautionary steps had been undertaken. Thus, when a 30-year probability estimate for a given segment is above .2, the question probably should not be "How soon will the quake actually arrive (with accompanying acrimonious debate among scientists and consequent misinformation disseminated in the press)?" But rather, since it is highly likely that it will arrive sometime in the humanly foreseeable future, "What are we going to do about it?"

This dual perspective poses a dilemma that weighs heavily on most of the scientists involved in earthquake prediction. As Ellsworth put it: "In this field we find ourselves trying to maintain a delicate equilibrium between research at the frontiers of our understanding of the phenomenon and its application to public policy." The USGS, after completing the Working Group report for 1990, decided to confront the threat of misinformation head on. Rather than leaving the scientific report (properly and documentarily dense, though well structured and clearly written) for the press and public to interpret and summarize for themselves, the USGS produced a full-color, 24-page pamphlet in English, Spanish, Chinese, and braille. When the scientific report was released, 41 newspapers in the region provided the more accessible pamphlet as an insert in their Sunday editions, leading to the distribution of 2.5 million copies. As is only reasonable given the intended readership, the pamphlet emphasized preparedness, but it also included a lucid section titled "Why a Major Earthquake is Highly Likely."

Using clear language and illustrative figures—and absent the tendency of scientific publications to overly qualify and understate important conclusions—the pamphlet would seem to be a major step in dealing with the dilemma seismologists face. Since so much of the important seismological research in America moves through a federal agency, the USGS, the sense of responsibility for the public interest is inherent, especially since refereed journals and the other mechanisms of peer review and scientific method remain in place to monitor validity. Notwithstanding the determination to transmit the most constructive message to the public, however, the dilemma remains. In the real world—until science can predict earthquakes within a much shorter time frame and with much more certainty—politicians and people will continue to act out these uncertainties in socially predictable ways.

GEOPHYSICS IN LABORATORY EARTH

Solar and lunar cycles are predictable. Ever since Copernicus and especially Newton, scientists—by constructing descriptions from observed phenomena and deriving from them the laws of nature and the laws of physics—may be said to have *explained* these cycles with their theories. As Bolt puts it, "A strong theoretical basis is usually needed to make reliable predictions such as in the prediction of the phases of the moon or the results of a chemical reaction" (Bolt, 1988, p. 159). By contrast, the seismic cycle does not represent a coherent explanation of known phenomena with fully identifiable causes; rather it seems to describe what appear to be categorical time periods. Look at seismicity in a region, on a fault, on a plate boundary, or in historic earthquake catalogs, and a general pattern seems to emerge. But the general pattern that arises from the conjunction of the earthquake cycle—generally accepted—and the elastic rebound theory does not mean that all earthquakes share enough underlying aspects to be encompassed by a descriptive model like the characteristic earthquake hypothesis.

The USGS does not evoke the specter of a monolithic bureaucracy. Ideas and theories bounce around and collide in hallways and through modems like molecules in a confined gas. As an example, Heaton is unpersuaded by the premise that many earthquakes repeat regularly: "It has been extremely difficult to test the characteristic earthquake model." He doesn't challenge the actual preinstrumental data in the nineteenth-century catalogs or the geological evidence Sieh and others have unearthed. But he does believe that, "Unfortunately, many questionable assumptions must be made when interpreting either of these types of data." Citing one example, he points out that, if the Loma Prieta earthquake had been prehistoric, searches such as Sieh's would probably not have found evidence for it since there was no trace close enough to the surface to dig up, the fault having been confined to between 5 and 18 kilometers underground. And while he concedes that the first Working Group report in 1988 "predicted" Loma Prieta in the sense that the Santa Cruz segment was assigned the highest probability of rupture, he says most of his colleagues found the character of the earthquake quite surprising. The fault rupture in 1989 did not duplicate the 1906 event and thus "raised many questions about the validity of the characteristic earthquake model for this stretch of the fault."

Heaton is challenging the very notion of a characteristic earthquake, in part because of the paucity of direct data and the many assumptions and inferences involved. He believes that "it will be many generations before we have collected enough reliable data to observe the repetition of rupture along given fault strands." Even before the October 1992 Parkfield revelations (to be discussed in the final section), Heaton noted that this fault also may have fractured differently at different times in the past. Such dilemmas have "caused much debate within the seismological community. . . . The Parkfield earthquake is now almost 4 years beyond its mean recurrence interval, giving plenty of time for scientists to raise serious questions about the validity of the Parkfield Prediction Experiment and the overall validity of the characteristic earthquake model." These and other problems have impelled him to take a different tack toward prediction, focusing on the dynamics of the faulting process.

It would be misleading, however, to suggest that the seismological community looking at recurrence models is taking different approaches altogether, since as Ellsworth points out, he and others are actively trying to establish or disprove the characteristic earthquake hypothesis with hard evidence. Ellsworth emphasizes that, "Our goal, whether we study the small earthquakes recorded by modern seismographs, historical earthquakes, or paleoearthquakes, is the same: To understand the underlying physics of the earthquake source."

The Stress Dilemma

A first step in this direction is to confront more directly what Heaton calls "the stress dilemma." Geologists are obviously concerned with the pressure and forces that might prevail at various depths beyond direct measurement. Bars (and their aggregation into thousands, kilobars) provide a convenient scale of measure, since a bar is equivalent to 98.697 percent of the atmospheric pressure at the earth's surface. (More precisely, a bar is the force of 15 pounds per square inch, or 10^6 dynes per square centimeter, where a dyne is the force needed to propel 1 gram 1 centimeter per second.) Calculating the apparent force exerted by the overlying rock (whose mass is generally known), scientists tell us that pressure obviously increases with depth. In the 5- to 15-kilometers-deep, shallow-focus zone where most earthquakes commence, a straightforward calculation of pressure yields about 1.5 to 5 kilobars. This is what is known as confining pressure, says Heaton, and it sets a general limit to the earthquake forces seismologists are curious about: "Almost any type of rock at [those] pressures can withstand shear [or tearing] stresses of at least one half of the confining pressure, before it yields. . . . Faults obviously yield during earthquakes," he reasons, and thus "we expect the shear stress to be on the order of 1 to 2 kilobars," that is, 1000 to 2000 times the atmospheric pressure at the surface. Whence, then, the dilemma?

The stress released during an earthquake is somewhat hard to actually measure, but a fairly close approximation is possible, and Heaton points out that these stress changes are two full orders of magnitude smaller, typically about 15 bars. "And it's a good thing," he continues, because if you had a full kilobar of stress drop on a fault rupture that was 15 kilometers in diameter, the activity at the surface would be devastating. While seismologists cannot say beyond a first order precisely what an earthquake is, Heaton knows what it is *not*: "An earthquake is not taking a piece of rock up to its ultimate strength and letting it totally fail." Such a scenario provides an unlikely stage on which the drama of evolution could have been played. The fact that earthquake stress drops appear to be much smaller "is why there is life on Earth. Otherwise, everything would be dead."

Heaton illustrates this prophesy with what physics tells us about a world where the smaller forces (that seem to be observed) were not the operative ones. The fault would slip 100 meters (compared to around 2 meters, for example, at Loma Prieta). The ground itself during this tear would move about 10 meters per second. Heaton hypothesizes: "If your building was strong enough to survive this motion—say a bomb shelter—then certainly you would be killed by a wall running into you at 25 miles an hour. . . . Nothing would survive the experience. The point is, that although earthquakes seem violent to humans they actually involve stress changes that are small (typically about 15 bars) compared with the confining stress at the depth of earthquakes." The measurements of accumulating strain in the crust over several decades also support this premise, says Heaton: "The rate at which shear strain builds on the San Andreas fault is about 2×10^{-7} per year, or less than 1 bar of shear stress per decade."

This is the first part of the dilemma: There would appear to be much more stress at the depths where earthquakes originate than is released during the event. But the essential stress number seismologists want to know is the amount at which the rock actually fails, and thus a second part of the stress dilemma must be explained: Many distinct lines of evidence suggest rocks are shearing at stresses much lower than would theoretically prevail at such depths. One such indicator comes from what is known about friction. If the shear stress was anywhere near 1 kilobar, laboratory experiments indicate, the rocks would melt during the rupture and enormous quantities of heat would accompany the event. But "despite extensive research on this problem, no such heat flow anomalies have been detected," said Heaton. Another piece of data comes from direct observation, for example, at a hole some 3.5 kilometers deep excavated at Cajon Pass, adjacent to the San Andreas. In 1987 Mark Zoback (of Stanford University) and others found the shear stresses at this depth to be relatively low, probably less than several hundred bars, said Heaton.

Heaton poses the essential question these anomalies raise: "Clearly there is a problem. . . . It is impossible that stress on the San Andreas fault steadily builds until a yield stress of more than a kilobar is reached. Why are faults so weak? In laboratory experiments, we have yet to observe yield in rocks at such low shear stresses and such high confining pressures." Possible answers to this conundrum take the form of two candidate solutions, each of which, emphasizes Heaton, seriously complicates the simple picture assumed for the characteristic earthquake model. First, the confining pressure in these environments may actually be less than that calculated at the engineer's desk, because of the effects of fluids under high pressure that may be present. Second, his own work and that of others on the physics of dynamic rupture suggest that earthquakes can occur at relatively low stresses.

Early in the careers of many of these scientists, only about a decade after plate tectonics began to take hold, seismology appeared to be in the grip of another revolution, based on the dilatancy-fluid diffusion model. In the early 1970s Scholz and many others reported laboratory experiments that appeared to explain some idiosyncracies in 1969 Russian seismological data. As Bolt explains, "It has been found that, under some circumstances, wet rocks under shear strain *increase* their volume rather than decrease it. This increase in volume . . . is called *dilatancy.* There is evidence that the volume increase from pressure arises from opening and extending the many microcracks in the rocks. The ground water that then moves into the microcracks is much less compressible than air, so the rocks no longer close easily under pressure" (Bolt, 1988, p. 65). While this observation has not proven to be the breakthrough in prediction that it once appeared it might be, most geologists believe that the pressure effects of liquids are a crucial ingredient in local faulting. Said Heaton: "Understanding the role of fluids in fault zones is an important problem for which we currently have few certain answers."

However, he added, important evidence that fluids are present in fault zones comes from straightforward geological analyses (like those conducted by Rick Sibson and colleagues in New Zealand) of quakes millions and billions of years old whose faults have been buried but later uplifted and revealed, which contain mineral veins that were hydrothermally deposited. Physics demonstrates that a column of liquid reaching to earthquake-focus depths exerts a hydrostatic pressure, but it is smaller than the lithostatic pressure exerted by a comparable column of rock. Ergo you have buoyancy, precluding the possibility that fluids migrate downward. Rather, he continues, we hypothesize that fluids migrate upward from a source deep in the crust, probably due to the metamorphism of hydrated minerals. Upward-rushing fluids would encounter fairly large pressure gradients at such depths, and Heaton does not rule out the possibility that such a powerful confluence of forces could trigger an earthquake by decreasing the effective confining stress. If this were so, he continues, it would be necessary to understand the plumbing of a fault zone in order to predict when earthquakes occur. Heaton's USGS colleague James Byerlee and Harvard's Jim Rice have looked into this fluid problem extensively.

Grappling with the stress dilemma puts geophysicists into a kind of would-be Walt Disney state of mind, because if only they could be present like some darting, impervious dervish of an animated creature in the midst of a dynamic rupture, many feel confident one or several more Rosetta stones would be revealed. In fact, the armamentarium of instruments sunk into and placed about the surface at Parkfield is the closest attempt yet mounted to descend into precisely that sort of adventure. Meanwhile, they must content themselves with the laboratory simulations, based on the best mix of theory and intuition. But these experiments are hounded by three crucial issues of scaling: Whether simulations can actually match, or at least transfer to, dynamic subterranean conditions; whether the compressed time of laboratory experiments distorts processes occurring over much longer periods; and also whether the lower range of seismic phenomena, about which there is more (and more accessible) data, is qualitatively different from what may be happening in M 6 or larger earthquakes.

"After all," observed McNutt, "earthquakes occur all the time." Heaton has already mentioned that the 40 or so daily earthquakes presently cataloged in California are indicative less of total seismicity than of the limits of the instruments. Adds Ellsworth: "So far as we know, earthquakes continue down in size indefinitely. There is some debate about this in the community, as to whether they bottom out at around magnitude zero. But it's a logarithmic scale, obviously. There is evidence of much smaller earthquakes that are seen in very near proximity of sensors that are put in deep bore holes." So, reasons McNutt, "predicting *when* they occur is not the question so much as predicting when they're going to stop, since that makes the difference between the big and the little earthquakes." To answer *this* question, replied Heaton, we need to understand rupture dynamics.

Rupture Dynamics—Propagating Pulses of Slip?

When an earthquake will stop, then, is the focus of Heaton's scrutiny, as much as when will it commence. "My question is not when these earthquakes will occur (they happen all the time), but instead, which of these will be a big earthquake? There is far more to the physics than simply initiating rupture. . . . Big earthquakes are ones in which the rupture area is large. Thus, predicting which of the many small earthquakes will be a big one is a matter of predicting which rupture will propagate a large distance." Looked at as a question of energetics, the problem is almost a tautology. "As long as the elastic energy released by sliding exceeds the work necessary to slide the fault surfaces past each other, the rupture will continue to propagate. . . . If the earthquake releases more energy than it absorbs, then it continues to run." Once the effort to overcome frictional resistance absorbs a crucial amount of energy, in essence diverting it from the work required to keep the slip propagating, "then it stops." That condition, where it stops, concludes Heaton, is really controlled by dynamic frictional sliding.

Heaton's enthusiasm comes from his belief that traditional models of friction during slip are inadequate. Slipping friction is exceedingly important, he insists. It can control whether a rupture continues to propagate or dies out, and it is often assumed to be a constant value. The stress dilemma reminds us that the coefficient of friction for a given material is a function of its confining pressure, but—in the local environment of the rupture—this confining pressure is not known for certain, nor are its complexities fully understood by geophysicists, though Terry Tullis from Brown University has thrown much light on this area. Heaton also believes that the slipping friction may actually be a complex function of the slip history itself. The complexity he refers to could arise from several phenomena: The fluids that may be present, from dilatancy, and from dynamic stress variations that would accompany the elastic sound waves that must also be present. It also begins to suggest a solution to the stress dilemma. Given these considerations, he explains, it is not surprising that ruptures can propagate with relatively low driving shear stresses.

The important point, Heaton emphasizes, is that the sliding friction is a complex phenomenon, not entirely understood, and certainly not some simple number in a table in an engineer's handbook. Add to this *x* factor the predicate that the underlying forces are tectonic, straining the materials in ways that are understood also only very generally. Once a slip begins, these two phenomena interact to permit the slip to propagate or not. This is a feedback system that is likely to be unstable, he says; that is, dynamic friction and the slip may be expected to vary (perhaps chaotically) as the rupture propagates along the fault. Notwithstanding the many uncertainties of tectonic drive and dynamic slip, Heaton has developed a candidate slip model, for which there is an elegant and simple analog first conceived by Jim Brune, who directs the seismological lab at the University of Nevada-Reno. "Carpet installers have long realized that it is much easier to move a rug by introducing a wrinkle than it is to slide the entire surface. . . . The theoretical calculations for slip in a continuum show that if a pulse of slip with low frictional stress is introduced into a medium, it will tend to propagate. . . . These theoretical solutions have only recently been recognized by seismologists," said Heaton, but are now under study.

By reexamining the seismogram records of major earthquakes, such as those in California at Imperial Valley in 1979 (M 6.5) and Morgan Hill in 1984 (M 6.2), Heaton and his colleague Stephen Hartzell saw that many ruptures were like a stony zipper, being opened and then rezipped from the rear, with only a very small portion of the fault actually slipping at any brief isolated moment. Their model accounts not only for the few direct observations by people but for some of the anomalies of the stress dilemma as well. Heaton's major premise—not hard for many to grant—is that dynamic friction is complex and variable, that is, heterogeneous throughout the local environment of the fault. Heterogeneous slip distributions indicate, he reasons, that dynamic variations in dynamic friction are to be expected. Given this premise, he believes that (for any of a number of possible reasons, already explained) dynamic friction on the fault is low near the crack front. The rupture begins to radiate, but frictional resistance, which increases at the outer edges of the slip, slows the slip velocity toward the rear of the opening. With less force propelling the rupture from the back, the slip "heals itself," and thus the actual slipping area that proceeds down the fault takes the form of a pulse, whose front has had time to move only a small distance forward. The pulse propagates only as far down the fault line as variations in the dynamic friction that is encountered will permit.

This model provides Heaton with what he believes is the final nail in the coffin. The implications of dynamic friction unpredictability as indicated by heterogeneous slip behavior pose major problems for the characteristic earthquake model, concludes Heaton: "If these earthquakes were to repeat themselves over several recurrence cycles, then large slips would accumulate in some regions, but other regions would have very little slip. This is clearly an untenable situation. It suggests that future earthquakes will have different slip distributions, and have large slips in regions of small slip during past earthquakes. In this case the earthquakes would not be characteristic." Rather than look for a specific recurrence interval, Heaton believes that "predicting the time of a large earthquake requires that we must be able to predict the nature of this dynamic friction," about which much is unknown, though clearly it "has interesting and complex physics." A predictable and mathematical model of recurrence times is not feasible here, because the actual cause of earthquakes is not in any sense linear. According to Heaton and his colleagues' models, "very long and very short recurrence intervals may be observed for any section of the fault. The key to knowing when a given section will rupture is the ability to predict the nature of dynamic rupture. This may be a very difficult task," he concedes, "perhaps impossible." But Heaton seems to prefer searching the earth itself for direct clues rather than cruising the earthquake catalog for meaningful statistical patterns. If no such underlying pattern exists—as he clearly believes—then successful models will be founded on basic geophysics, and both Laboratory Earth as well as the labs where the physical processes are probed through simulation are where the secrets are buried.

THE PARKFIELD "OBSERVATORY"

As this piece is being written, 1993 has just begun, ringing down the curtain on the 10-year window predicted for the Parkfield quake. When the Parkfield Working Group met in June on the morning of the Landers quake (two of its members were later to write), "the long wait for the Parkfield earthquake could lead to a high degree of frustration in the Working Group. However, perhaps because of the significant gains that have and remain to be made, this attitude did not prevail at the meeting" (Michael and Langbein, 1992, p. 9). They invoked the words of G. K. Gilbert, who, 85 years earlier, writing just a year after the great San Francisco quake, had said:

> It is the natural and legitimate ambition of a properly constituted geologist to see a glacier, witness an eruption, and feel an earthquake. The glacier is always there ready, awaiting his visit; the eruption has a course to run, and alacrity only is needed to catch its more important phases; but the earthquake, unheralded and brief, may elude him through his entire lifetime.

Gilbert's words were more wistful than dejected, as he in fact had observed earthquakes. And thus it should be emphasized that the science going on at Parkfield, which—after years of development—now might fairly be regarded as a National Earthquake Prediction Laboratory, is not overly reliant on *when* the next Parkfield quake will arrive. To be sure, the extensive attention to measuring the rupture will provide irrefutable "facts" that the characteristic hypothesis must be able to accommodate. Ellsworth feels that characteristic earthquakes are but one possible phenomenon and in fact thinks that Heaton's zipper model could well be fortified or disconfirmed by the Parkfield quake. And, most importantly, the many phenomena that, almost since the birth of seismology, have been reported as "short-term precursors" of major earthquakes can be evaluated with more scientific forethought and validity than ever before.

Wayne Thatcher, a USGS seismologist, clarified the scientific goals of the Parkfield experiment several years ago. Since "no single, simple set of events preceding earthquakes" has been identified, it cannot be claimed (of any precursory signals that might be identified at Parkfield, after the quake) that such a set "would universally precede earthquakes. However, given the strongly focused nature and high sensitivity of the Parkfield monitoring networks, there can be little doubt that new and unexpected features of the earthquake mechanism will be uncovered, and that significant constraints will be placed on the mechanics of the precursory process" (Thatcher, 1988, p. 78).

A catalog of these experiments indicates the leading candidates for a precursor. The subject of precursors, says Joann Stock of Caltech, provokes heated controversy and could be explored at great length, but the phenomena involved only *may* correlate with earthquakes and, if so, nobody knows exactly *how*. If there is such a thing it will be recorded at Parkfield with much greater forethought and precision than any of the serendipitous and therefore hard-to-rely-on discoveries in the past that have preceded other earthquakes. The incessant flood of data is being continuously recorded in real time and analyzed, much of it with computer software designed to detect anomalies and to notify some of the many scientists who are tethered to the experiment by beepers. Six principal networks are in place:

- *Seismicity*. More than 20 advanced seismometers are located within 25 kilometers of Parkfield (three of them buried in boreholes). They provide continuous, high-gain, high-frequency seismic information for quakes down to the incredibly small, M 0.25 (about as much energy as it takes to light a parking lot). Machines to record larger M 3–4 shocks, called accelerometers, also are in place. All of these are hooked up to a complex data-processing array, providing an unprecedented fine-grained picture of seismic waves in the area.
- *Creep*. Thirteen creepmeters continuously monitor the slip at the surface, sending an update every 10 minutes to Menlo Park. The movement of the surface has been analyzed and recorded intensively, almost since the 1966 event.
- *Strain*. As the crust is being deformed, an array of strainmeters and dilatometers at various depths are detecting "principal strains, shear

strains, directions of maximum shear, areal strain, and various other strain parameters," said Bakun. Tilt, one of the more difficult changes to record because of solid-earth tides and the instability of near-surface materials, is still part of the search, with four closely spaced shallow borehole tiltmeters near Parkfield.

- *Water wells.* Eighteen wells, installed at 13 sites near the fault, are sampled and measured continuously for changes in the level or chemistry of the groundwater. One well near the expected epicenter should be able to detect pore pressure changes and dilatancy if and when it occurs.
- *Magnetic field.* Magnetometers are in place to detect changes in the earth's magnetic field due to stress changes in the crust.
- *Electrical resistivity and radio frequency currents.* Among the most often cited precursory phenomena are electrical signals recorded just before quakes, and the fluid changes that many think precede rupture could well explain changes in electrical earth currents. These may actually provide the most useful results, says Stock, and they have been the subject of much attention in Russia and China.

On October 20, 1992, this array of measuring devices picked up an M 4.7 earthquake, a potential foreshock to the anticipated event and large enough to trigger the automatic warning system that had been established at Parkfield to provide California's Office of Emergency Services an opportunity to prepare people for an impending quake. The system was more or less automatic. That is, seismologists had agreed that M 4.5 shocks (which occur generally every 4 to 5 years) once in about every three times reliably precede shocks of M 6 or more. Thus, they had agreed years before, in establishing the generic system at Parkfield, that a shock of that size would trigger a warning in the context of this history, to wit: There was a 37 percent chance of the quake arriving in the next 72 hours.

This "warning system" clearly reveals the foreshock dilemma. With most larger quakes, seismograms reveal a series of ruptures near the hypocenter, leading up to and falling away from the mainshock over time. In hindsight, the dilemma disappears, as the mainshock nearly always stands out clearly as the largest event, the unambiguous spike in the graph. But in present time the dilemma makes even the most seasoned seismologist nervous because *only time will tell* whether, say, an M 4.7 event is an isolated quake or among an escalating series of shocks that could grow into a much larger earthquake, even "The Big One." In October it turned out to be part of a smaller series, and the long-awaited Parkfield characteristic earthquake remains a no-show. To the pack of

A pair of streams that have been offset by right-lateral slip on the San Andreas fault (lineament extending from left to right edge of photograph). View northeastward across fault toward the Temblor Range. (Photograph by Sandra Schultz Burford, USGS. Reprinted from Wallace, 1990, p. 38.)

Some of the damage caused by the October 17, 1989, earthquake in the Marina district of San Francisco. Liquefaction of the ground there caused most of the damage, including this tilted building and the burned apartments in the foreground, where nearly one-quarter of a city block was destroyed by fire. (Photograph by Charles E. Meyer, USGS. (Photograph by Charles E. Meyer, USGS)

Yellowish thin film of almost pure buckyballs (C_{60}) shows concentric rings due to optical interference. The film was sublimed out of fullerene-containing soot made by researchers at the IBM Almaden Research Center who vaporized graphite rods. They heated the soot to 500° C under a quartz substrate that captured the escaping molecules. (Courtesy of IBM Almaden Research Center.)

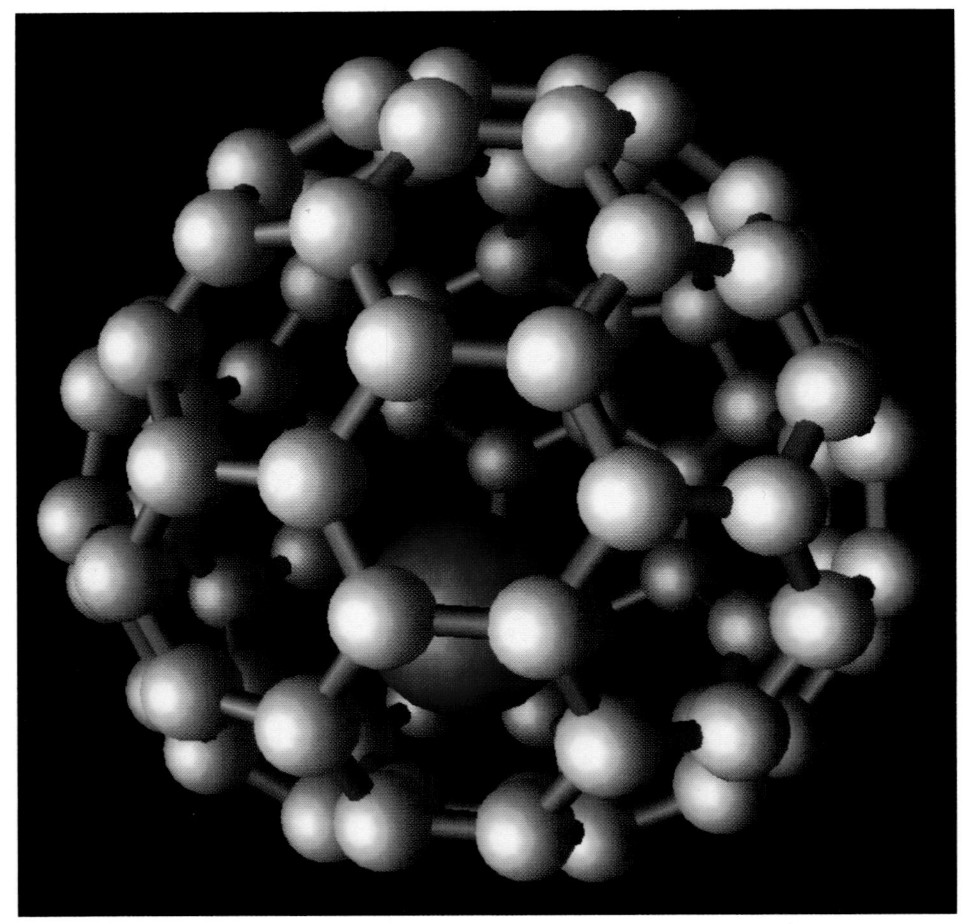

One lanthanum atom in a C-82 cage. This computer-generated image shows how a single atom of the metal lanthanum (red) is imprisoned inside a carbon-82 cage. Such exotic molecules, called metallofullerenes, are one of the exciting, potentially useful byproducts of the recent discovery of fullerenes, a whole family of round, hollow carbon molecules that can be stuffed, covered, or packed together with other atoms to make entirely new materials. This metallofullerene was made by scientists at the IBM Research Division's Almaden Research Center in San Jose. (Courtesy of IBM Almaden Research Center.)

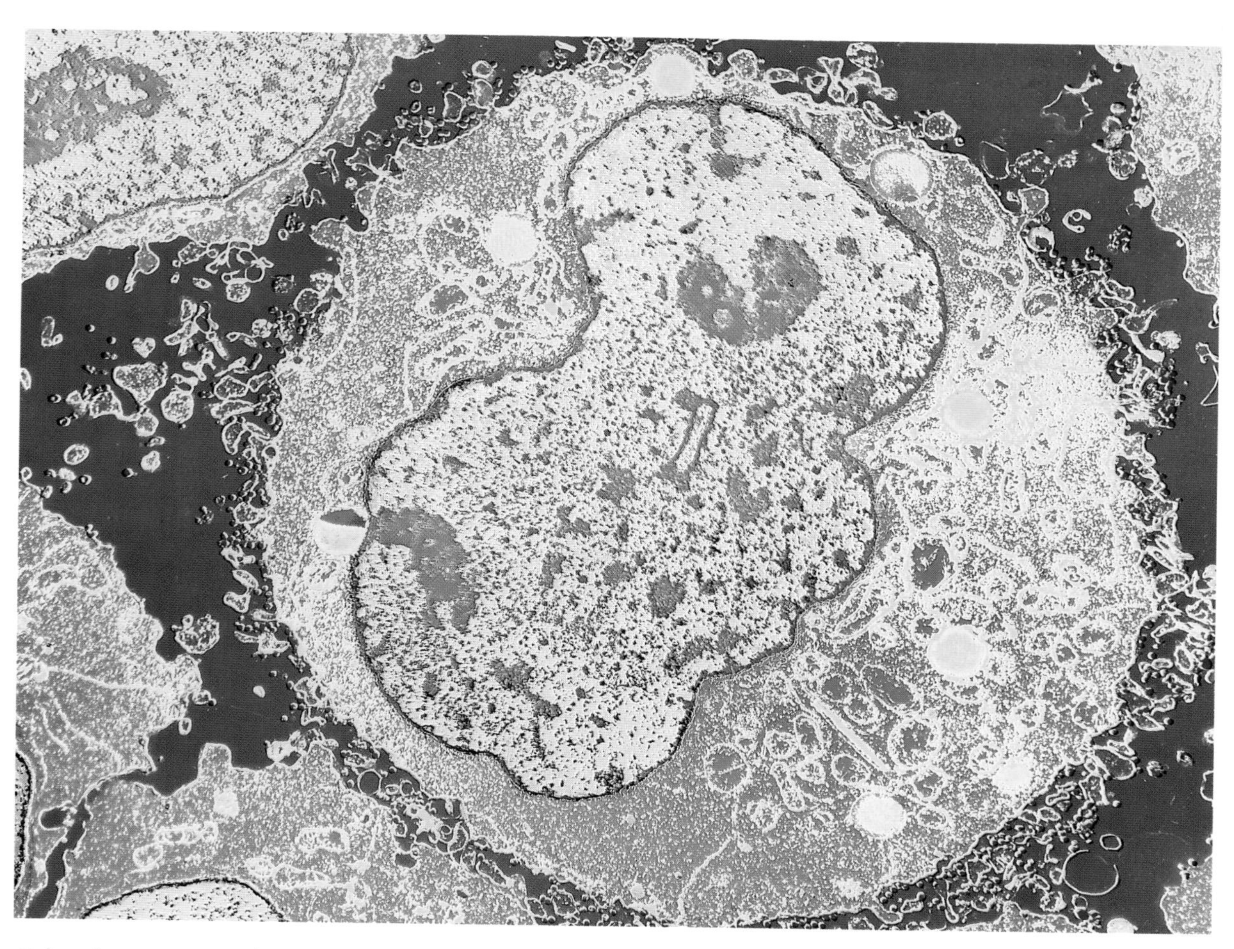

Colored transmission electron micrograph of a T-lymphocyte cell infected with Human Immunodeficiency virus (HIV), the causative agent of AIDS. At center, the pale pink nucleus of the T-cell can be seen. An infected T-cell typically has a lumpy appearance with irregular surface protrusions. The small red spherical virus particles on the cell surface are in the process of budding away from the cell membrane. The virus has instructed the cell to reproduce more viruses. Depletion of the number of T-cells through HIV infection is the main reason for the destruction of a person's immune system in the disease of AIDS. Magnification: x6150. (Credit: NIBSC/Science Photo Library.)

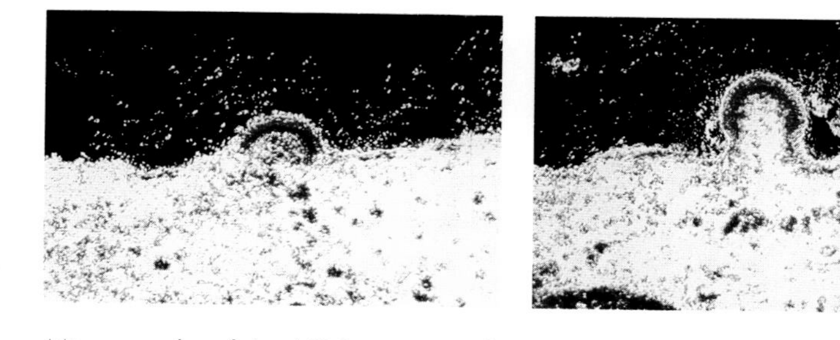

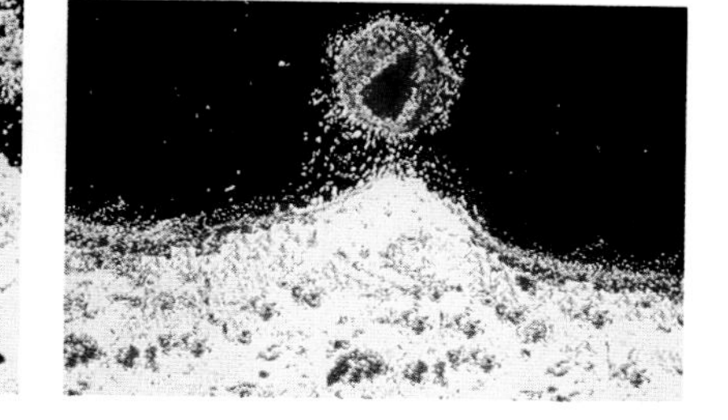

Micrographs of the AIDS virus as it leaves an infected lymphocyte. (Credit: Charles Daguet/Pasteur Institute/Petit format/Science Source.)

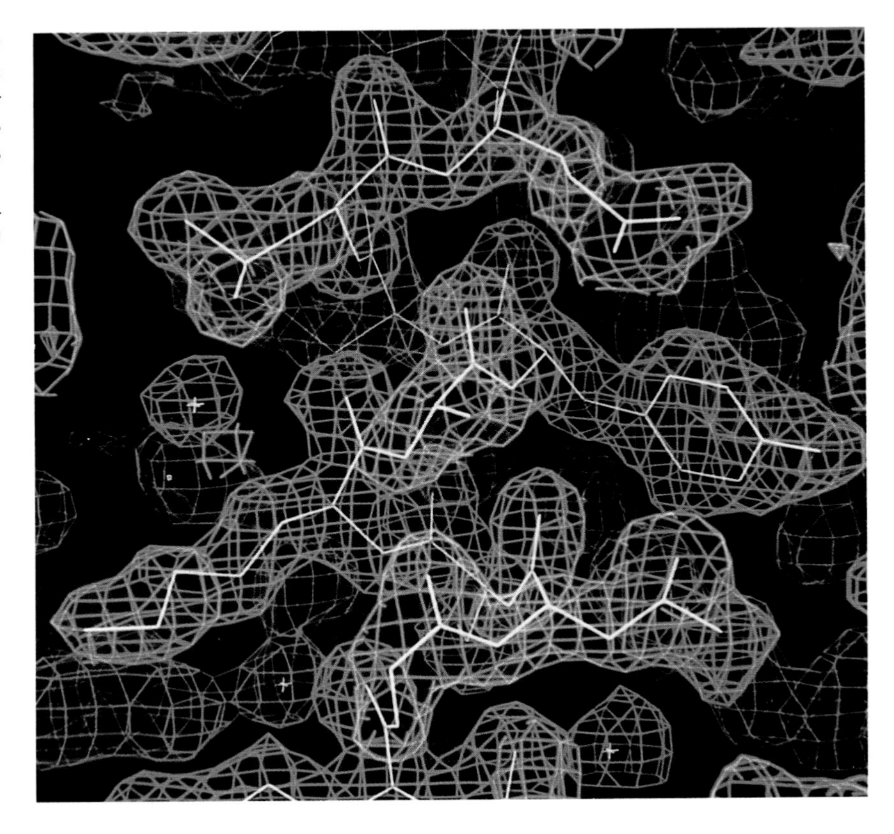

The electron density map of a leucine zipper, in blue, is superimposed on the molecular model of same, in yellow. The good fit indicates an accurate model. (Courtesy of Tom Alber, University of California-Berkeley.)

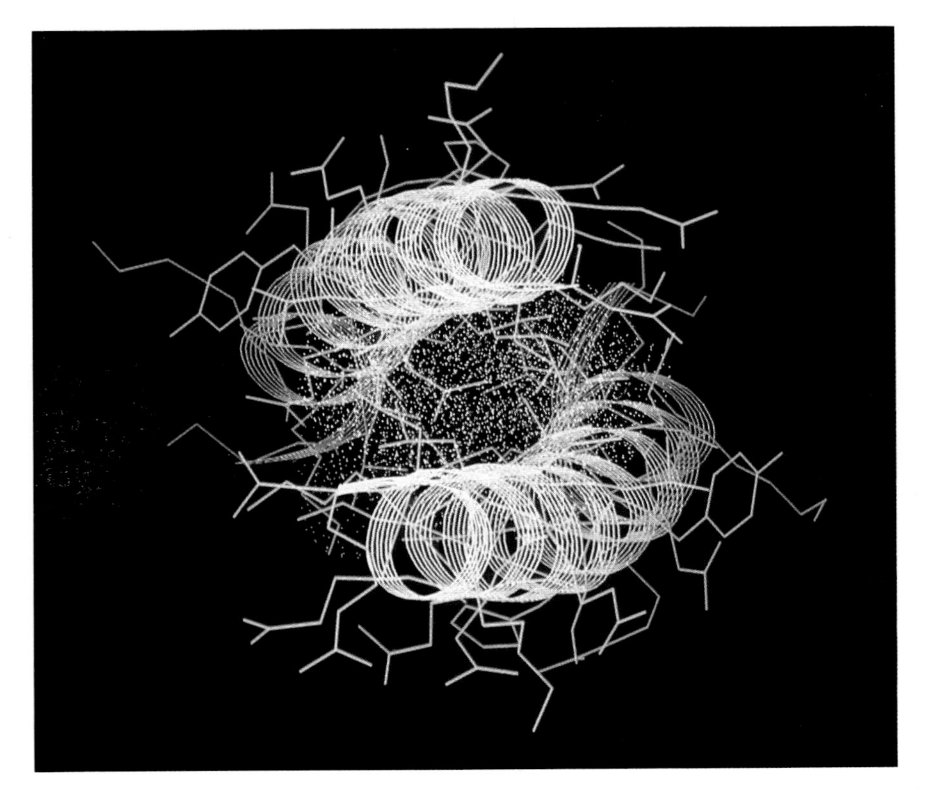

End-on view of a leucine zipper. The structure contains two 30-amino-acid alpha-helices. They coil around each other to form one-quarter turn of a super-helix. The blue lines represent the molecular structure and the red stippling represents the surface of the leucine side chains of the two coils, which fill the space between them. The purple springs are a schematic representation of the individual alpha helices. (Courtesy of Tom Alber, University of California-Berkeley.)

national journalists who descended on Parkfield, rummaging around for comprehension as well as flamboyant quotes, the event provided another opportunity to becloud the public's appreciation of the subtleties, and the success, of modern American seismology.

In point of fact, the Parkfield Working Group in June had real cause to celebrate the project's success, calling it an excellent observatory site for testing monitoring methods and expressing the belief that it had become, in reality, a National Earthquake Prediction Observatory. Citing a number of significant scientific and sociological results, they believe it should be staffed, funded, and supported as such. An instrument cluster is now being installed in the San Francisco Bay area, based on empirical experience gained at Parkfield. Greater subtlety in the relationship of wave structure to fault behavior has been developed from the experiment and recognized to be operative on other faults in California. The USGS pamphlet distributed to millions after Loma Prieta was modeled on a mailing that had been sent to Parkfield residents. And the alert structure being refined (through experiences like that in October 1992) at Parkfield provides an invaluable model to public planners.

The Future

Civilization was built atop a planet that, until the nineteenth century, even scientists believed to be permanent in form and recent in origin. And though its mass is conserved, the plate tectonics paradigm portrays the earth as a cauldron boiling with internal turmoil, whose surface volcanoes and earthquakes remind us what a heat engine our planet actually is. The tale told here—of dueling models, warring hypotheses—is testament to the intellectual excitement that fuels modern Earth science. Al Lindh and Bill Ellsworth believe that Parkfield and the search for pattern have already borne rich fruit. Probably not, say Jim Brune and Tom Heaton, bristling with their own belief that earthquake genesis is random and inherently unpredictable. And Kerry Sieh continues to break paths and upset applecarts (and the occasional colleague) with one intriguing find after another; just recently he unearthed records of previously unknown faults running right under the heart of Los Angeles. In 1993 theories of earthquake genesis and prediction are still embryonic and fairly rough, but it is precisely this inviting frontier—waiting to be explored and mapped—and the vital interests of millions of people that fire the energy, enthusiasm, and commitment of this band of modern scientists engaged in nothing less than the effort to better decipher the working drawings of humanity's one common hearth and home.

BIBLIOGRAPHY

Ad Hoc Working Group on the Probabilities of Future Large Earthquakes in Southern California. 1992. Future Seismic Hazards in Southern California; Phase I: Implications of the 1992 Landers Earthquake Sequence. California Divison of Mines and Geology, Sacramento, November.

Agnew, D. C., and W. L. Ellsworth. 1991. Earthquake prediction and long-term hazard assessment. Reviews of Geophysics (Supplement to U.S. National Report to International Union of Geodesy and Geophysics 1987–1990). (April):877–889.

Agnew, D. C., and K. E. Sieh. 1978. A documentary study of the felt effects of the great California earthquake of 1857. Bulletin of the Seismological Society of America 68(6):1717–1729.

Bakun, W. H., and A. G. Lindh. 1985. The Parkfield, California, earthquake prediction experiment. Science 229:619–624.

Bolt, B. A. 1988. Earthquakes. W.H. Freeman, New York.

Brune, J., S. Brown, and P. Johnson. 1991. Rupture mechanism and interface separation in foam rubber models of earthquakes: A possible solution to the heat flow paradox and the paradox of large overthrusts. In Proceedings of the International Workshop on New Horizons in Strong Motion, Santiago, June 4–7, 1991. Technophysics, special issue (1993): "New Horizons in Strong Motion: seismic studies and engineering practice, " Ed., Fernando Lunt, Vol. 218, nos. 1–3, pp. 59–67.

DeMets, C., R. G. Gordon, S. Stein, and D. F. Argus. 1987. A revised estimate of Pacific-North America motion and implications for western North America plate boundary zone tectonics. Geophysical Research Letters 14(9):911–914.

Ellsworth, W. L. 1990. Earthquake history, 1769–1989. Pp. 153–181 in The San Andreas Fault System, California. Robert E. Wallace, ed. U.S. Government Printing Office, Washington, D.C.

Ellsworth, W. L., A. G. Lindh, W. H. Prescott, and D. G. Herd. 1981. The 1906 San Francisco earthquake and the seismic cycle. In Earthquake Prediction—An International Review, Maurice Ewing Series 4. American Geophysical Union, Washington, D.C.

Heaton, T. H. 1990. Evidence for and implications of self-healing pulses of slip in earthquake rupture. Physics of the Earth and Planetary Interiors 64:1–20.

Heaton, T. H. 1990. The calm before the quake? Nature 343:511–512.

Heppenheimer, T. A. 1990. The Coming Quake: Science and Trembling on the California Earthquake Frontier. Paragon House, New York.

Lay, T. 1992. Theoretical seismology. In Encyclopedia of Earth System Science, Vol. 4, pp. 153–173. Academic Press, New York.

McPhee, J. 1993. Assembling California. Farrar, Strauss and Giroux, New York.

Michael, A. J., and J. Langbein. 1992. Parkfield: Learning While Waiting. Summary of the 1992 Santa Cruz Review Meeting. USGS Working Paper, November 24 draft.

Rice, J. 1991. Fault stress states, pore pressure distributions, and the weakness of the San Andreas fault. Pp. 475–503 in Fault Mechanics and Transport Properties in Rocks. B. Evans and T. F. Wong, eds. Academic Press, New York.

Richter, C. F. 1958. Elementary Seismology. W.H. Freeman, San Francisco.
Scholz, C. H. 1990. The Mechanics of Earthquakes and Faulting. Cambridge University Press, Cambridge.
Shimizaki, K., and T. Nakata. 1980. Time-predictable recurrence model for large earthquakes. Geophys. Res. Lett. 7:279–82.
Sibson, R., F. Robert, and F. Poulson. 1988. High-angle reverse faults, fluid pressure cycling, and mesothermal gold deposits. Geology 16:551–555.
Sieh, K. 1984. Lateral offsets and revised dates of large prehistoric earthquakes at Pallett Creek, Southern California. Journal of Geophysical Research 89:7641–7670.
Sieh, K., M. Stuvier, and D. Brillinger. 1989. A more precise chronology of earthquakes produced by the San Andreas fault in Southern California. Journal of Geophysical Research 94(B1):603–623.
Thatcher, W. 1988. Scientific goals of the Parkfield Earthquake Prediction Experiment. Earthquakes and Volcanoes 20(2):78.
U.S. Geological Survey. 1988. Configuration and Uses of Existing Networks. Publ. No. 1031. United States Government Printing Office, Washington, D.C.
U.S. Geological Survey. 1989. Lessons Learned from the Loma Prieta, California, Earthquake of October 17, 1989. Publ. No. 1045. United States Government Printing Office, Washington, D.C.
U.S. Geological Survey. 1992. Goals, Opportunities, and Priorities for the USGS Earthquake Hazards Reduction Program. Publ. No. 1079. United States Government Printing Office, Washington, D.C.
U.S. Geological Survey. 1988. Parkfield: The prediction and the promise. Earthquakes and Volcanoes 20(2). United States Government Printing Office, Washington, D.C.
U.S. Geological Survey. 1989. The Loma Prieta earthquake of October 17, 1989. Earthquakes and Volcanoes 21(6). United States Government Printing Office, Washington, D.C.
Wallace, R. E., ed. 1990. The San Andreas Fault System, California. U.S. Geological Survey Professional Paper 1515. U.S. Government Printing Office, Washington, D.C.
Working Group on California Earthquake Probabilities. 1988. Probalities of Large Earthquakes Occurring in California on the San Andreas Fault. U.S. Geological Survey Open-File Report 88–398. U.S. Government Printing Office, Washington, D.C.
Working Group on California Earthquake Probabilities. 1990. Probabilities of Large Earthquakes in the San Francisco Bay Region, California. U.S. Geological Survey Circular 1053. U.S. Government Printing Office, Washington, D.C.

THE MATHEMATICAL MICROSCOPE

Waves, Wavelets, and Beyond

by Barbara Burke

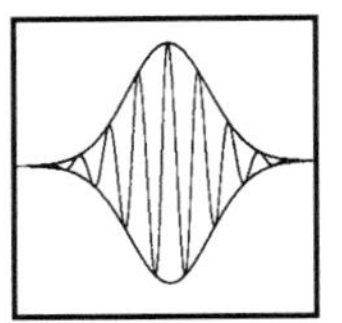

Mathematician Michael Frazier of Michigan State University was educated in the tradition that maintains that "real" mathematics by "real" mathematicians is and should be useless. "I never expected to do any applications—I was brought up to believe I should be proud of that," he says. "You did pure harmonic analysis for its own sake, and anything besides that was impure, by definition." But in the summers of 1990 and 1991 he found himself using a mathematical construction to pick out the pop of a submarine hull from surrounding ocean noise.

In St. Louis, Victor Wickerhauser was using the same mathematics to help the Federal Bureau of Investigation store fingerprints more economically, while at Yale University Ronald Coifman used it to coax a battered, indecipherable recording of Brahms playing the piano into yielding its secrets. In France, Yves Meyer of the University of Paris-Dauphine found himself talking to astronomers about how they might use these new techniques to study the large-scale structure of the universe.

Over the past decade a number of mathematicians accustomed to the abstractions of pure research have been dirtying their hands—with great enthusiasm—on a surprising range of practical projects. What these tasks have in common is a new mathematical language, its alphabet consisting of identical squiggles called wavelets, appropriately stretched, squeezed, or moved about.

A whole range of information—your voice, your fingerprints, a snapshot, x-rays ordered by your doctor, radio signals from outer space, seismic waves—can be translated into this new language, which emerged independently in a number of different fields, and in fact was only recently understood to be a single language. In many cases this transformation into wavelets makes it easier to transmit, compress, and analyze information or to extract information from surrounding "noise"—even to do faster calculations.

In their initial excitement some researchers thought wavelets might virtually supplant the much older and very powerful mathematical language of Fourier analysis, which you use every time you talk on the telephone or turn on a television. But now they see the two as complementary and are exploring ways to combine them or even to create more languages "beyond wavelets."

Different languages have different strengths and weaknesses, points out Meyer, one of the founders of the field: "French is effective for analyzing things, for precision, but bad for poetry and conveying emotion—perhaps that's why the French like mathematics so much. I'm told by friends who speak Hebrew that it is much more expressive of poetic images. So if we have information, we need to think, is it best expressed in French? Hebrew? English? The Lapps have 15 different words for snow, so if you wanted to talk about snow, that would be a good choice."

Some information processing is best done in the language of Fourier; other with wavelets; and yet other tasks might require new languages. For the first time in a great many years—almost two centuries, if one goes back to the very birth of Fourier analysis—there is a choice.

A MATHEMATICAL POEM

Although wavelets represent a departure from Fourier analysis, they are also a natural extension of it: the two languages clearly belong to the same family. The history of wavelets thus begins with the history of Fourier analysis. In turn the roots of Fourier analysis predate Fourier himself (and much of what is now called Fourier analysis is due to his successors). But Fourier is a logical starting point; his influence on mathematics, science, and our daily lives has been incalculable, if to many people invisible. Yet he was not a professional mathematician or scientist; he fit these contributions into an otherwise very busy life.

His father's twelfth child, and his mother's ninth, Joseph Fourier was born in 1768 in Auxerre, a town roughly halfway between Paris and Dijon. His mother died when he was nine and his father the following year. Although two younger siblings were abandoned to a foundling hospital after their mother's death, Fourier continued school and in 1780 entered the Royal Military Academy of Auxerre, where at age 13 he became fascinated by mathematics and took to creeping down at night to a classroom where he studied by candlelight.

Fourier's academic success won the favor of the local bishop. But when at the end of his studies his application to join the artillery or the army engineers was rejected, he entered the abbey of St. Benoît-sur-Loire. (The popular story that he was rejected by the army because he was not of noble birth—and therefore ineligible "even if he were a second Newton"—is questioned by at least two of Fourier's contemporaries.)

The French Revolution erupted before Fourier took his vows. At first indifferent, he became increasingly committed to the cause of establishing "a free government exempt from kings and priests"[1] and in 1793 joined the revolutionary committee of Auxerre.

Twice he was arrested, once in the bloody days shortly before the fall of Robespierre and again the following year, on charges of terrorism. In defending himself Fourier pointed out that during the Terror no one in Auxerre was condemned to death; a friend related that once, to prevent a man he believed to be innocent from arrest and the guillotine, Fourier invited the agent charged with the arrest to lunch at an inn and, "having exhausted every means of retaining his guest voluntarily," left the room on a pretext, locked the door, and ran to warn the suspect, returning later with excuses.

After several years teaching in Paris, Fourier next accompanied Napoleon to Egypt, serving as permanent secretary of the Institute of Egypt that Napoleon had set up, in part to study Egypt's past and natural history. Upon Fourier's return to France, Napoleon appointed him prefect of the department of Isère. He served as prefect, living in Grenoble, for 14 years, earning a reputation as an able administrator; he was responsible for the draining of some 20,000 acres of swamps that

Mathematical Analysis

. . . mathematical analysis . . . defines all observable relationships, measures time, space, force, temperature. This difficult science grows slowly but once ground has been gained it is never relinquished. . . .

Analysis brings together the most disparate phenomena and discovers the hidden analogies which unify them. If material escapes us like air or light because of its extreme fineness, if bodies are placed far from us in the immensity of space, if man desires to know the aspect of the heavens for times separated by many centuries, if the effects of gravity and temperature occur in the interior of the solid earth at depths which will remain forever inaccessible, yet mathematical analysis can still grasp the laws governing the phenomena. Analysis makes them actual and measurable and seems to be a faculty of human reason meant to compensate for the brevity of life and the imperfection of our senses. . . . —Joseph Fourier, *La Théorie Analytique de la Chaleur*

had caused annual epidemics of fever. After Waterloo, denied a government pension because he had served under Napoleon, he found a safe haven in the Bureau of Statistics in Paris, and in 1817 (after an initial rebuff by King Louis XVIII) he was elected to the Academy of Sciences.

Despite his administrative duties—and his isolation from Paris for many years—Fourier managed to pursue his scientific and mathematical interests. Victor Hugo called him a man "whom posterity has forgotten,"[2] but Fourier's name is as familiar to countless scientists, mathematicians, and engineers as the names of their own children. This fame rests on ideas he set forth in a memoir in 1807 and published in 1822 in his book, *La Théorie Analytique de la Chaleur* (*The Analytic Theory of Heat*).

Physicist James Clark Maxwell called Fourier's book "a great mathematical poem," but the description does not begin to give an idea of its influence. In the seventeenth century Isaac Newton had a new insight: that forces are simpler than the motions they cause, and the way to understand the natural world is to use differential and partial differential equations to describe these forces—gravity, for example.

Newton's differential equation showing how the gravitational pull between two objects is determined by their mass and the distance between them replaced countless observations, and predictive science became possible. Theoretically possible, at least. Solving differential equations—actually predicting where we will be taken by forces that themselves depend at each moment on our changing position—is not easy. As Fourier himself wrote in *La Théorie Analytique de la Chaleur*, although the equations that describe the propagation of heat have a very simple form, "existing methods do not give any general way to integrate them; as a result it is impossible to determine the values of the temperature after a given period of time. This numerical interpretation . . . is nevertheless essential. . . . As long as we cannot achieve it . . . the truth that we wish to discover is as thoroughly hidden in the formulas of analysis as in the physical question itself." [3]

Some 150 years after Newton, Fourier provided a practical way to solve numerically a whole class of such equations, linear partial differential equations. His ideas dominated mathematical analysis for 100 years and had surprising ramifications even for number theory and probability. Outside mathematics their influence is difficult to exaggerate. Virtually every time scientists or engineers model systems or make predictions, they use Fourier analysis. Fourier's ideas have also found applications in linear programming, in crystallography, and in countless devices from telephones to radios and hospital x-ray machines; they are, in mathematician T. W. Körner's words, "built into the commonsense of our society."

A RABBLE OF FUNCTIONS

There are two parts to Fourier's contribution: first, a mathematical statement (actually proved later by Dirichlet), and, second, showing why this statement is useful.

The mathematical statement is that any periodic (repeating) function can be represented as a sum of sines and cosines (see Figure 7.1). Roughly what this means is that any periodic curve, no matter how irregular (the output of an electrocardiogram, for example), can be expressed as the sum, or superposition, of a series of perfectly regular sine and cosine curves, of different frequencies. The irregular curve and the sum of sines and cosines are two different representations of the same object in different "languages."

Even a jagged line can be represented as a Fourier series. The trick is to multiply the sines and cosines by a coefficient to change their *amplitude* (the height of their waves) and to shift them so that they either add or cancel (changing the *phase*). One can also treat nonperiodic functions this way, using the *Fourier transform* (see Box on p. 202).

Fourier himself found this statement "quite extraordinary," and it met with some hostility. Mathematicians were used to functions that when graphed took the form of regular curves; the function $f(x) = x^2$, for example, produces a well-behaved, symmetrical parabola. (A function gives a rule of changing an arbitrary number into something else; $f(x) = x^2$ says to square any number x; if $x = 2$, then $f(x) = 4$.)

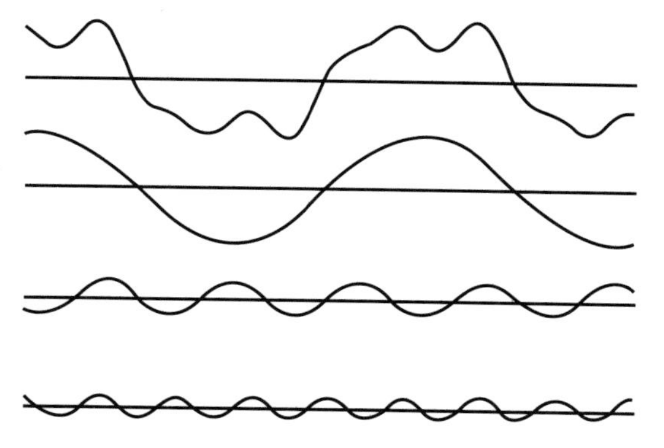

FIGURE 7.1 *In 1807 Fourier showed that any function (even an irregular one) can be expressed as the sum of a series of sines and/or cosines. The function at the top is a combination of the three functions below: sin (x), 0.3 sin (3x) and 0.2 cos (5x).*

The idea that any arbitrary curve could be expressed as a series of sines and cosines and thus treated as a function came as a shock and contributed to a profound and sometimes disturbing change in mathematics; mathematicians spent much of the nineteenth century coming to terms with just what a function was.

"We have seen a rabble of functions arise whose only job, it seems, is to look as little as possible like decent and useful functions," wrote French mathematician Henri Poincaré in 1889. "No more continuity, or perhaps continuity but no derivatives. . . . Yesterday, if a new function was invented it was to serve some practical end; today they are specially invented only to show up the arguments of our fathers, and they will never have any other use."[4]

Ironically, Poincaré himself was ultimately responsible for showing that seemingly "pathological" functions are essential in describing nature (leading to such fields as chaos and fractals), and this new direction for mathematics proved enormously fruitful, giving new vigor to a discipline that some had found increasingly anemic, if not moribund. In 1810 the French astronomer Jean-Baptiste Delambre had issued a report on mathematics expressing the fear that "the power of our methods is almost exhausted,"[5] and some 30 years earlier Lagrange, comparing mathematics to a mine whose riches had all been exploited, wrote that it was "not impossible that the mathematical positions in the academies will one day become what the university chairs in Arabic are now."[6]

"Looking back," writes Körner in his book *Fourier Analysis*, we can see Fourier's memoir "as heralding the surge of new mathematical methods and results which were to mark the new century."

THE EXPLANATION OF NATURAL PHENOMENA

The German mathematician Carl Jacobi wrote that Fourier believed the chief goal of mathematics to be "the public good and the explanation

The Fourier Transform

The Fourier transform is the mathematical procedure by which a function is split into its different frequencies, like a prism breaking light into its component colors. But the Fourier transform goes further and tells both how much of each frequency the function contains (the amplitude of the frequency) and the phase of the signal at each frequency (the extent to which it is shifted with respect to a chosen origin).

The term also describes the result of that operation: the Fourier transform of a particular function (that varies with time) is a new function (that varies with frequency). A Fourier series is a special case of Fourier transform, representing a periodic, or repeating, function.

For functions that vary with time, such as sound recorded as changes in air pressure, or fluctuations in the stock market, frequency is generally measured in hertz or cycles per second. For functions that vary with space, "frequency" is often related to the inverse of a distance. For instance, the Fourier transform of a fingerprint will have large values near the "frequency" 15 ridges per centimeter.

The Fourier series of a periodic function of period 2π is given by the formula

$$f(x) = \frac{a_0}{2} + (a_1 \cos x + b_1 \sin x) + (a_2 \cos 2x + b_2 \sin 2x) + \ldots \qquad (1)$$

The Fourier coefficients, a_n and b_n, tell how much a signal contains of each frequency n: its amplitude at frequency n is the square root of $(a_n^2 + b_n^2)$.

Calculating coefficients of the Fourier series of a function $f(x)$ involves integral calculus:

$$a_n = \frac{1}{\pi}\int_{-\pi}^{\pi} f(x) \cos nx \, dx \quad \text{and} \quad b_n = \frac{1}{\pi}\int_{-\pi}^{\pi} f(x) \sin nx \, dx$$

(This means that you multiply the function $f(x)$ by the appropriate sine or cosine and integrate, measuring the area enclosed by the resulting curve; the result is divided by π. To multiply the function by a sine or cosine, the value of each point on the function is multiplied by the value of the corresponding point on the sine or cosine.)

The phase can, in principle, be calculated from the coefficients. If you plot the point a_n, b_n on a coordinate system, the amplitude is the length of the line from the origin to that point, while the phase is the angle formed by that line and the positive x axis (Figure 7.2).

The function can be reconstructed from the coefficient using formula (1) above; the sine or cosine at each frequency is multiplied by its coefficient, and the resulting functions are added together, point by point; the first term is divided by 2.

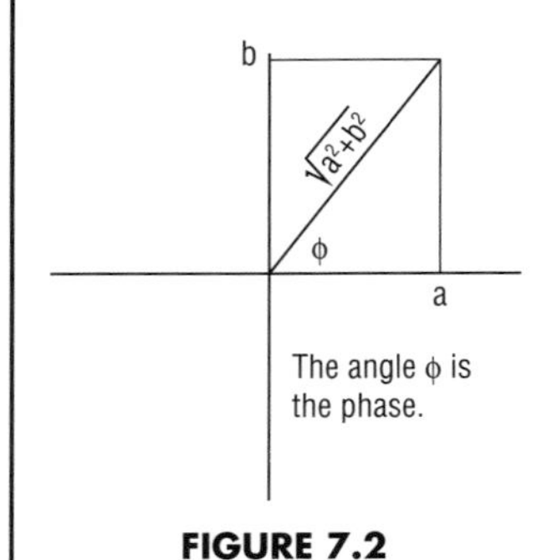

The angle ϕ is the phase.

FIGURE 7.2

of natural phenomena," and Fourier showed how his mathematical statement could be used to study natural phenomena such as heat diffusion, by turning a difficult differential equation into a series of simple equations.

Suppose, for example, we want to predict the temperature at time t of each point along a metal bar that has been briefly heated at one end. We start by establishing the initial temperature, at time zero, which we consider a function of distance along the bar. (This is why Fourier needed a technique that would work with all functions, even irregular or discontinuous ones: he couldn't expect the initial temperature to be so obliging as to take the form of a regular curve.)

When that function is translated into a Fourier series a remarkable thing happens: the intractable differential equation describing the evolution of the temperature *decouples,* becoming a series of independent differential equations, one for the coefficient of each sine or cosine making up the function. These equations tell how each Fourier coefficient varies as a function of time. The equations, moreover, are very simple—the same as the equation that gives the value of a bank account earning compound interest (negative interest in this case).

One by one, we simply plug in the coefficients describing the temperature at time zero and crank out the answers; these are the Fourier coefficients of the temperature at time t, which can be translated back into a new function giving the new temperature at each point on the bar. The procedure is no harder than the one banks use to compute the balance in their clients' accounts each month.

Essentially we have made a little detour in Fourier space, where our calculations are immensely easier—as if, faced with the problem of multiplying the Roman numerals LXXXVI and XLI, we translated them into Arabic numerals to calculate $86 \times 41 = 3526$, and then translated the answer back into Roman numerals: LXXXVI × XLI = MMMDXXVI.

The techniques Fourier invented have of course had an impact well beyond studies of heat or even solutions to differential equations. Real data tend to be very irregular: consider an electrocardiogram or the readings of a seismograph. Such signals often look like "complicated arabesques," to use Yves Meyer's expression—tantalizing curves that contain all the information of the signal but that hide it from our comprehension. Fourier analysis translates these signals into a form that makes sense.

In addition, in many cases the sines and cosines making up a Fourier series are not simply a mathematical trick to make calculations easier; they correspond to the frequencies of the actual physical waves making up the signal. When we listen to music or conversation, we hear changes in air pressure caused by sound waves—high sounds having high frequency and low sounds having lower frequency. (In fact, a piano can perform a kind of Fourier analysis: a loud sound near a piano with the damper off will cause certain strings to vibrate, corresponding to the different frequencies making up the sound.)

Similarly—although this was not known in Fourier's time—radio waves, microwaves, infrared, visible light, and x-rays are all electromagnetic waves differing only in frequency. Being able to break down sound waves and electromagnetic waves into frequencies has myriad uses, from tuning a radio to your favorite station to interpreting radiation from distant galaxies, using ultrasound to check the health of a developing fetus, and making cheap long-distance telephone calls.

With the discovery of quantum mechanics, it became clear that Fourier analysis is the language of nature itself. On the "physical space" side of the Fourier transform, one can talk about an elementary particle's position; on the other side, in "Fourier space" or "momentum space," one can talk about its momentum or think of it as a wave. The modern realization that matter at very small scales behaves differently from matter on a human scale—that at small scales we cannot have precise knowledge of both sides of the transform at once, we cannot know simultaneously both the position and momentum of an elementary particle—is a natural consequence of Fourier analysis.

BEING ACADEMIC OR BEING REAL

While irregular functions can be expressed as the sum of sines and cosines, those sums usually are infinite. Why translate a complex signal into an endless arithmetic problem, calculating an infinite number of coefficients, and summing an infinite number of waves? Fortunately, a small number of coefficients is often adequate. From the heat diffusion equation, for example, it is clear that the Fourier coefficients of high-frequency sines and cosines rapidly get exceedingly close to zero, and so all but the first few frequencies can safely be ignored. In other cases engineers may assume that a limited number of calculations give a sufficient approximation until proved otherwise.

In addition, engineers and scientists using Fourier analysis often don't bother to add up the sines and cosines to reconstruct the signal—they "read" Fourier coefficients (at least the amplitudes; phases are more difficult) to get the information they want, rather the way some musicians can hear music silently, by reading the notes. They may spend hours on end working quite happily in this "Fourier space," rarely emerging into "physical space."

But the time it takes to calculate Fourier coefficients *is* a problem. In fact, the development of fast computers and fast algorithms has been crucial to the pervasive, if quasi-invisible, use of Fourier analysis in our daily lives in connection with today's digital technology.

The basis for digital technology was given by Claude Shannon, a mathematician at Bell Laboratories whose *Mathematical Theory of Communications* was published in 1948; while he is not well known among the general public, he has been called a "hero to all communicators."[7] Among his many contributions to information theory was the sampling theorem (discovered independently by Harry Nyquist and others). This theorem proved that if the range of frequencies of a signal measured in hertz (cycles per second) is n the signal can be represented with complete accuracy by measuring its amplitude $2n$ times a second.

This result, a direct consequence of Fourier analysis, is simple to state and not very difficult to prove, but it has had enormous implications for the transmission and processing of information. It is not necessary to reproduce an entire signal; a limited number of samples is enough.

Since the range of frequencies transmitted by a telephone line is about 4000 hertz, 8000 samples per second are sufficient to reconstitute your voice when you talk on the telephone; when music is recorded on a compact disc, about 44,000 samples a second are used. Measuring the amplitude more often, or trying to reproduce it continuously, as with old-fashioned records, does not gain anything.

Another consequence is that, in terms of octaves, more samples are needed in high frequencies than in low frequencies, since the frequency doubles each time you go up an octave: the range of frequency between the two lowest As on a piano is only 28 hertz, while the range of frequency between the two highest As is 1760 hertz. Encoding a piece of music played in the highest octave would require 3520 samples a second; in the lowest octave, 56 would be enough.

The sampling theorem opened the door to digital technology: a sampled signal can be expressed as a series of digits and transmitted as a series of on-and-off electrical pulses (creating, on the other hand, round-off errors). Your voice can even be shifted temporarily into different frequencies so that it can share the same telephone line with many other voices, contributing to enormous savings. (In 1915 a 3-minute call from coast to coast cost more than $260 in today's dollars.) In 1948 Shannon and his colleagues Bernard Oliver and John Pierce expected digital

The Fast Fourier Transform

An algorithm is a recipe for doing computations. When schoolchildren learn to "carry," "borrow," or multiply two-digit numbers without a calculator, they are learning algorithms. Fast algorithms are mathematical shortcuts for dealing with large computations. The fast Fourier transform, published by J. W. Cooley and T. W. Tukey in 1965, is a prime example. It cuts the number of computations from n^2 in standard Fourier analysis to $n \log n$. When n is big, this makes a substantial difference: if $n = 1{,}000{,}000$, then $n^2 = 1{,}000{,}000{,}000{,}000$, but $n \log n = 6{,}000{,}000$ (log 1,000,000 is 6 since $10^6 = 1{,}000{,}000$). The logarithm is roughly the number of digits, so it grows slowly.

The larger n is, the more impressive the difference becomes. If n is a billion and a computer can complete a billion calculations a second, this cuts computing time from approximately 32 years to 9 seconds. With the FFT one can compute π to a billion digits in about 45 minutes; without it the job would take almost 10,000 years.

transmission to "sweep the field of communications;"[8] the revolution came, if later than they had expected.

Fueling this revolution was the fast Fourier transform (see Box on this page), a mathematical trick that catapulted the calculation of Fourier coefficients out of horse-and-buggy days into supersonic travel. With it, calculations could be done in seconds that previously were too costly to do at all. "It's the difference," Michael Frazier says, "between being academic and being real."

This fast algorithm, known as the FFT, requires computers in order to be useful. "Once the method was established it became clear that it had a long and interesting prehistory going back as far as Gauss," Körner writes. "But until the advent of computing machines it was a solution looking for a problem." On the other hand, the gain in speed from the FFT is greater than the gain in speed from better computers; indeed, significant gains in computer speed have come from such fast algorithms built into computer hardware.

DRIVING A CAR HALF A BLOCK

It could be argued that the fast Fourier transform was too successful. "Because the FFT is very effective, people have used it in problems where it is not useful—the way Americans use cars to go half a block," says Yves Meyer. "Cars are very useful, but that's a misuse of the car. So the FFT has been misused, because it's so practical."

The problem is that Fourier analysis does not work equally well for all kinds of signals or for all kinds of problems. In some cases, scientists using it are like the man looking for a dropped coin under a lamppost, not because that is where he dropped it but because that's where the light is.

Fourier analysis works with linear problems. *Nonlinear* problems tend to be much harder, and the behavior of nonlinear systems is much less predictable: a small change in input can cause a big change in output. The law of gravity is nonlinear and using it to predict the very long term behavior of even three bodies in space is wildly difficult, perhaps impossible; the system is too unstable. (Engineers make clever use of this instability when sending space probes to distant planets: NASA's Pioneer and Voyager spacecraft were both aimed at Jupiter in such a way that Jupiter's gravity accelerated the probes and bent their paths, sending them on to Saturn.)

"It is sometimes said," quips Körner, "that the great discovery of the nineteenth century was that the equations of nature were linear, and the great discovery of the twentieth century is that they are not."

Engineers faced with a nonlinear problem often resort to the rough-and-ready expedient of treating it like a linear problem and hoping the answer won't be too far off. For example, the engineers charged with defending Venice from the high waves that flood it every year, forcing Venetians to make their way across the Piazza San Marco on sidewalks set on stilts, want to predict the flood waters far enough in advance so that eventually they could raise inflatable dikes to protect the city. Since they can't solve the nonlinear partial differential equation that determines the behavior of the waves (an equation involving winds, position of the moon, atmosphere pressure, and so on), they simply reduce it to a linear equation and solve it with Fourier analysis. Despite some progress, they are still taken by surprise by sudden rises in water level of up to a meter.

In information processing Fourier analysis has other limits as well: it is poorly suited to very brief signals or to signals that change suddenly and unpredictably. A Fourier transform makes it easy to see how much of each frequency a signal contains but very much harder to figure out when the various frequencies were emitted or for how long. It pretends, so to speak, that any given instant of the signal is identical to any other, even if the signal is as complex as a Bach prelude or changes as dramatically as the electrocardiogram of a fatal heart attack.

Mathematically this is correct. The same sines and cosines can represent very different moments in a signal because they are shifted in phase so that they amplify or cancel each other. A B-flat that appears, from the Fourier transform, to be omnipresent in a prelude may in fact appear only intermittently; the rest of the time it is part of an elaborate juggling act. But as physicist J. Ville wrote in 1948, "If there is in this concept a dexterity that does honor to mathematical analysis, one can't hide the fact that it also distorts reality."[9] One moment of a prelude does not sound like another; the flat line of an electrocardiogram that announces death is not the same as the agitated lines produced by a beating heart.

The time information is not destroyed by the Fourier transform, just well hidden, in the phases. But the fact that information about any point in a signal is spread out over the entire transform—contained in all the frequencies—is a serious drawback in analyzing signals or functions. Brief changes often carry the most interesting information in a signal; in medical tests, for example, being able to detect them could make the difference between a correct or an incorrect diagnosis. In theory, the phases containing this time information can be calculated from the Fourier coefficients; in practice, calculating them with adequate precision is virtually impossible.

In addition, the lack of time information makes Fourier transforms vulnerable to errors. "If when you record an hour-long signal you have an error in the last five minutes, the error will corrupt the whole Fourier transform," Meyer points out. And phase errors are disastrous; if you make the least error in phase, you can end up with something that has nothing to do with your original signal.

To get around this problem, the *windowed Fourier transform* was created in the 1940s. The idea is to analyze the frequencies of a signal segment by segment; that way, one can at least say that whatever is happening is happening somewhere in a given segment. So while the Fourier transform uses sines and cosines to analyze a signal, windowed Fourier uses a little piece of a curve. This curve serves as the "window," which remains fixed in size for a given analysis; inside it one puts oscillations of varying frequency.

But these rigid windows force painful compromises. The smaller your window, the better you can locate sudden changes, such as a peak or discontinuities, but the blinder you become to the lower-frequency components of your signal. (These lower frequencies won't fit into the little window.) If you choose a bigger window, you can see more of the low frequencies but the worse you do at "localizing in time."

So Yves Meyer, who as a harmonic analyst was well aware of the power and limitations of Fourier analysis, was interested when he first heard of a new way to break down signals, or functions, into little waves—*ondelettes* in French—that made it possible to see fleeting changes in a signal without losing track of frequencies.

TALKING TO HEATHENS

"I got involved almost by accident," recalls Meyer. "I was a professor at the Ecole Polytechnique, where we shared the same photocopy machine with the department of theoretical physics. The department chairman liked to read everything, know everything; he was constantly making photocopies. Instead of being exasperated when I had to wait, I would chat with him while he made his copies.

"One day in the spring of 1985 he showed me an article by a physicist colleague of his, Alex Grossmann in Marseille, and asked whether it interested me. It involved signal processing, using a mathematical technique I was familiar with. I took the train to Marseille and started working with Grossmann."

Often pure math "trickles down" to applications, but this was not the case for wavelets, Meyer added. "This is not something imposed by the mathematicians; it came from engineering. I recognized familiar mathematics, but the scientific movement was from application to theory. The mathematicians did a little cleaning, gave it more structure, more order."

Structure and order were needed; the predecessors of today's wavelets had grown in a topsy-turvy fashion, to the extent that in the early days wavelet researchers often found themselves unwittingly recreating work of the past. "I have found at least 15 distinct roots of the theory, some going back to the 1930s," Meyer said. "David Marr, who worked on artificial vision and robotics at MIT, had similar ideas. The physics community was intuitively aware of wavelets dating back to a paper on renormalization by Kenneth Wilson, the Nobel prize winner, in 1971."

But all these people in mathematics, physics, vision, and information processing didn't realize they were speaking the same language, partly for the simple reason that they rarely spoke to one another but also partly because the early work existed in such disparate forms. (Grossmann had in fact spoken about wavelets to other people in Meyer's field, but they "didn't make the connection," he said. "With Yves it was immediate, he realized what was happening.")

Wavelet researchers sometimes joke that the main benefit of wavelets is to allow them to have wavelet conferences. Behind that joke lies the reality that the modern coherent language of wavelets has provided an unusual opportunity for people from different fields to speak and work together, to everyone's benefit.

"Under normal circumstances the fields are pretty much water-tight one to the other," Grossmann said. "So one of the main reasons that many people find this field very interesting is that they have found themselves outside of their usual universe, talking to heathens of various kinds. Anybody who is not in one's little village is a heathen by definition, and people are always surprised to see— 'look, they have two ears and a single nose, just like us!' That has been a very pleasant experience for everyone."

"THIS MUST BE WRONG. . . ."

Tracing the history of wavelets is almost a job for an archeologist, but let's take as a starting point Jean Morlet, who developed them independently in prospecting for oil for the French oil company Elf-Aquitaine. (It was he who baptized the field, originally using the term "wavelets of constant shape" to distinguish them from other "wavelets" in geophysics.) A standard way to look for underground oil is to send vibrations into the earth and analyze the echoes that return. Doing this, Morlet became increasingly dissatisfied with the limits of Fourier analysis; he wanted to be able to analyze brief changes in signals. To do this he figured out both a way to decompose a signal into wavelets and a way to reconstruct the original signal.

"The thing that is surprising is how very far Jean went all by himself, with no formal baggage. He had a lot of intuition to make it work without knowing why it worked," says Marie Farge of the Ecole Normale Supérieure in Paris, who uses wavelets to study turbulence.

But when Morlet started showing his results to others in the field, he was told that "this must be wrong, because if it were right, it would be known." Convinced that his wavelets were important—and aware that he didn't understand why they worked—Morlet spoke to a physicist at the Ecole Polytechnique who sent him, in 1981, to see Alex Grossmann in Marseille.

"Jean was sent to me because I work in phase space quantum mechanics," Grossmann said. "Both in quantum mechanics and in signal processing you use the Fourier transform all the time—but then somehow you have to keep in mind what happens on both sides of the transform.

When Jean arrived, he had a recipe, and the recipe worked. But whether these numerical things were true in general, whether they were approximations, under what conditions they held, none of this was clear."

The two spent a year answering those questions. Their approach was to show mathematically that when wavelets represent a signal, the amount of "energy" of the signal (a measure of its size) is unchanged. This means that one can transform a signal into wavelet form and then get exactly the same signal back again—a crucial condition. It also means that a small change in the wavelet representation produces a correspondingly small change in the signal; a little error or change will not be blown out of proportion.

The work involved a lot of experimenting on personal computers. "One of the many reasons why the whole thing didn't come out earlier is that just about this time it became possible for people who didn't spend their lives in computing to get a little personal computer and play with it," Grossmann says. "Jean did most of his work on a personal computer. Of course, he could also handle huge computers, that's his profession, but it's a completely different way of working. And I don't think I could have done anything if I hadn't had a little computer and some graphics output."

A MATHEMATICAL MICROSCOPE

Wavelets can be seen as an extension of Fourier analysis. As with the Fourier transform, the point of wavelets is not the wavelets themselves; they are a means to an end. The goal is to turn the information of a signal into numbers—coefficients—that can be manipulated, stored, transmitted, analyzed, or used to reconstruct the original signal (see Box on p. 212; Figure 7.3).

The basic approach is the same. The coefficients tell in what way the analyzing function (sines and cosines, a Fourier window or wavelets) needs to be modified in order to reconstruct the original signal. The idea underlying the calculation of coefficients is the same (although in practice the mathematical details vary). For his wavelets Morlet even used the Gaussian, or bell-shaped, function often used in windowed Fourier analysis. But he used it in a fundamentally different way. Instead of filling a rigid window with oscillations of different frequencies, he did the reverse. He kept the number of oscillations in the window constant and varied the width of the window, stretching or compressing it like an accordion or a child's slinky (see Figure 7.4). When he stretched the wavelet, the oscillations inside were stretched,

The Wavelet Transform

The wavelet transform decomposes a signal into wavelet coefficients. Each wavelet is multiplied by the corresponding section of the signal. Then one integrates (measures the area enclosed by the resulting curve). The result is the coefficient for that particular wavelet.

Essentially, the coefficient measures the correlation, or agreement, between the wavelet (with its peaks and valleys) and the corresponding segment of the signal. "With wavelets, you play with the width of the wavelet in order to catch the rhythm of the signal," Meter says. "Strong correlation means that there is a little piece of the signal that looks like the wavelet."

Constant stretches give wavelet coefficients with the value zero. (By definition, a wavelet has an integral of zero—half the area enclosed by the curve of the wavelet is above zero and half is below. Multiplying a wavelet by a constant changes both the positive and the negative components equally, so the integral remains zero.)

Wavelets can also be made that give coefficients of zero when they meet linear and quadratic stretches and even higher polynomials. The more zero coefficients, the greater the compression of the signal, which makes it cheaper to store or transmit, and can simplify calculations. A typical signal may have about 100,000 values, but only 10,000 wavelets are needed to express it; parts of the signal that give coefficients of zero are automatically disregarded.

Using wavelets that are "blind" to linear and quadratic stretches, and higher polynomials, also makes it easier to detect very irregular changes in a signal. Such wavelets react violently to irregular changes, giving big coefficients that stand out against the background of very small coefficients and zero coefficients indicating regular changes.

decreasing their frequency; when he squeezed the wavelet, the oscillations inside were squeezed, resulting in higher frequencies.

As a result, wavelets adapt automatically to the different components of a signal, using a big "window" to look at long-lived components of low frequency and progressively smaller windows to look at short-lived components of high frequency. The procedure is called multiresolution; the signal is studied at a coarse resolution to get an overall picture and at higher and higher resolutions to see increasingly fine details. Wavelets have in fact been called a "mathematical micro-

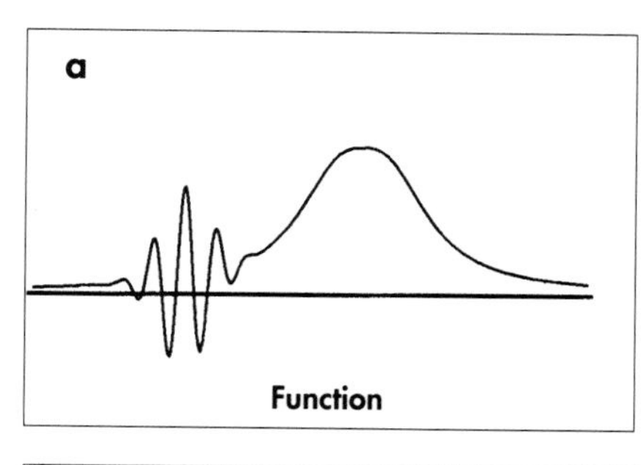

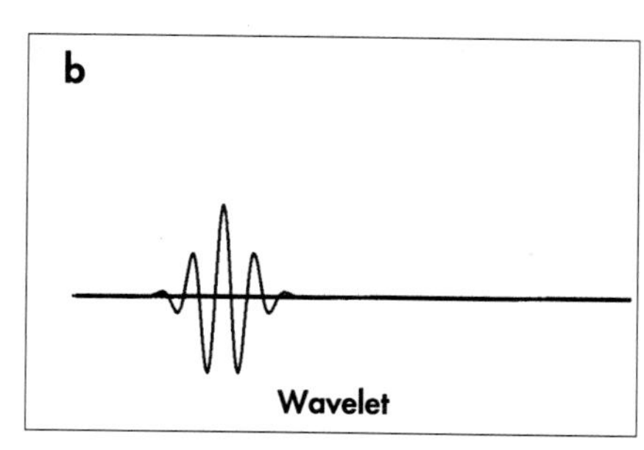

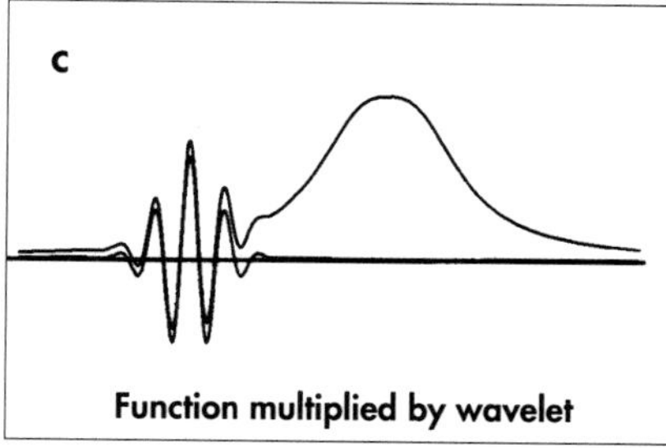

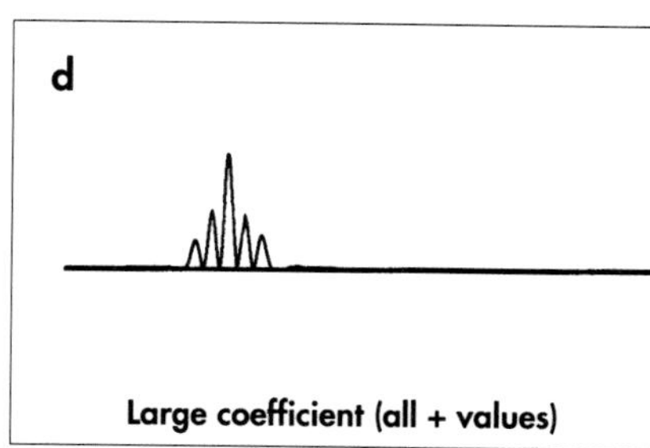

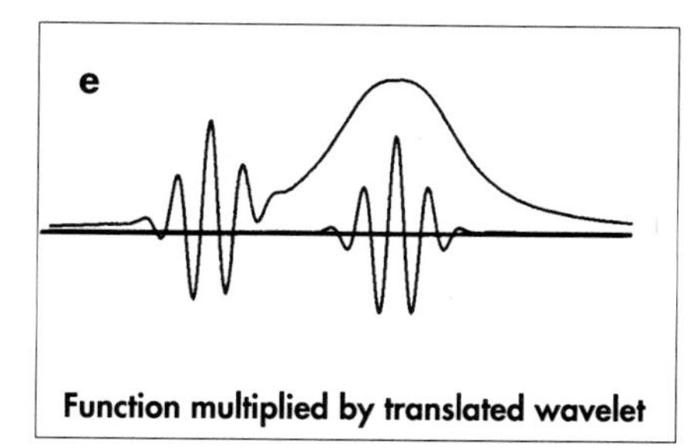

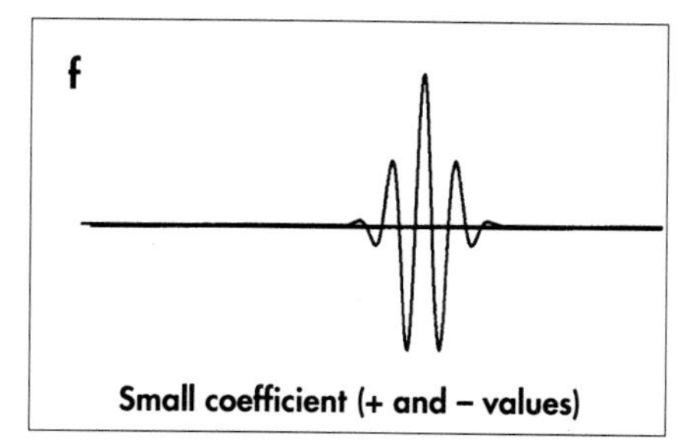

FIGURE 7.3 *In the wavelet transform a wavelet is compared successively to different sections of a function. The product of the section and the wavelet is itself a curve; the area under that curve is the wavelet coefficient. Sections of the function that look like the wavelet give big coefficients, as seen in panels c and d. (Two negative functions multiplied together give a positive function.) Slowly changing sections produce small coefficients, as seen in panels e and f. (Courtesy of John Hubbard, Cornell University.)*

scope"; compressing wavelets increases the magnification of this microscope, enabling one to take a closer and closer look at small details in the signal.

And unlike a Fourier transform, which treats all parts of a signal equally, wavelets only encode changes in a function. That is, unchanging stretches of a signal give coefficients with the value zero, which can be ignored. This makes them good for "seeing" changes—peaks in a signal, for example, or edges in a picture—and also means that they can be effective for compressing information.

"Wavelet analysis is a way of saying that one is sensitive to changes," Meyer says. "It's like the way we respond to speed." We don't feel the speed of a train as long as the speed is constant, but we notice when it speeds up or slows down.

MOTHER OR AMOEBA?

In wavelet analysis a function is represented by a family of little waves: what the French call a "mother" wavelet, a "father," and a great many babies of various sizes. But the French terminology "shows a scandalous misunderstanding of human reproduction," objects Cornell mathematician Robert Strichartz. "In fact the generation of wavelets more closely resembles the reproductive life style of an amoeba."

To make baby wavelets one clones *la mère* and then either stretches or compresses the new wavelets; in mathematical jargon they are "dilated." These new wavelets can then be shifted about or "translated."

But the father function plays an important role. If you were looking at changes in temperature, you might be interested in broad changes over millions of years (the ice ages, for example), fluctuations over the past hundred years, or changes between day and night. If your real interest were the effect of climate on wheat production in the nineteenth century, you might look at temperatures on the scale of a year, a decade, or possibly a century; you wouldn't bother with changes on the scale of a

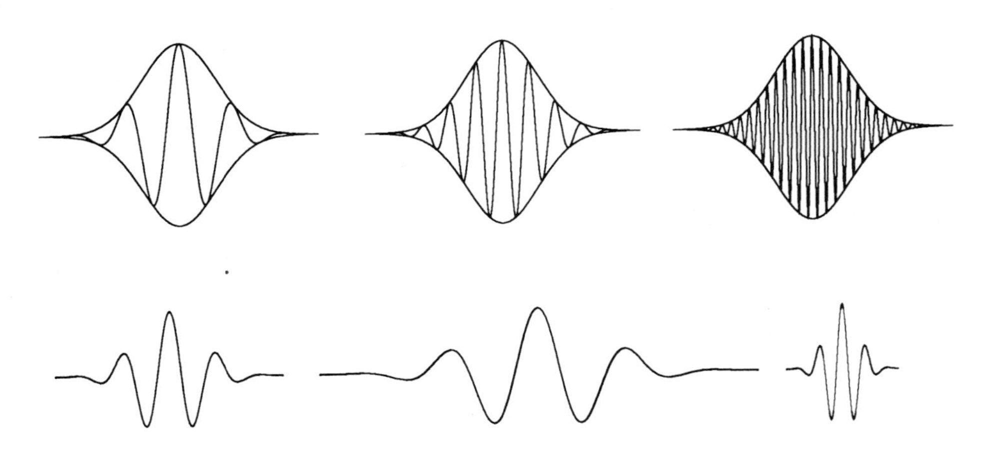

FIGURE 7.4 *Top row: In windowed Fourier analysis the size of the window is fixed and the number of oscillations varies. A small window will be "blind" to low frequencies, which are too large to fit into the window. If a large window is used, information about a brief change will be submerged in the information about the entire section of the signal corresponding to the window. Bottom row: In wavelet analysis, a "mother wavelet" (left) is stretched or compressed to change the window size, making it possible to analyze a signal at different scales. The wavelet transform has been called a "mathematical microscope": wide wavelets give an overall picture, while smaller and smaller wavelets are used to "zoom in" on small details. (Courtesy of John Hubbard, Cornell University.)*

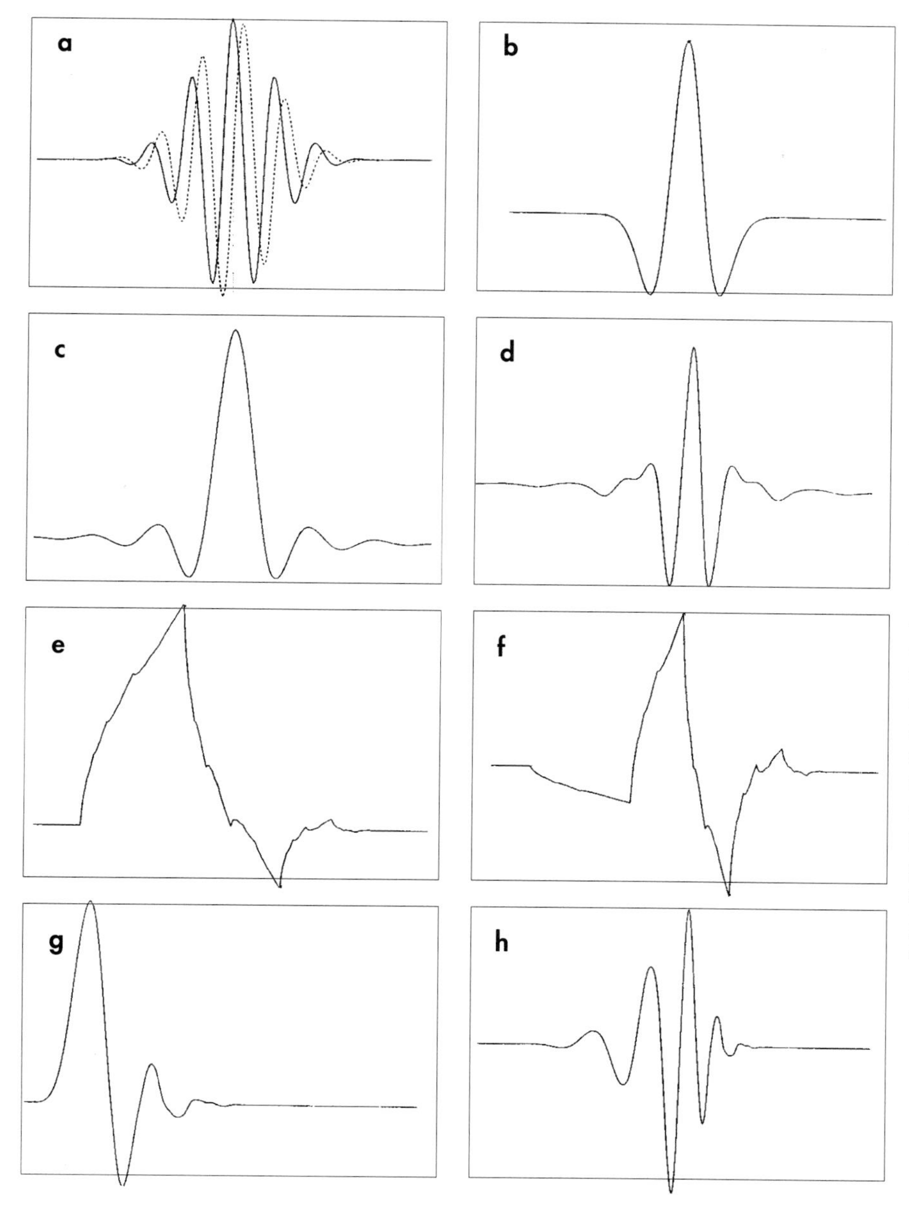

FIGURE 7.5 *A small sample of wavelets and scaling functions (father functions). Wavelets shown in panels a and b are used in continuous representations; no father function is needed. The Morlet wavelet in panel a is complex valued; the dotted line gives the imaginary part. The "Mexican hat" shown in panel b is the second derivative of a Gaussian function (bell curve). The other wavelets are orthogonal, and are associated with a scaling function: c, Battle-Lemarié scaling function; d, Battle-Lemarié wavelet corresponding to c; e, Daubechies scaling function, order 2; f, Daubechies wavelet, order 2; g, Daubechies scaling function, order 7; h, Daubechies wavelet, order 7. The "order" of a wavelet is a measure of its regularity or smoothness; the Daubechies wavelet in panel h is more regular than the one in panel f. (Courtesy of Marie Farge and Eric Goirand.)*

thousand years, much less a million. The father function (now more often referred to as the "scaling function") gives the starting point (see Figure 7.5).

To do this, you construct a very rough approximation of your signal, using only father functions. Imagine covering your signal with a row of identical father functions. You then multiply each one by the corresponding section of the signal and integrate (measure the space under the resulting curve). The resulting numbers, or coefficients, give the information you would need to reconstitute a very rough picture of your function.

Wavelets are then used to express the details that you would have to add to that first rough picture in order to reconstitute the original signal. At the coarsest resolution, perhaps a hundred fat wavelets are lined up next to each other on top of the signal. The wavelet transform (see Box on p. 212) assigns to each one a coefficient that tells how much of the function has the same frequency as the fat wavelets.

At the next resolution the next generation of wavelets—twice as many, half as wide, and with twice the frequency—is put on top of the signal, and the process is repeated. Each step gives more details; at each step the frequency doubles and the wavelets are half as wide. (Typically, up to five different resolutions are used.)

If the procedure is compared to expressing the number 78/7 in decimal form (11.142857142 . . .), then the number 11 corresponds to the information given by the father functions, and the first set of wavelets would encode the details 0.1; the next, skinnier wavelets would encode 0.04, the next wavelets 0.002, the next wavelets 0.0008, and so on. The farther you go in layers of detail the more accurate the approximation will be, and each time you will be using a tool of the appropriate size. At the end, your signal has been neatly divided into different-frequency components—but unlike Fourier analysis, which gives you global amounts of different frequencies for the whole signal, wavelets give you a different frequency breakdown for different times on your signal.

To reconstruct the signal, you add the original rough picture of the function and all the details, by multiplying each coefficient by its wavelet or father function and adding them all together, just as to reconstruct the number 78/7 you add the number 11 given by the father and all the details: 0.1, 0.04, 0.002, and so on. Of course, the number will still be in decimal form; in contrast, when you reconstruct a signal from a father function, wavelets, and wavelet coefficients, you switch back to the original form of representation, out of "wavelet space."

AVOIDING REDUNDANCY

When Meyer took the train to Marseille to see Grossmann in 1985 the idea of multiresolution existed—it originated with Jean Morlet—but wavelets were limited and sometimes difficult to use, compared with the choices available today (which are still changing rapidly). For one thing computing wavelet coefficients was rather slow. For another the wavelet transforms that existed then were all continuous. Imagine a wavelet slowly gliding along the signal, new wavelet coefficients being computed as it moves. (The process is repeated at all possible frequencies or scales; instead of brutally changing the size of the wavelets by a factor of 2, you stretch or compress it gently to get all the intermediate frequencies.)

In such a *continuous* representation, there is a lot of repetition, or redundancy, in the way information is encoded in the coefficients. (The number of coefficients is in fact infinite, but in practice "infinite may mean 10,000, which is not so bad," Grossmann says.) This can make it easier to analyze data, or recognize patterns. A continuous representation is *shift invariant*: exactly where on the signal one starts the encoding doesn't matter; shifting over a little doesn't change the coefficients. Nor is it necessary to know the coefficients with precision.

"It's like drawing a map," says Ingrid Daubechies, professor of mathematics at Princeton and a member of the technical staff at AT&T Bell Laboratories, who has worked with wavelets since 1985. "Many men draw these little lines and if you miss one detail you can't find your way. Most women tend to put lots of detail—a gas station here, a grocery store there, lots and lots of redundancy. Suppose you took a bad photocopy of that map, if you had all that redundancy you still could use it. You might not be able to read the brand of gasoline on the third corner but it would still have enough information. In that sense you can exploit redundancy: with less precision on everything you know, you still have exact, precise reconstruction."

But if the goal is to compress information in order to store or transmit it more cheaply, redundancy can be a problem. For those purposes it is better to have a different kind of wavelet, in an *orthogonal* transform, in which each coefficient encodes only the information in its own particular part of the signal; no information is shared among coefficients (see Box on p. 218).

At the time, though, Meyer wasn't thinking in terms of compressing information; he was immersed in the mathematics of wavelets. A few years before it had been proved that it is impossible to have an orthogonal representation with standard windowed Fourier analysis; Meyer was convinced that orthogonal wavelets did not exist either (more precisely, infinitely differentiable orthogonal wavelets that soon get close to zero on either side). He set out to prove it—and failed, in the summer of 1985, by constructing precisely the kind of wavelet he had thought didn't exist.

Orthogonality

The practical consequences of orthogonality for wavelets are both good and bad. Orthogonal wavelets encode information efficiently, and their coefficients can be calculated very rapidly. But the coefficients can be hard to interpret and the representation of the signal is not "shift invariant"—where you start encoding a signal makes a difference.

Mathematically, orthogonality refers to a geometrical relationship. Over the past 150 years or so mathematicians have come to think of functions geometrically, as single points in an infinite dimensional space. You can define a point on a line with one number, a point in a place with two numbers, and a point in three-dimensional space with three numbers, but to define an entire function you need to know all its values: you need infinitely many numbers. A function—a wavelet, for example—becomes a single point in an infinite dimensional space.

Mathematicians can't really picture infinite dimensional space, but they learn to use their intuition about ordinary three-dimensional space to think about it. To get an idea of what orthogonality means geometrically, think of ordinary space. Any point in space can be defined by a vector going from zero to that point. Vectors are orthogonal if they form right angles with each other. In three-dimensional space, no more than three vectors can be mutually perpendicular, and in the plane only two. But in an infinite-dimensional space, you can choose infinitely many vectors, each of which is perpendicular to all the others.

Families of orthogonal wavelets form such systems: the vectors formed by one mother wavelet and all its "babies" (its "translates" and "dilates") are all perpendicular to each other. The speed of computing orthogonal wavelet coefficients is a consequence of this geometry (as is the speed of calculating Fourier coefficients; sines and cosines also form an orthogonal basis).

That smooth orthogonal wavelets exist is not obvious; Yves Meyer, who found that they did, originally set out to prove the opposite. The Haar function (1909), together with its translates and dilates, forms an orthogonal system, but some wavelet researchers do not consider it a true wavelet because it is so jerky:

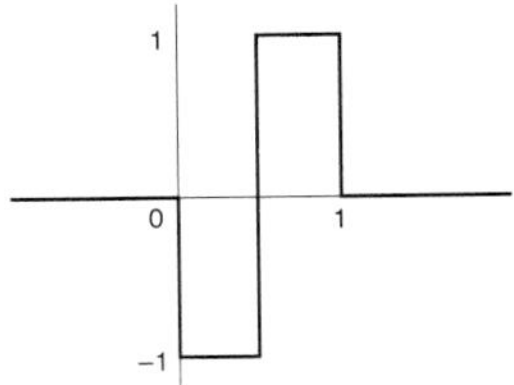

An intermediary construction between the Haar function and Meyer's wavelets was made by J.O. Strömberg (1981); his functions were orthogonal but not infinitely differentiable (not perfectly smooth).

UNIFICATION

The following year, in the fall of 1986, while Meyer was giving a course on wavelets at the University of Illinois at Urbana, he received several telephone calls from a persistent 23-year-old graduate student in computer vision at the University of Pennsylvania in Philadelphia.

Stéphane Mallat (now at the Courant Institute of Mathematical Sciences in New York) is French and had been a student at the Ecole Polytechnique, one of France's prestigious *grandes écoles*, when Meyer taught there, but the two hadn't met. "The system at Ecole Polytechnique is very rigid and very elitist, the students have a rank when they graduate," Meyer said. "According to their rank they are directed to this or that profession. Mallat had found this system absurd and had decided that he would do what he wanted." So after graduating he did something that, from a French perspective, was extraordinary: abandoning the social and professional advantages of being a *polytechnicien* (the first 150 in each graduating class are even guaranteed a salary for life), he left France for the United States.

"Working in the United States, for Stéphane Mallat, was starting over from zero," Meyer said. "No one there knows what Ecole Polytechnique is, they couldn't care less, he was in the same situation as a student from Iran, for example. . . . He's completely original, in his behavior, his way of thinking, his way of progressing in his career."

Mallat had heard about Meyer's work on wavelets from a friend in the summer of 1986 while he was vacationing in St. Tropez; to him it sounded suspiciously familiar. So on returning to the United States, he called Meyer, who agreed to meet him at the University of Chicago. The two spent 3 days holed up in a borrowed office ("I kept telling Mallat that he absolutely had to go to the Art Institute in Chicago, but we never had time," Meyer says) while Mallat explained that the multiresolution Meyer and others were doing with wavelets was the same thing that electrical engineers and people in image processing were doing under other names.

"This was a completely new idea," Meyer said. "The mathematicians were in their corner, the electrical engineers were in theirs, the people in vision research like David Marr were in another corner, and the fact that a young man who was then 23 years old was capable of saying, you are all doing the same thing, you have to look at that from a broader perspective—you expect that from someone older."

In 3 days the two worked out the mathematical details; since Meyer was already a full professor, at his insistence the resulting paper,

"Multiresolution Approximation and Wavelets," appeared under Mallat's name alone. The paper made it clear that work that existed in many different guises and under many different names—the pyramid algorithms used in image processing, the quadrature mirror filters of digital speech processing, zero-crossings, wavelets—were at heart all the same. For using wavelets to look at a signal at different resolutions can be seen as applying a succession of filters: first filtering out everything but low frequencies, then filtering out everything but frequencies twice as high, and so on. (And, in accordance with Shannon's sampling theorem, wavelets automatically "sample" high frequencies more often than low frequencies, since as the frequency doubles the number of wavelets doubles.)

The realization benefited everyone. A whole mathematical literature on wavelets existed by 1986, some of it developed before the word wavelet was even coined; this mathematics could now be applied to other fields.

"All those existing techniques were tricks that had been cobbled together; they had been made to work in particular cases," says Marie Farge. "Mallat helped people in the quadrature mirror filters community, for example, to realize that what they were doing was much more profound and much more general, that you had theorems, and could do a lot of sophisticated mathematics."

Wavelets got a big boost because Mallat also showed how to apply to wavelet multiresolution fast algorithms that had been developed for other fields, making the calculation of wavelet coefficients fast and automatic—essential if they were to become really useful. And he paved the way for the development by Daubechies of a new kind of regular orthogonal wavelet that was easier and faster to use: wavelets with "compact support." Such wavelets have the value zero everywhere outside a certain interval (between –2 and 2, for example).

EVADING INFINITY

Daubechies, who is Belgian, was trained as a mathematical physicist; she had worked with Grossmann in France on her Ph.D. research and then spent 2 years in the United States working on quantum mechanics. She is the recipient of a 5-year MacArthur fellowship.

"Ingrid's role has been crucial," Grossmann said. "Not only has she made very important contributions, but she has made them in a form that was legible, and usable, to various communities. She is able to speak to engineers, to mathematicians; she is trained as a physicist and one sees her training in quantum mechanics."

Daubechies had heard about Meyer and Mallat's multiresolution work very early on. "Yves Meyer had told me about it at a meeting. I had been thinking about some of these issues, and I got very interested," she said. The orthogonal wavelets Meyer had constructed trail off at the sides, never actually ending; this meant that calculating a single wavelet coefficient was a lot of work.

"I said why can't we just start from the fact that we want those numbers and that we want a scheme that has these properties [and] proceed from there," Daubechies said. "That's what I did. . . . I was extremely excited; it was a very intense period. I didn't know Yves Meyer so very well at the time. When I had the first construction, he had gotten very excited, and somebody told me he had given a seminar on it. I knew he was a very strong mathematician and I thought, oh my God, he's probably figuring out things much faster than I can. . . . Now I know that even if that had been true, he would not have taken credit for it, but it put a very strong urgency on it; I was working very hard. By the end of March 1987 I had all the results."

Together, multiresolution and wavelets with compact support formed the wavelet equivalent of the fast Fourier transform: again, not just doing calculations a little faster, but doing calculations that otherwise very likely wouldn't be done at all.

HEISENBERG

Multiresolution and Daubechies's wavelets also made it possible to analyze the behavior of a signal in both time and frequency with unprecedented ease and accuracy, in particular, to zoom in on very brief intervals of a signal without becoming blind to the larger picture.

But although one mathematician hearing Daubechies lecture objected that she seemed "to be beating the uncertainty principle," the Heisenberg uncertainty principle still holds. You cannot have perfect knowledge of both time and frequency. Just as you cannot know simultaneously both the position and momentum of an elementary particle (if you could hold an electron still long enough to figure out where it was, you would no longer know how fast it would have been going if you hadn't stopped it). The product of the two uncertainties (or spreads of possible values) is always at least a certain minimum number.

One must always make a compromise; knowledge gained about time is paid for in frequency, and vice versa. "At low frequencies I have wide wavelets, and I localize very well in frequency but very badly in time. At very high frequency I have very narrow wavelets, and I localize very well in time but not so well in frequency," Daubechies said.

This imprecision about frequency results from the increasing range of frequencies at high frequencies: as we have seen, frequency doubles each time one goes up an octave. This widening spread of frequencies can be seen as a barrier to precision, but it's also an opportunity that engineers have learned to exploit. It is the reason why the telephone company shifts voices up into higher frequencies, not down to lower ones, when it wants to fit a lot of voices on one line: there's a lot more room up there. It also explains the advantage of fiber optics, which carry high-frequency light signals, over conventional telephone wires.

PUTTING WAVELETS TO WORK

Wavelets appear unlikely to have the revolutionary impact on pure mathematics that Fourier analysis has had. "With wavelets it is possible to write much simpler proofs of some theorems," Daubechies said. "But I know of only a couple of theorems that have been proved with wavelets, that had not been proved before." But the characteristic features of wavelets make them suited for a wide range of applications.

Because wavelets respond to change and can narrow in on particular parts of a signal, researchers at the Institut du Globe in Paris are using them to study the minuscule effect on the speed of the earth's rotation of the El Niño ocean current that flows along the coast of Peru. British scientists are using wavelets to study ocean currents around the Antarctic, and researchers and mechanics are exploring their use in detecting faults in gears by analyzing vibrations.

Multiresolution lends itself to a variety of applications in image processing. One can imagine transmitting pictures electronically quickly and cheaply by sending only a coarse picture, calling up a more detailed picture only when needed. Mathematician Dennis Healy, Jr., and radiologist John Weaver of Dartmouth College are exploring the use of wavelets for "adaptive" magnetic resonance imaging, in which higher resolutions would be used selectively, depending on the results already found at coarser scales. (Since a half-hour magnetic resonance imaging exam costs $500 to $1000 or more, anything that reduces the time spent is of obvious interest.)

Multiresolution is also useful in studying the large-scale distribution of matter in the universe, which for years was thought to be random but which is now seen to have a complicated structure, including "voids" and "bubbles."[10] Wavelets have enabled astronomers at the Observatoire de la Côte d'Azur in Nice to identify a subcluster at the center of the Coma supercluster, a cluster of about 1400 galaxies. Subsequently, that subcluster was identified as an x-ray source. "Wavelets were like a telescope pointing to the right place," Meyer said. And at the Centre de Recherche Paul Pascal in Pessac (near Bordeaux), Alain Arnéodo and colleagues have exploited "the fascinating ability of the wavelet transform to reveal the construction rule of fractals"[11].

In addition, it can be instructive to compare wavelet coefficients at different resolutions. Zero coefficients, which indicate no change, can be ignored, but nonzero coefficients indicate that something is going on—whether an abrupt change in the signal, an error, or noise (an unwanted signal that obscures the real message). If coefficients appear only at fine scales, they generally indicate the slight but rapid variations characteristic of noise. "The very fine scale wavelets will try to follow the noise," Daubechies explains, while wavelets at coarser resolutions are too approximate to pick up such slight variations.

But coefficients that appear at the same part of the signal at all scales indicate something real. If the coefficients at different scales are the same size, it indicates a jump in the signal; if they decrease, it indicates a singularity—an abrupt, fleeting change. It is even possible to use scaling to sharpen a blurred signal. If the coefficients at coarse and medium scales suggest there is a singularity, but at high frequencies noise overwhelms the signal, one can project the singularity into high frequencies by restoring the missing coefficients—and end up with something better than the original.

CUT THE WEEDS AND SPARE THE DAISIES

Wavelets also made possible a revolutionary method for extricating signals from pervasive white noise ("all-color," or all-frequency, noise), a method that Meyer calls a "spectacular application" with great potential in many fields, including medical scanning and molecular spectroscopy.

An obvious problem in separating noise from a signal is knowing which is which. If you know that a signal is smooth—changing slowly—and that the noise is fluctuating rapidly, you can filter out noise by averaging adjacent data to kill fluctuations while preserving the trend. Noise can also be reduced by filtering out high frequencies. For smooth signals, which change relatively slowly and therefore are mostly lower frequency, this will not blur the signal too much.

But many interesting signals (the results of medical tests, for example) are not smooth; they contain high-frequency peaks. Killing all high frequencies mutilates the message—"cutting the daisies along with the weeds," in the words of Victor Wickerhauser of Washington University in St. Louis.

FIGURE 7.6 *The wavelet transform can detect the edges of an image at several scales. Among all edges of panel a, a computer program has selected the "important" ones. Panel b displays these edges at a fine scale. Panel c is reconstructed from edges selected at different scales, with an algorithm developed by Mallat and Zhifeng Zhang. The edge selection removes the noise and small irregular structures. The skin is now smoother.*

A simple way to avoid this blind slaughter has been found by a group of statisticians. David Donoho of Stanford University and the University of California at Berkeley and his colleague Iain Johnstone of Stanford had proved mathematically that if a certain kind of orthogonal basis existed, it would do the best possible job of extracting a signal from white noise. (A *basis* is something with which you can represent any possible function in a given space; each mother wavelet provides a different basis, for example, since any function can be represented by it and its translates and dilates.)

This result was interesting but academic since Donoho and Johnstone did not know whether such a basis existed. But in the summer of 1990, when Donoho was in St. Flour, in France's Massif Central, to teach a course in probability, he heard Dominique Picard of the University of Paris-Jussieu give a talk on the possibility of using wavelets in statistics. After discussing it with her and with Gérard Kerkyacharian of the University of Picardy in Amiens, "I realized it was what we had been searching for a long time," Donoho recalled. "We knew that if we used wavelets right, they couldn't be beaten."

The method is simplicity itself: you apply the wavelet transform to your signal, throw out all coefficients below a certain size, at all frequencies or resolutions, and then reconstruct the signal. It is fast (because the wavelet transform is so fast), and it works for a variety of kinds of signals. The astonishing thing is that it requires no assumptions about the signal. The traditional view is that one has to know, or assume, something about the signal one wants to extract from noise—that, as Grossmann put it, "if there is absolutely no *a priori* assumption you can make about your signal, you may as well go to sleep. On the other hand, you don't want to put your wishes into your algorithm and then be surprised that your wishes come out."

The wavelet method stands this traditional wisdom on its head. Making no assumptions about the signal, Donoho says, "you do as well as someone who makes correct assumptions, and much better than someone who makes wrong assumptions." (Furthermore, if you *do* know something about the signal, you can adjust the coefficient threshold and get even better results.)

The trick is that an orthogonal wavelet transform makes a signal look very different while leaving noise alone. "Noise in the signal becomes noise in the wavelet transform and it has about the same size at every resolution and location," Donoho says. (That all orthogonal representations leave noise unchanged has been known since the 1930s.) So while noise masks the signal in "physical space," the two become disentangled in "wavelet space."

In fact, Donoho said, a number of researchers—at the Massachusetts Institute of Technology, Dartmouth, the University of South Carolina, and elsewhere—independently discovered that thresholding wavelet coefficients is a good way to kill noise. "We came to it by mathematical decision theory, others simply by working with wavelet transforms and noticing what happened," he said.

Among those was Mallat, who uses a somewhat different approach. Donoho's method works for a whole range of functions but isn't necessarily optimal for each. When it is applied to blurred images, for example, it damages some of the edges; the elimination of small coefficients creates ripples that can be annoying. Mallat and graduate student Wen Liang Hwang developed a way to avoid this by computing the wavelet transform of the signal and selecting the points where the correlation between the curve and the wavelet is greatest, compared to nearby points. These maximum values, or wavelet *maxima*, are kept if the points are thought to belong to a real edge and discarded if they are thought to correspond to noise. (That decision is made automatically, but it requires more calculations than Donoho's method; it is based on the existence and size of maxima at different resolutions.)

WAVELETS DON'T EXIST . . .

Although Donoho and Johnstone's technique is simple and automatic, wavelets aren't always foolproof. With orthogonal wavelets it can matter where one starts encoding the signal: shifting over a little can change the coefficients completely, making pattern analysis hazardous. This danger does not exist with continuous wavelets, but they have their own pitfalls; what looks like a correlation of coefficients (different coefficients "seeing" the same part on the signal) may sometimes be an artifact introduced by the wavelets themselves. "It's the kind of thing where you can shoot yourself in the foot without half trying," Grossmann said.

Generally, using wavelets takes practice. "With a Fourier transform, you know what you get," Meyer says. "With a wavelet transform you need some training in order to know what you get. I have a report from EDF [Electricité de France] giving conclusions of engineers about wavelets—they say they have trouble with interpretation." Part of that difficulty may be fear of trying something new; Gregory Beylkin of the University of Colorado at Boulder reports that one student, who learned to use wavelets before he knew Fourier analysis, experienced no difficulty. Farge has had similar experiences, but Meyer thinks the problem is real.

Because Fourier analysis has existed for so long, and most physicists and engineers have had years of training with Fourier transforms, interpreting Fourier coefficients is second nature to them. In addition, Meyer points out, Fourier transforms aren't just a mathematical abstraction: they have a physical meaning. "These things aren't just concepts, they are as physical, as real, as this table. But wavelets don't exist in nature; that's why it is harder to interpret wavelet coefficients," he said.

Curiously, though, both our ears and our eyes appear to use wavelet techniques in the first stages of processing information. The work on "wavelets" and hearing goes back to the 1930s, Daubechies said. "They didn't talk about wavelets—they talked about constant q filtering, in which the higher the frequency, the better resolution you have in time." This is unlikely to lead to new insights about hearing or vision, she said, but it could make wavelets effective in compressing information. "If our ear uses a certain technique to analyze a signal, then if you use that same mathematical technique, you will be doing something like our ear. You might miss important things, but you would miss things that our ear would miss too."

COMPRESSING INFORMATION

One way to cope with an ever-increasing volume of signals is to widen the electronic highways—for example, by moving to higher frequencies. Another, which also reduces storage and computational costs, is to compress the signal temporarily, restoring it to its original form when needed.

In fact, only a small number of all possible "signals" are capable of being compressed, as the Russian mathematician Andrei Kolmogorov pointed out in the 1950s. A compressible signal can by definition be expressed by something shorter than itself: one sequence of digits (the signal) is encoded by a shorter sequence of digits (e.g., a computer program).

It is easy to see that, using any given language (such as the computer language Pascal), the number of short sequences is much smaller than the number of long sequences: most long sequences cannot be encoded by anything shorter than themselves. (Even a highly efficient encoding scheme like a library card catalog cannot cope with an infinite number of books; eventually, the only way to distinguish one book from another would be to print the entire book in the card catalog.) Like Heisenberg with his uncertainty principle, Kolmogorov has set an absolute limit that mathematicians and scientists cannot overcome, however clever they are. Accepting some loss of information makes more compression possible, but a limit still remains.

In practice, however, many signals people want to compress have a structure that lends itself to compression; they are not random. For instance, any point in such a signal might be likely to be similar to points near it. (In a picture of a white house with a blue door, a blue point is likely to be surrounded by other blue points, a white point by other white points.) Wavelets lend themselves to such compression; because wavelet coefficients indicate only changes, areas with no change (or very small change) are automatically ignored, reducing the number of figures that have to be kept to encode the information.

So far, Daubechies says, image compression factors of about 35 or 40 have been achieved with wavelets with little loss. That is, the information content of the compressed image is about 1/35th or 1/40th the information content of the original. But wavelets alone cannot achieve those compression factors; an even more important role is played by clever quantization methods, mathematical ways of giving more weight in the encoding to information that is important for human perception (edges, for example) than information that is less so.

"If you just buy a commercially available image compressor you can get a factor of 10 to 12, so we're doing better than that," Daubechies said. "However, people in research groups who fine-tune the Fourier transform techniques in commercial image compressors claim they can also do something on the order of 35. So it's not really clear that we can beat the existing techniques. I do not think that image compression—for instance, television image compression—is really the place where wavelets will have the greatest impact."

But the fact that wavelets concentrate the information of a signal in relatively few coefficients makes them good at detecting edges in images, which may result in improved medical tests. Healy and Weaver have found that with wavelets they can use magnetic resonance imaging to track the edge of the heart as it beats by sampling only a few coefficients. And wavelet compression is valuable in speeding some calculations. In the submarine detection work that Frazier and colleague Jay Epperson did for Daniel H. Wagner Associates, they were able to compress the original data by a factor of 16 with good results.

Ways to compress huge matrices (square or rectangular arrays of numbers) have been developed by Beylkin, working with Ronald Coifman and Vladimir Rokhlin at Yale. The matrix is treated as a picture to be compressed; when it is translated into wavelets, "every part of the matrix that could be well represented by low-degree polynomials will have very small coefficients—it more or less disappears," Beylkin says. Normally, if a matrix has n^2 entries, then almost any computation requires at least n^2 calculations and sometimes as many as n^3. With wavelets one can get by with n calculations—a very big difference when n is large.

Talking about numbers "more or less" disappearing, or treating very small coefficients as zero, may sound sloppy but it is "very powerful, very important"—and must be done very carefully, Grossmann says. It works only for a particular large class of matrices: "If you have no *a priori* knowledge about your matrix, if you just blindly use one of those things, you can expect complete catastrophe."

Just how important these techniques will prove to be is still up in the air. Daubechies predicts that "5, certainly 10 years from now you'll be able to buy software packages that use wavelets for doing big computations, in simulations, in solving partial differential equations."

Meyer is more guarded. "I'm not saying that algorithmic compression by wavelets is a dead end; on the contrary, I think it's a very important subject. But so far there is very little progress; it's just starting." Of the matrices used in turbulence, he said, only one in 10 belongs to the class for which Beylkin's algorithm works. "In fact, Rokhlin has abandoned wavelet techniques in favor of methods adapted to the particular problem; he thinks that every problem requires an ad hoc solution. If he is right, then Ingrid Daubechies is wrong, because there won't be 'prefabricated' software that can be applied to a whole range of problems, the way prefabricated doors or windows are used in housing construction."

TURBULENCE AND WAVELETS

Rokhlin works on turbulent flows in connection with aerodynamics; Marie Farge in Paris, who works in turbulence in connection with weather prediction, remains confident that wavelets will prove to be an effective tool. She was working on her doctoral thesis when she heard about wavelets from Alex Grossmann in 1984.

"I was very much excited—in turbulence we have needed a tool like wavelets for a long time," she said. (Much later she learned that some turbulence researchers in the former Soviet Union, in Perm, had been working with similar techniques completely independently since 1976.)

"When you look at turbulence in Fourier space," Farge explains, "you see cascades of energy, where energy is transferred from one wavenumber [frequency] to another. But you don't see how those cascades relate to what is happening in physical space; we had no tool that let us see both sides at once, so we could say, *voilà*, this cascade corresponds to this interaction. So when Alex showed me that wavelets were objects that allowed one to unfold the representation both in physical space and in scale, I said to myself, this is it, now we're going to get somewhere.

"I invited him to speak in a seminar and told everyone in the turbulence community in Paris to come. I was shocked by their reaction to his talk. 'Don't waste your time on that,' they told me. Now some of the people who were the most skeptical are completely infatuated and insist that everyone should use wavelets. It's just as ridiculous. It's a new tool, and one cannot force it into problems as shapeless as turbulence if it isn't calibrated first on academic signals that we know very well. We have to do a lot of experiments, get a lot of practice, develop methods, develop representations."

She uses orthogonal wavelets, or related wavelet packets, for compression but continuous wavelets for analysis: "I would never read the coefficients themselves in an orthogonal basis; they are too hard to read." With continuous wavelets she can take a one-dimensional signal that varies in space, put it into wavelet space, and get a two-dimensional function that varies with space and scale: she can actually see, on a computer screen or a printout what is happening at different scales at any given point.

Orthogonal wavelets also give time-scale information, of course (although in a rougher form, since one doubles the scale each time, ignoring intermediate scales). The difference is largely one of legibility. Once, Meyer says, Jean Jacques Rousseau invented a musical notation based on numbers rather than notes on a staff, only to be told that it would never catch on, that musicians wanted to see the shape and movement of music on the page, to see the patterns formed by notes.

The coefficients of orthogonal wavelets correspond to Rousseau's music by numbers; continuous wavelets to the musical notation we know.

Farge compares the current state of turbulence research to "prescientific zoology." Many observations are needed, she says, to see what structures in turbulence are dynamically important and to try to recreate the theory in terms of their interactions. Possible candidates are ill-defined creatures called "coherent structures" (a tornado, for example, or the vortex that forms when you drain the bath). She uses wavelets to isolate them and to see how many exist at different scales or whether a single structure exists at a whole range of scales.

Identifying the dynamically important structures would tell researchers "where we should invest lots of calculations and where we can skimp," Farge said. For studying turbulence requires calculations that defy the most powerful computers. The Reynolds number for interactions of the atmosphere—a measure of its turbulence—ranges from 10^9 to 10^{12}; direct computer simulations of turbulence can now handle Reynolds numbers on the order of 10^2 or 10^3.

But so far the results have been disappointing, Meyer says: "There should be something between turbulence and wavelets, everyone thinks so, but so far no one has a real scientific fact to offer." Certainly wavelets do not offer an easy trick for solving nonlinear equations (such as the Navier-Stokes equation used to describe turbulent flows), in the way that Fourier turned many linear equations into cookbook problems.

"Wavelets are structurally a little better adapted to nonlinear situations," such as those found in turbulence, Meyer said. "But is something that is better in principle actually better in practice?" He is disturbed that when wavelets are used in nonlinear problems they "are used in a neutral way; they are always the same wavelets—they aren't adapted to the problem. . . . It is in this sense that there is perhaps a doubt about using wavelets to solve nonlinear problems. What can one hope for from methods that don't take the particular problem into account? At the same time, there are general methods in science. So one can give a different answer depending on one's personality."

FINGERPRINTS AND HUNGARIAN DANCES

One contribution of wavelets, Farge says, is that they have "forced people to think about what the Fourier transform is, forced them to think that when they choose a type of analysis they are in fact mixing the signal and the function used for the analysis. Often when people use the same technique for several scientific generations, they become blind to it."

As work with wavelets progressed, it became clear that if Fourier analysis had limitations, wavelet analysis did also. As David Marr wrote in *Vision*, "Any particular representation makes certain information explicit at the expense of information that is pushed into the background and may be quite hard to recover." Very regular, periodic signals are more easily recognized, and more efficiently encoded, by a Fourier transform than by wavelets, for example.

So Coifman, Meyer, and Wickerhauser developed an information-compression scheme to take advantage of the strengths of both Fourier and wavelet methods: the "Best Basis" algorithm.

In Best Basis a signal enters a computer like a train entering a switchyard in a train station. The computer analyzes the signal and decides what basis could encode it most efficiently, with the smallest possible amount of information. At one extreme it might send the signal to Fourier analysis (for signals that resemble music, with repeating patterns). At the other extreme it might send it to a wavelet transform (irregular signals, fractals, signals with small but important details). Signals that don't fall clearly into either group are represented by "wavelet packets" that combine features of both Fourier analysis and wavelets.

Loosely speaking, a wavelet packet is the product of a wavelet by a wiggle, an oscillating function. The wavelet itself can then react to abrupt changes, while the wiggle inside can react to regular oscillations. "The idea is that you are introducing a new freedom," Meyer said. Since the choice of wiggles is infinite, "it gives a family that is very rich."

Working with the FBI, Wickerhauser and Coifman applied Best Basis to the problem of fingerprint compression, and in a test conducted by the FBI's Systems Technology Unit it outperformed other methods. (Because Best Basis was being patented, Wickerhauser said, the FBI did not adopt it but instead custom-made a similar technique.) So far, the wavelet technique is intended only to compress fingerprints for storage or transmission, reconstructing them before identification by people or machines. But the FBI plans to hold a competition for automatic identification systems. "Those who understand how to use wavelet coefficients to identify will probably win, on speed alone if nothing else, because the amount of data is so much less," Wickerhauser said.

In studies with military helicopters, Best Basis has been used to simplify the calculations needed to decide, from radar signals, whether a possible target is a tank or perhaps just a boulder. In trials the Best Basis algorithm could compress the 64 original numbers produced by the radar system to 16 and still give "identical or better results than the original 64, especially in the presence of noise," Wickerhauser said.

But probably the most unusual use of Best Basis has been in removing noise from a battered recording of Brahms playing his own work, recorded in 1889 on Thomas Edison's original phonograph machine, which used tinfoil and wax cylinders to record sound. The Yale School of Music had entrusted it to Coifman after all else had failed.

"Brahms was recorded playing his music for Edison," Wickerhauser said. "It was played on the radio sometime in the 1920s—possibly using a wooden needle, and was recorded off the radio. Then it was converted to a 78 record. That was the condition in which Yale had it—beaten to death."

Coifman's approach was to say that noise can be defined as everything that is not well structured and that "well structured" means easily expressed, with very few terms, with something like the Best Basis algorithm. So the idea is to use Best Basis to decompose the signal and to remove anything that is left over. The result was not musical—no one hoped for that, from a recording that contained perhaps 30 times as much noise as signal—but they were able to identify the music as variations on Hungarian dances. The project, with Jonathan Berger of the Yale School of Music, is still going on. "We don't know yet how far we can restore it," Coifman said.

BEYOND WAVELETS

For some purposes, however, Best Basis is not ideal. Because it treats the signal as a whole, it has trouble dealing with highly nonstationary signals—signals that change unpredictably. To deal with such signals Stéphane Mallat and Zhifeng Zhang have produced a more flexible system, called Matching Pursuits, which finds the best match for each section of the signal, out of "dictionaries" of waveforms.

"Instead of trying to globally optimize the match for the signal, we're trying to find the right waveform for each feature," Mallat said. "It's as if we're trying to find the best match for each 'word' in the signal, while Best Basis is finding the best match for the whole sentence."

Depending on the signal, Matching Pursuits uses one of two "dictionaries": one that contains wavelet packets and wavelets, another that contains wavelets and modified "windowed Fourier" waveforms. (While in standard windowed Fourier the size of the window is fixed and the number of oscillations within the window varies, in Matching Pursuits the size of the window is also allowed to vary.) First, the appropriate dictionary is scanned and the "word" chosen that best matches the signal; then that "word" is subtracted out, and the best match is chosen for the remaining part of the signal, and so on. (To some extent, the dictionaries can produce words on command, modifying waveforms to provide a better fit.)

Because the waveforms used are not orthogonal to each other, the system is slower than Best Basis: n^2 calculations compared to $n \log n$ for Best Basis. On the other hand, the lack of orthogonality means that it doesn't matter where on the signal you start encoding. This makes it better suited for pattern recognition, for encoding contours and textures, for example.

But the quest for new ways to encode information is far from over. "When you speak, you have a huge dictionary of words and you pick the right words so that you can express yourself in a few words. If your dictionary is too small, you'll need a lot of words to express one idea," Mallat said. "I think the challenge we are facing right now is, when you have a problem, how are you going to learn the right representation? The Fourier transform is a tool, and the wavelet transform is another tool, but very often when you have complex signals like speech, you want some kind of hybrid scheme. How can we mathematically formalize this problem?"

In some cases the task goes beyond mathematics; the ultimate judge of effective compression of a picture, or of speech, is the human eye or ear, and developing the right mathematical representation is often intimately linked to human perception. Information is not all equal. Even a young child can draw a recognizable outline of a cat, for example, while the very notion of a drawing without edges is perplexing, like the Cheshire cat who vanished, leaving only his smile. Other differences are less well understood. People have no trouble differentiating textures, for example, while "after 20 years of research on texture, we still don't really know what it is mathematically," Mallat said.

Wavelets may help with this, especially since some wavelet-like techniques are used in human vision and hearing, but any illusions researchers may have had that wavelets will solve all problems unsuited to Fourier analysis have long since vanished; the field has become wide open.

"Wavelets have gone off in many directions; it becomes a little bit of a scholastic question, what you call wavelets," Grossmann says. "Some of the most interesting very recent things would technically not be called wavelets, the scale is introduced in a somewhat different way—but who cares?"

It may not be the least of the contributions made by wavelets that they have inspired both a closer and a broader look at mathematical languages for expressing information: a more judicious look at Fourier analysis, which was often used reflexively ("The first thing any engineer does when he gets hold of a function is to Fourier transform it—it's an automatic reaction," one mathematician said) and a more free-ranging look at what else might be possible.

"When you have only one way of expressing yourself, you have limits that you don't appreciate," Donoho says. "When you get a new way to express yourself, it teaches you that there could be a third or a fourth way. It opens up your eyes to a much broader universe."

NOTES

1. Herivel (1975), p.27 (letter to Villetard, 1795; original in the archives of the Académie des Sciences, Institut de France, Paris).
2. Hugo, Les Misérables, edition folio, vol. 1, p.185
3. Fourier (1888), p. 9
4. Körner (1988), p. 42; original in "La logique et l'intuition dans la science mathématique et dans l'enseignement," in Oeuvres de Henri Poincaré, vol. 11.
5. Delambre (1810), p. 125
6. Lagrange (1882). Oeuvres de Lagrange, vol. 13, p. 368
7. Pierce and Noll (1990), p. 55
8. Pierce and Noll (1990), p. 79
9. Meyer (1993), p. 86
10. Farge (1993), p. 195
11. Meyer (ed., 1992), p. 286

REFERENCES

Cousin, V. 1831. Notes biographiques pour faire suite à l'éloge de M. Fourier (Biographical notes to follow praise of Mr. Fourier), Fourier file AdS, archives, Académie des Sciences, Institut de France, Paris.

Delambre, Jean-Baptiste. 1810. Rapport historique sur le progrès des sciences mathématiques depuis 1789 (Historical report on the progress of mathematical sciences since 1789), part of Rapport à l'Empereur sur le progrès des sciences, des lettres et des arts depuis 1789, published by Belin, Paris, 1989.

Encyclopedia Britannica, 11th ed. 1972. Harmonic analysis. Vol. 11, pp. 106–107.

Farge, M., J.C.R. Hunt and J.C. Vassilicos (eds). 1993. Wavelets, Fractals, and Fourier Transforms. Clarendon Press, Oxford.

Fourier, J. 1822. Théorie analytique de la chaleur (published in Oeuvres de Fourier, Vol. 1, Gauthier-Villars et fils, Paris 1888).

Healy, D., Jr., and J. B. Weaver. 1992. Two Applications of Wavelets Transforms in Magnetic Resonance Imaging. IEEE Transactions on Information Theory, Vol. 38, No. 2, March 1992.

Herivel, J. 1975. Joseph Fourier — the Man and the Physicist. Clarendon Press, Oxford.

Körner, T.W. 1988. Fourier Analysis. Cambridge University Press, Cambridge.

Lagrange, J.-L. Oeuvres de Lagrange, Gauthier-Villars, Paris, 1867–1892.

Marr, D. 1982. Vision. W. H. Freeman, New York.

Meyer, Y. (ed). 1992. Wavelets and Applications: Proceedings of the International Conference, Marseille, France. Masson and Springer-Verlag.

Meyer, Y. 1993. Wavelets: algorithms & applications. Society for Industrial and Applied Mathematics, Philadelphia. (This is an English translation of Les Ondelettes Algorithmes et Applications, Meyer, Y., Armond Colin, Paris, 1992.)
Pierce, J. R., and A. M. Noll. 1990. Signals — The science of telecommunications. Scientific American Library, New York.
Poincaré, H. 1956. Oeuvres de Henri Poincaré, vol. 11, Gauthier-Villars, Paris.
Strichartz, R. 1993. How to make wavelets, American Mathematical Monthly, Vol. 100, no. 6, June-July 1993, pp. 539–556.
Strömberg, J.-O. 1981. A modified Franklin system and higher-order spline systems on R^n as unconditional bases for Hardy spaces. Conference on Harmonic Analysis in Honor of Antoni Zygmund, vol. II, pp. 475–494; W. Beckner, A. Caldéron, R. Fefferman, and P. Jones, eds. University of Chicago, 1983.

RECOMMENDED READING

Books Accessible to the General Public

Gratten-Guinness, I., and J.R. Ravetz. 1972. Joseph Fourier, 1768–1830. MIT Press, Cambridge, Mass.
Herivel, John. 1975. Joseph Fourier—The Man and the Physicist. Clarendon Press, Oxford.
Körner, T.W. 1988. Fourier Analysis. Cambridge University Press, Cambridge.
(This is a book for mathematicians, but it includes delightful sections that require no mathematical background, such as the description of the laying of transatlantic cables, where "half a million pounds was being staked on the correctness of the solution of a partial differential equation.")
Pierce, John R., and A. Michael Noll. 1990. Signals—The Science of Telecommunications. Scientific American Library, New York.

A Sampling of Technical Books

Benedetto, J., and M. Frazier (eds.) 1993. Wavelets: Mathematics and Applications. CRC Press, Boca Raton.
Chui, C.K. (ed.). 1992. Wavelets-A Tutorial in Theory and Applications. Academic Press, Boston.
Daubechies, I. 1992. Ten Lectures on Wavelets. Society for Industrial and Applied Mathematics, Philadelphia.
Farge, M., J.C.R. Hunt, and J.C. Vassilicos (eds.). 1993. Wavelets, Fractals, and Fourier Transforms. Clarendon Press, Oxford.
Körner, T.W. 1988. Fourier Analysis. Cambridge University Press, Cambridge.
Meyer, Y. 1990. Ondelettes et Opérateurs. Three volumes. Hermann, Paris. (English translation published by Cambridge University Press in 1992.)
Meyer, Y. 1992 Wavelets and Operators, Cambridge University Press, Cambridge and New York. (This is a translation of Ondelettes et Operateurs, Hermann, Paris, 1990.)
Meyer, Y. 1993. Wavelets: algorithms & applications. Society for Industrial and Applied Mathematics, Philadelphia.
Ruskai, B. (ed.) 1992. Wavelets and Their Applications. Jones and Bartlett, Boston.

A FAMILY AFFAIR
The Top Quark and the Higgs Particle

by T. A. Heppenheimer

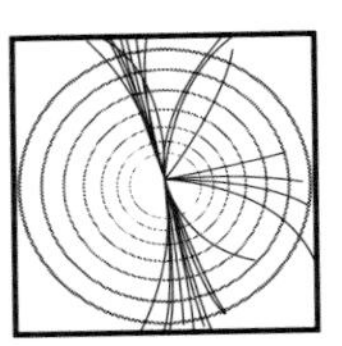

There was a time, just before World War II, when physicists had reason to believe that they had penetrated the deepest issues in their field. Werner Heisenberg, Erwin Schroedinger, Max Born, and Niels Bohr had taken the lead in inventing quantum mechanics, offering insights so powerful that graduate students were solving problems that had bedeviled savants for decades. Studies of atoms held the focus of concern, and here too there was fundamental advance.

People knew that an atom consisted of a nucleus surrounded by electrons. By elaborating this concept, Bohr had given a theoretical explanation for the regularities seen within the periodic table of elements, the foundation of chemistry. The nucleus, for its part, consisted of protons and neutrons. In addition, Britain's P. A. M. Dirac had recently given a powerful theory that described the electron.

The electron has a negative electric charge, and Dirac's theory predicted that there should also be positively charged electrons. In 1932 Carl

Anderson, an American, found them in cosmic rays; these particles became known as positrons. Here was a genuine advance, for this represented the first prediction of a new particle's existence, entirely from theory. The general view at the time was that a Dirac-like theory would soon come forth to describe the proton and neutron. At that point, physicists would have dug down to bedrock in their search for nature's ultimate secrets.

Today, after six decades and many more disappointed hopes, the field of physics is in a rather similar situation. Once again we have a set of powerful theories—the Standard Model, which offers great predictive power. Indeed, it not only accounts for all physical experiments performed to date, it has even shown its power by once again successfully predicting the existence of new particles. Yet today's researchers are not satisfied. Important features of the Standard Model remain unconfirmed; not all of its predictions have yet been borne out. Furthermore, even if one accepts it without reservation, it raises a number of new questions, which lie within the reach of experiments. In pursuing these matters, today's physicists are setting an agenda for the coming century.

The road to the Standard Model has not been smooth. It began just after the war, as the federal government allocated funds for construction of an increasingly powerful series of particle accelerators. In studies of the atom's nucleus, they quickly replaced the older technique of relying on observations of cosmic rays. The new accelerators produced beams of particles, such as electrons or protons, that were particularly intense. They also offered high energy, and experimenters could control this energy, turning it up and down.

The standard technique was to direct such a beam onto a target made of some material and observe the debris that came out as particles with that beam "split the atom"—or, more properly, shattered some of the target's nuclei. New types of particles might lie within those sprays of nuclear debris, and experimenters were not disappointed. Indeed, during the years after 1950, they found themselves with an embarrassment of riches. There appeared to be not one or two but rather dozens of new particles. So far as anyone could tell, they might all be as fundamental as the proton or the electron.

These new particles generally were very short lived. Unlike electrons and even positrons, they did not form well-defined tracks within a detector. Instead, they decayed into other particles, in as little as 10^{-16}

seconds. Still, it was quite possible to describe them and particularly to note the energy required for their formation. As one proceeded with an experiment, gradually increasing the energy in an accelerator's beams, there would be characteristic values of energy where the incidence of these decay products would markedly increase. That meant that new types of particles were forming at those specific energies, even if they had very short lifetimes.

During the 1950s, these experimental results put physics in a quandary. Its leaders had sought to define theoretical physics as the deepest of the sciences, able to make fundamental and far-reaching predictions on the basis of pure thought. But the work of the accelerator labs was turning particle physics into merely a descriptive enterprise. Like botanists who catalog new species of plants without knowing how they had evolved, physicists could do little more than describe the new particles they were discovering. No theory had predicted their existence; indeed, their existence had come as a major surprise. The bright hope of the 1930s appeared premature, if not completely out of reach. Clearly, if physicists indeed were to get down to bedrock, they would have to do a lot more digging.

The first important step came after 1960. By then the roster of new particles was sufficiently numerous to bear some resemblance to the roster of chemical elements, for both numbered in the dozens. A century earlier, Russia's Dmitri Mendeleev had brought order to chemistry by organizing the known elements into his periodic table. This had proven to be more than an exercise in taxonomy; it had pointed toward an underlying order, for the periodicity in this table demanded an explanation based on fundamental principles. Similarly, in 1960, physicists faced the issue of building a counterpart to the periodic table, able to describe their particles in terms of their own underlying principles.

Two theorists, Murray Gell-Mann at the California Institute of Technology and Yuval Ne'eman, an Israeli, emerged as the Mendeleev of particle physics. Working independently, both of them showed that one could produce a periodic table of the particles by relying on a structure taken from abstract algebra, the group SU(3). This not only organized the known particles, it also brought the prediction of a new one, the omega-minus. In a spectacular confirmation of these ideas, researchers at Brookhaven National Laboratory, in 1964, found the omega-minus. This showed that there indeed was a pattern and order within the system of known particles. Their properties were not mutually independent; instead they were linked.

But the work of Gell-Mann and Ne'eman resembled that of

Mendeleev in another way, which showed that much work still lay ahead. These physicists' use of SU(3), like Mendeleev's periodic table, gave order and predictability, without disclosing the underlying source of the order. For Mendeleev the source had proved to lie in the structure of electron orbits within atoms. For physicists in the 1960s, the order that showed itself in SU(3) suggested strongly that the particles possessed some kind of deeper physical structure, able to stand as this underlying source. In seeking this structure, Gell-Mann again came to the forefront along with George Zweig at the European Center for Nuclear Research (CERN).

In developing his theory, Gell-Mann had considered whether he might be able to describe the particles as being built up from more fundamental units. He found that this was possible—and immediately rejected the idea, for the new particles, the building blocks, would have fractional electric charges such as 1/3 and 2/3. No one had ever seen such particles; all the known ones had charges of either 0 or 1. But in 1963 he decided that this might not be a problem after all. These fractionally charged entities might be trapped inside the particles they formed, trapped so strongly that they could never be seen.

Gell-Mann called them quarks. He invented the term out of whimsy, as a reaction against pretentious terminology, but soon found that James Joyce had used this word in *Finnigans Wake.*

> Three quarks for Muster Mark!
> Sure he hasn't got much of a bark.
> And sure any he has it's all beside the mark.

Gell-Mann's theory indeed called for three quarks, which should combine to yield all the particles organized under SU(3). Still, a major problem lay in his suggestion that they could not be seen, even in principle. Such a viewpoint smacked of theology rather than physics, for while theologists indeed begin by asserting that one cannot see the matters that concern them, physics had always been an experimental science par excellence. The quark concept offered a compelling theory, but to win widespread acceptance it would have to pass the test of observation and experiment.

The work that would provide this confirmation got under way during 1967 at the Stanford Linear Accelerator Center (SLAC). It involved scattering electrons off protons, at increasingly high energies. Initially, at modest energies, the electrons simply bounced away, somewhat like bullets recoiling from impacts with a cannonball. But as the energies increased, the data took on new forms. Now it indicated that the electrons were no longer ricocheting in a simple fashion, scattering off the protons as a whole. Instead, they were deflecting off tiny constituent objects that lay *within* each proton. This was understandable if those objects were quarks, and with this Gell-Mann's theory received a major boost.

While this work was progressing, other theorists were making headway as well. The studies of SU(3), and of quarks, had involved particles that feel the strong force. This is one of the fundamental forces of physics; it holds atomic nuclei together. Another force, equally fundamental, is the weak force, which mediates the production of energy in the sun. Here the road to understanding lay in unification, in creating a theory that would join the weak force to that of electromagnetism.

Unification was an old theme in physics. Scotland's James Clerk Maxwell had pursued it a century earlier, giving equations that unified the electric and magnetic fields. Maxwell's equations had then gone on to achieve the status of a cornerstone of physics. Of course, Maxwell had known nothing of subsequent discoveries: the electron, Einstein's relativity theory, quantum mechanics. Hence, an important project for physicists had featured creation of a theory that indeed would accommodate these findings; Dirac's work had stood as a major contribution. The full theory, known as quantum electrodynamics, reached fruition around 1950.

With this achievement in hand, theorists turned to the weak force. The view was that at sufficiently high energies it too would unify, joining with the electromagnetic force to yield a single "electroweak" force. The separate nature of these forces then would become apparent only at lower energies. It took over a decade before a suitable electroweak theory was in hand, but in 1971 this effort also achieved success. The main contributors were Steven Weinberg, Sheldon Glashow, Abdus Salam, and Gerard 't Hooft.

Hence, by 1972, particle physics had turned around. A decade earlier it had been experiment rich and theory poor, beset by a plethora of newly discovered particles that resisted understanding. In 1972 the field was theory rich and experiment poor. Theories were in hand for quarks, the strong force, and the electroweak force. The problem now was to test them by conducting appropriate experiments. If these theories passed those tests, they could offer a new set of foundations for theoretical physics.

Electroweak theory was the first to face this challenge. It predicted the existence of a new particle, the Z, that could give rise to characteristic events that might occur within a particle accelerator. For instance, a single electron might materialize in a detector, with no other particle in view. The Z itself would remain unseen in such an encounter, but the electron could show its presence. During 1973 and 1974, groups working at CERN in Switzerland and at Fermilab near Chicago both announced success in observing such events. Here was solid evidence both for the Z and for the electroweak theory.

In November 1974 it was the turn of the quark theorists. Gell-Mann's original theory, in 1963, had postulated three types of quark: Up, down, and strange. The up and down quarks sufficed to build protons and neutrons, the stuff of atomic nuclei and hence of the world of matter that we live in. The third or strange quark had its own roles. In combination with up or down quarks or with other strange quarks, it yielded the so-called strange particles. These had been seen in cosmic-ray studies during the postwar years, receiving that adjective because they had unusually long lifetimes before they decayed. Such particles could be seen in accelerators as well; the omega-minus, which had clinched Gell-Mann's early work with SU(3), was made up of three strange quarks.

However, this was not the end of the matter. Working from theory, Sheldon Glashow proposed in 1964 that there should be a fourth quark, which he called "charm." He declared that these four quarks should group into two families. Up and down would form the first, charm and strange the second.

In 1974 it was as true as ever that one could not see any type of quark in isolation, by knocking it out of a particle. In Glashow's words, "You can't even pull one out with a quarkscrew." But the theory predicted that charmed quarks should join with others to form new types of particles, having observable characteristics. At Brookhaven and SLAC, groups headed by Samuel Ting and Burton Richter, respectively, went on to find them.

The two labs had very different equipment. Brookhaven relied on a 1960-vintage accelerator, the Alternating Gradient Synchotron, which was not well suited to the task of finding particles with charm. Indeed, Ting would work with it for months before he was sure what he had found. The accelerator at SLAC was much newer and far more capable; if Richter knew the energy of these particles, he could discover them in a day. Ting knew this, and, as his data took form during the summer and fall of 1974, he went to considerable effort to prevent a leak of information that might allow his rival to beat him to the discovery.

Ting was not looking for charm; rather, he was looking for new particles of a different type, and he hoped to find them by searching through a broad energy range, exploring this range using the Brookhaven accelerator. He was working with energies measured in billions of electron volts—GeV, in physicists' notation. (One GeV corresponds to slightly more than the energy required to form a proton or a neutron.) At an energy of 3.1 GeV he found a sharp peak in the data. Here was clear and dramatic evidence for a new particle, standing out as sharply as a bright spectral line.

Fortunately for Ting, Richter was working in a different energy range. But late in October, in reviewing recent data, several SLAC physicists noted some curious features in results taken near 3.1 GeV. To go back to that energy was no simple matter of turning a knob; it meant shutting down the accelerator and resetting the magnets along with an injection beam. Nevertheless, Richter's colleagues did just this and were quickly rewarded with the same high and narrow peak in the data that Ting had seen. Indeed, it was so narrow that they missed it the first time around. Both Ting and Richter could claim to have made this discovery, independently and nearly simultaneously. Ting called it the J particle. Richter named it the psi, and most other physicists, who did not care to take sides, called it the J/psi.

One question remained: What was it? Glashow quickly declared that it was indeed a type of charmed particle, which he called "charmonium." Here was a decisive confirmation of the quark theory, showing that this theory was sufficiently powerful to predict the existence of a new particle. Indeed, it could predict a whole new class of particles, containing one or more charmed quarks and assembled along the lines of the strange particles in Gell-Mann's SU(3) theory.

Nevertheless, this triumph raised a new question. It now was clear that nature indeed had created two families of quarks: The up and down in the first, strange and charm in the second. But if there were two families, might there not be three? Four? An arbitrarily large number? No known result constrained these possibilities; theory offered no guidance, while experimental findings were similarly unhelpful. It was all too reminiscent of the situation prior to World War II, when physicists seemed to stand on the verge of complete understanding—only to see these prospects wash away in a flood of unpredicted and unexpected new particles.

In the ongoing dialogue between theory and experiment, the theorists certainly had had their say. Their work now stood as well confirmed, capable of accounting for all experimental observations, contradicted by none. Amid this success, people took to referring to this body of theory as the Standard Model. Still, the experimental work had hardly begun.

The immediate agenda lay in search for direct observation of the heavy particles predicted by the electroweak theory. There were three of them, denoted Z, W^+, and W^-, and they supposedly came into play in reactions mediated by the weak force. The work at CERN and Fermilab, in 1973 and 1974, had found evidence for the Z that was convincing but indirect; it was as if you would infer that a burglar had been in your house by finding the window open and your credit cards missing. The new effort sought the equivalent of catching the burglar red-handed.

These particles resisted discovery because they are very massive. Einstein's famous formula, $e = mc^2$, shows that energy and mass are really two different forms of the same thing; hence, a unit of mass can also be regarded as a unit of energy. Preliminary experiments gave data that could predict the masses of the W and Z particles, from physical theory. The predicted masses were 79.5 and 90 GeV, respectively, corresponding to nearly a hundred times greater than a proton.

To search for these particles, and to study physics at these energies, physicists needed new tools. Conventional particle accelerators produced energetic beams and slammed them into fixed targets. These beams' energies had risen markedly through the postwar decades. In 1952 the Cosmotron, at Brookhaven, had been the first machine to achieve multiple GeV. Twenty years later the accelerator at Fermilab could put 200 GeV on target. Yet in searching for the W and Z particles, even this was not enough. The reason was that as the energy went up, the efficiency of the particle production went down. Less and less of the energy in the beam's particle was available to create new particles, for most of it was wasted in collisional debris.

Fortunately, there was an alternative. Instead of striking a fixed target, an accelerator might produce colliding beams. It would produce two beams, circulating in opposite directions and interpenetrating each other at selected locations. Most of the beams' particles would fly past each other and miss, but if the beams were both intense and narrowly focused, a useful number of these particles would collide head on. All their energy would then be available to drive reactions and to create new particles.

It helped that this colliding beam approach fitted in neatly with the engineering design of accelerators. The big ones, as at Fermilab, featured long strings of powerful magnets set end to end like a train of railroad cars, with the magnets curving gently to form a ring. The magnets stored the beam, sending it repeatedly through stations that added energy to the beam. This architecture lent itself readily to arrangements in which two such beams could counterrotate within the same ring, flying in opposite directions. Such an experiment would amount to running two Indianapolis 500 races simultaneously, one with cars circling the track clockwise and the other counterclockwise. In the collisions of particles within such demolition derbies, physicists now hoped to make their finding.

At SLAC, Burton Richter was a particular advocate of this approach. He had built SPEAR, the Stanford Positron-Electron Accelerating Ring, and had used it in finding his psi particle, which contained charm. The nature of SPEAR testified to further advances in physics, for it collided beams of electrons and positrons. The positron, far from being a curiosity as in the days of Carl Anderson, now was a mainstay of physics research. Indeed, in SPEAR the positrons were so numerous that they formed beams and served to discover particles such as the psi, which were newer still.

For higher energies yet, physicists could use antiprotons. Antiprotons, like positrons, are a form of antimatter. These particles have all the properties of their conventional counterparts with one exception: They have opposite electric charges. The positron, for one, has the same mass and spin as the electron; but whereas the electron carries negative charge, that of the positron is positive. Similarly, the antiproton has the same mass and spin as the proton, but the proton carries positive charge, whereas that of the antiproton is negative. Emilio Segre and Owen Chamberlain had discovered the antiproton in 1955, using the Bevatron accelerator at the University of California at Berkeley; it had been one of the prizes of the new era of particle physics. Like the positron, the antiproton too had graduated to become a tool of research. A beam of antiprotons, colliding head on with one of the protons, could offer energies sufficient to create W and Z particles.

The Fermilab accelerator had been using protons since day one; it was a natural candidate for conversion. However, rather than carry through a straightforward modification, the lab's directors proceeded with a far-reaching upgrade that would feature an entirely new ring of magnets. Rather than shooting a 200-GeV beam at a fixed target, the new system would produce beams with energies of 1000 GeV, a trillion electron volts, or TeV. Two such beams, countercirculating, would collide head on to produce a total energy of 2 TeV. Still, this project was not due for completion until 1985, which left European researchers with an opportunity.

The opportunity lay with an existing accelerator in CERN, the Super Proton Synchrotron (SPS). In 1976 it was operating more as a powerful counterpart of the Fermilab system, directing proton beams at fixed targets, with energy as great as 400 GeV. During that year, several physicists proposed to convert it to a colliding beam operation by making provision for a countercirculating beam of antiprotons. This conversion was more straightforward than the Fermilab upgrade, but it nevertheless took several years.

Finally, in July 1981 the machine was ready. It produced peak energies of 270 GeV in each beam, or 540 GeV in collisions. Initial operation involved low beam intensities, and these runs failed to produce W or Z particles. That was not surprising; the theory predicted that they would be rare, and the experimenters had only an approximate idea of the energy levels that would produce them. Until they zeroed in on the right energies, these particles would be rarer still.

Carlo Rubbia of Harvard University, who had proposed converting the SPS in 1976, now led the experimenters. They boosted the beam intensity by an order of magnitude, greatly increasing the collision rate, and proceeded with their search. In 1983 they gained full success, detecting the W^+ and W^- as well as the Z. Subsequent worked pinned down their masses: 80.15 GeV for the W particles, 91.187 GeV for the Z. This showed that in an important respect the electroweak theory was stronger than the quark theory. Quark theorists, such as Glashow, had predicted the existence of charm but had not been able to predict the mass of charmed particles. By contrast, electroweak theory had predicted masses as well as existence and with remarkably good accuracy.

Meanwhile, the matter of quark families was coming to the fore. In 1977 Leon Lederman, working at Fermilab, discovered a member of the third family, which received the name of the bottom quark. Its mass proved to be 4.7 GeV, which confirmed a pattern. Quarks of the second family, the charm and strange, had proven to be considerably more massive than their counterparts in the first family. Third-family quarks were heavier still. Indeed, subsequent work showed that the partner of the bottom quark—known quite logically as the top quark—certainly merited its name. Its mass indeed was at the top; searches up to 91 GeV failed to find it. That meant it would be even heavier than the W or Z particles, the heaviest yet found.

There still remained the question of how many such quark families existed in nature. In a virtuoso feat, investigators at SLAC and CERN proceeded to find a direct answer through studies of the Z. In doing so, they introduced a new concept: That studies of specific particles could shed light on fundamental problems that did not involve these particles directly. The multiplicity of families stood as an issue in quark theory; it did not appear within electroweak theory, which represented a separate

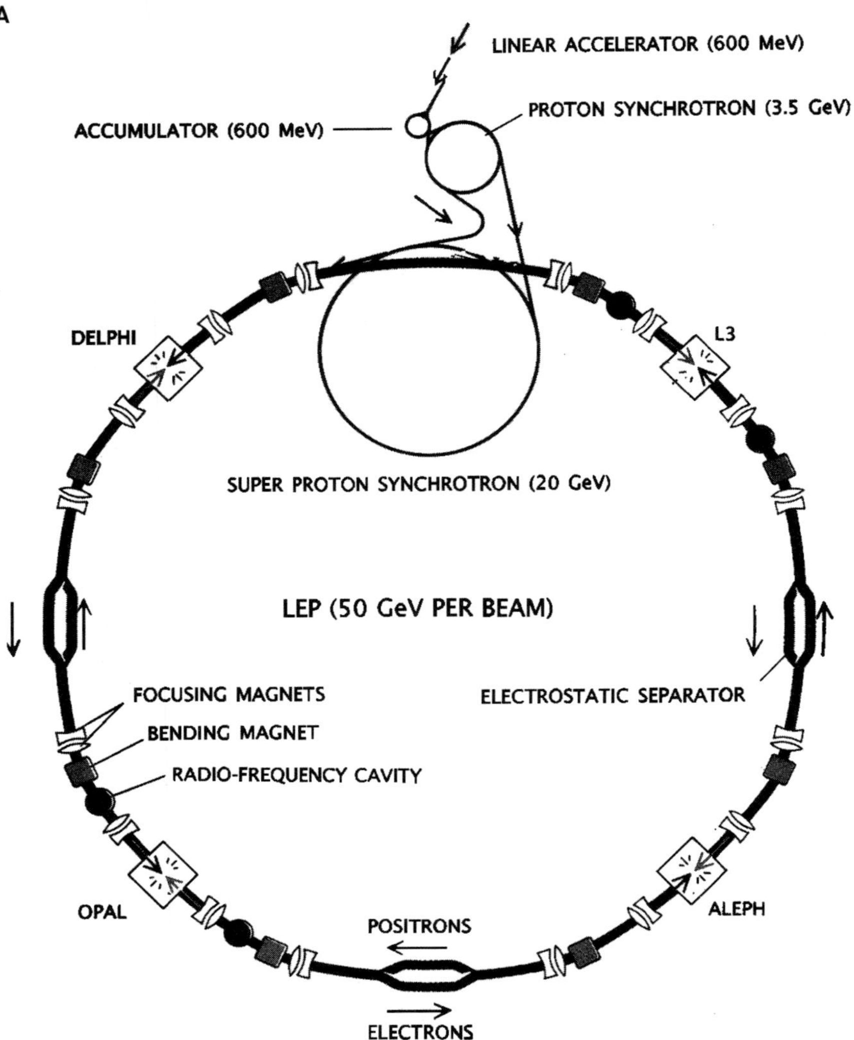

FIGURE 8.1 *Bird's-eye view of LEP storage ring shows how electron and positron bunches are accelerated by a series of smaller machines. Four bunches each of electrons and positrons circulate in opposite directions and collide inside the four gigantic detectors: Aleph, Delphi, L3, and Opal. The bunches also cross paths at four intermediate sites, where they are prevented from colliding by electrostatic separator plates. (From "The LEP Collider" by Stephen Myers and Emilio Picasso. Copyright © 1990 by Scientific American, Inc. All rights reserved.)*

and independent body of physical law. Yet the Z particle, featured in electroweak theory, proved to offer the key to the problem of quark families.

The reason lay in the uncertainty principle of quantum mechanics. This principle states that when a particle exists for a vanishingly short time—for the Z, some 10^{-25} seconds—there is a corresponding uncertainty in measuring its energy and therefore its mass. One might seek to determine the mass of a type of particle by making repeated measurements with great accuracy, but this principle would defeat the attempt. Any individual measurement might indeed show high precision, but no two measurements would agree. Different observations would show different values for the particle mass, with the ensemble of these values showing a bell-shaped curve, a distribution common in the field of statistics.

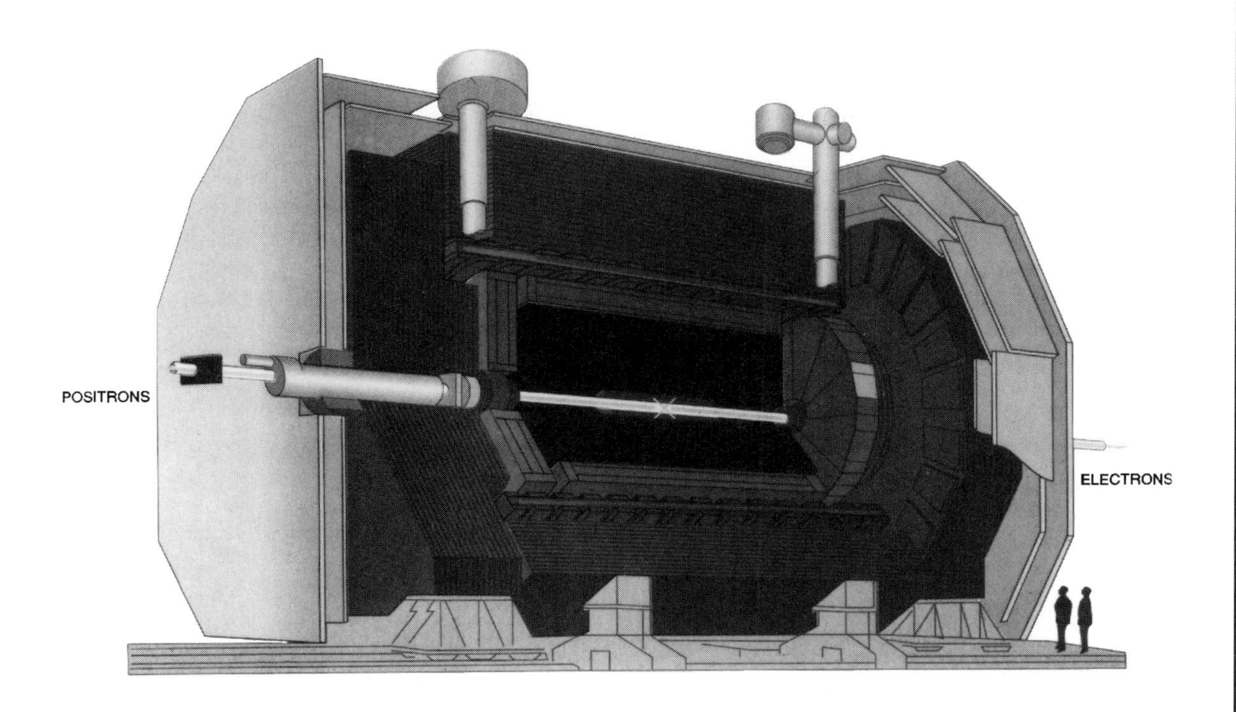

FIGURE 8.2 *Graphic presentation of events allows physicists to visualize the trajectories and energies of particles within the structure of the detector (in this case, the Aleph experiment). The diagram above shows the components of Aleph. (Reprinted, with permission, from Breuker et al., 1991. Copyright © 1991 by Scientific American, Inc. All rights reserved.)*

At SLAC and CERN, five groups carried out a related investigation. Rather than make direct measurements of the mass of the Z and observe the statistical scatter, they set their accelerator to operate at a succession of energies. The cited mass of the Z, 91.187 GeV, in fact was an average value; these groups sought to produce Z particles at energies ranging from 88 to 95 GeV. The rate of production varied with energy, peaking at 91.187; a plot of these variations, showing production rate as a function of energy, formed its own bell-shaped curve.

What did that have to do with the number of quark families? The width of this curve, and particularly the height of its peak, was related to the lifetime of the Z. That lifetime, in turn, depended on the number of different ways a Z could decay. Each such decay mode furnished a channel whereby the process of decay would create other particles. More channels meant a lower peak; fewer channels meant a higher peak. And the number of channels gave a direct determination of the number of quark families.

The work featured two new accelerators, the Stanford Linear Collider at SLAC and the Large Electron-Positron Collider (LEP) at CERN (see Figures 8.1, 8.2, and 8.3). The SLAC machine, an upgrade of the existing system, raised the collisional energy to 100 GeV, working with colliding beams of electrons and positrons (see Figure 8.4). However, the LEP was the workhorse of the effort, producing Zs at a rate

FIGURE 8.3 *LEP detector at CERN. Beams of particles fly in from left and right, with their collisions taking place within the detector itself. Shielding screens out less-penetrating particles, while penetrating ones, such as muons, reach the detector periphery.*

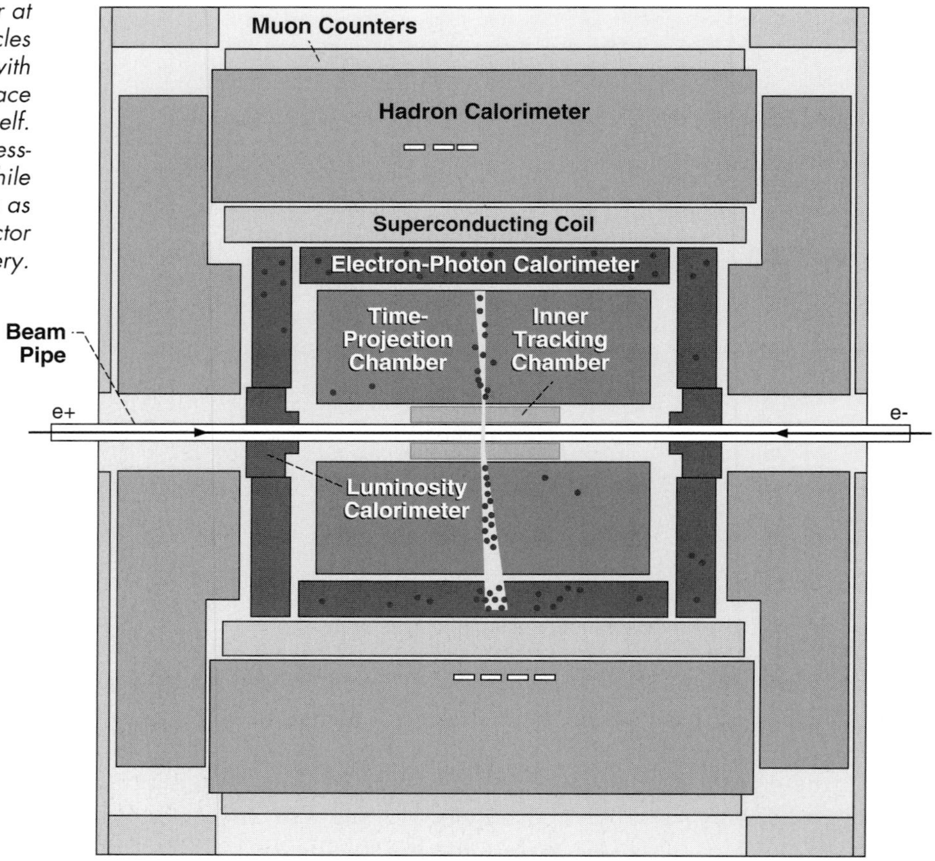

over a hundred times greater. During its construction, LEP had been the largest civil engineering project in Europe. It took form as a ring of magnets some 27 kilometers in circumference, and its beams had the intensity needed to make LEP a true Z factory. During the main experimental run, which lasted 4 months, LEP produced some 100,000 Zs.

Late in 1989 the directors of the five experimental groups pooled their data and announced the result: The number of quark families was 3.09 ± 0.09. Because that number had to be an integer, it actually was exactly 3 (see Figure 8.5). The third family, consisting of the bottom and the still-unseen top quark, would complete the roster.

With this new result now in hand, one might think that there is little left today for particle physicists, at a fundamental level. One more round of experiments might find the top quark, completing the list of elementary particles, and then everyone can go home. Such a view amounts to declaring that after 60 years we once again are where we thought we were in the time of Dirac, ready to wrap everything up with one more burst of creativity. In fact, a most important topic remains for study, one whose investigation has barely begun.

This is the Higgs particle, a theoretical concept proposed in 1964 by Peter W. Higgs of the University of Edinburgh. It has potentially far-reaching significance, for many theorists believe that in this particle lies the origin of mass. This origin, which would account for the masses of elementary particles, stands as a very deep issue. The Standard Model has little to say on the subject. It asserts that some particles are massless, such as the photon, which carries light. It states that the electron and

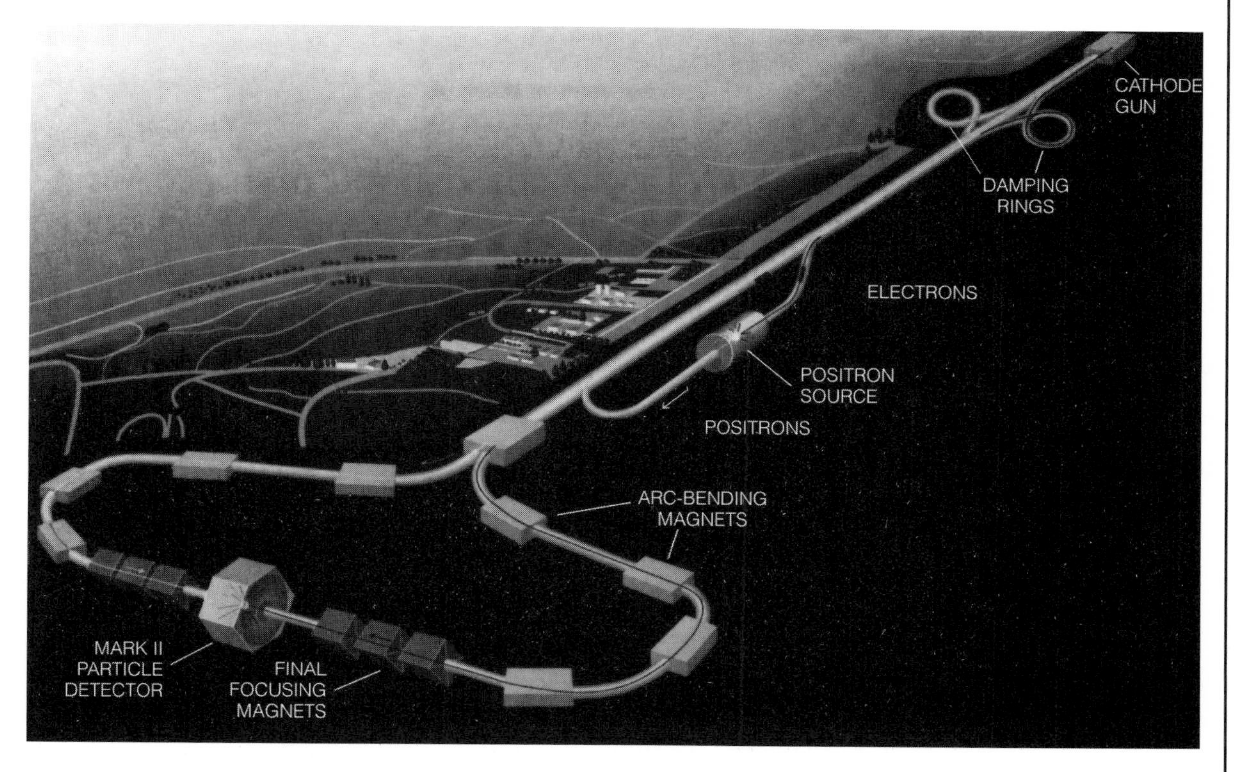

FIGURE 8.4 *Stanford Linear Collider speeds positrons and electrons along a 3-kilometer straightaway. The injector (top right) shoots electrons into a damping ring, which condenses them for later focusing. One bunch then enters the straightaway behind a bunch of positrons. The two bunches accelerate in tandem before entering separate arcs that focus and direct them to collision in the Mark II detector (bottom left). Meanwhile, the second bunch of electrons slams into a target, producing positrons (center). The positrons are returned to the front and are damped and stored. (From "The Number of Families of Matter" by Gary J. Feldman and Jack Steinberger. Copyright © 1991 by Scientific American, Inc. All rights reserved.)*

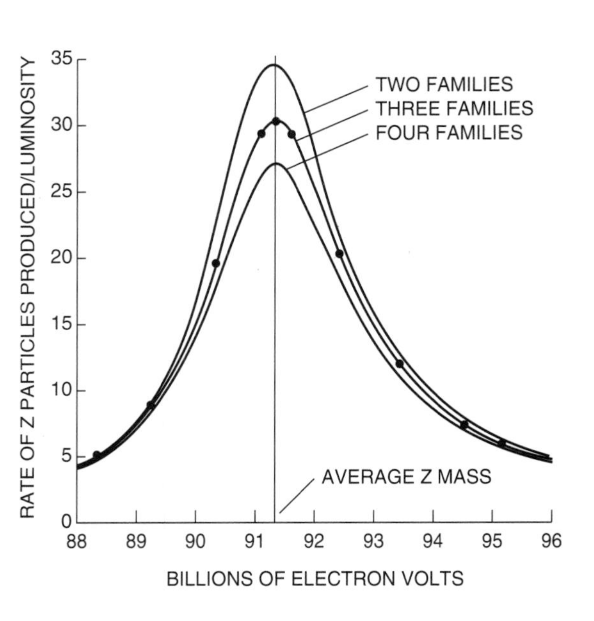

FIGURE 8.5 *Evidence for three families of quarks. Predicted curves for different numbers of families differ noticeably. The data points fall neatly on the three-family curve.*

positron are equal in mass. But it gives no account of the masses of quarks.

The concept of Higgs particles holds that they permeate empty space, like raindrops in the air during a thunderstorm. Other particles then gain mass by absorbing them; we could say that they eat the Higgs to gain weight. However, that is not to say that we understand this procedure, not by a long shot. "That hasn't explained anything," notes Kevin Einsweiler of Lawrence Berkeley National Laboratory. "There's no deep understanding contained in that. For instance, the top quark gets its mass because it couples very strongly to the Higgs. But that's not an explanation; it's a description of a mechanism of how it happens in the theory. The theory says nothing about what these couplings are or why the couplings of the top to the Higgs should be extremely large."

Hence, the concept of the Higgs particle offers a set of ideas that might lead to a true theory of mass, but that could easily be taken as a matter for some future day. In fact, the Higgs particle is central to electroweak theory. A key attribute of that theory is its ability to make precise calculations based on only a few parameters, calculating experimentally observable quantities to any desired accuracy. As Gerard 't Hooft first showed in 1971, it is the Higgs particle that gives electroweak theory this power. In the absence of the Higgs, for some problems it would give absurd results. Martinus Veltman, who worked with 't Hooft, gives an example: The scattering of one W particle off another. Veltman notes that, if one deletes the Higgs from the theory, at high energies the probability of the scattering could be calculated as greater than 1. "Such a result is clearly nonsense," he writes. "The statement is analogous to saying that even if a dart thrower is aiming in the opposite direction from a target, she will still score a bull's-eye."

Nevertheless, the Higgs particle quickly leads to difficulties. Gravity should couple with the Higgs particles that pervade free space, and this would cause the entire universe to curl up into something the size of a football. Needless to say, this is contrary to observation. To get around this, some physicists assume that in the absence of the Higgs the universe would curl up in the opposite way, again to the size of a football. The Higgs-induced curvature then cancels the "natural" curvature, yielding a universe that extends, as we see it, for billions of light-years.

This means that the Higgs concept is intimately tied up with cosmology, with the structure of the universe. It also leads to problems in cosmology for which the answers amount to little more than speculation. Nor are these issues trivial. Joseph Polchinski of the University of Texas, who has worked on closely related matters, notes that "there have been a lot of reasons offered" as to why the universe is not football size, "and most of them are obviously wrong. It's a very hard problem. It requires something new in physics." Steven Weinberg adds that this "may be the deepest problem we have."

The Higgs particle thus enters directly into a number of very significant issues: The physical basis for electroweak theory, the origin of mass, the large scale structure of the universe. No one has seen a Higgs, and an obvious problem would then be for experimenters to find it. At first blush this appears possible, for although the mass of the Higgs is ill defined, a variety of arguments lead to the conclusion that it should lie between 60 and 1000 GeV. That is not so much greater than the mass of the W and Z particles. And with the upgraded accelerator at Fermilab, the Tevatron, now reaching 1800 GeV in proton-antiproton collisions, the way would appear open to a direct attack on the problem.

However, other issues come into play, effectively ruling out the likelihood of a successful search with the present generation of accelerators. These are all of the colliding beam type, which give the most energy, and they come in two versions. There are positron-electron colliders, such as LEP; there also are proton-antiproton machines, such as Tevatron.

LEP today offers collisions up to 110 GeV and is to be upgraded to reach 200. But this type of machine loses efficiency at very high energy due to the problems of synchrotron radiation. That is a type of radiation emitted by electrons or positrons, as they whirl in a circle. One eases this problem by making the circle quite large; that is why LEP has a diameter greater than 8 kilometers. There nevertheless are limits to the amount of real estate that can be dedicated for the use of such a facility; there also are limits to the number of magnets that even a major project can afford.

LEP presses these limits, making it unlikely that a larger electron-proton collider will be built, at least in the near future. And if one were to try to raise the energy of LEP to 1000 GeV, one would find that energy drained off by synchrotron radiation.

One turns then to the alternative, in which the colliding particle beams feature protons and antiprotons. These particles are much heavier than electrons and positrons, and their high mass suppresses synchrotron radiation. However, at very high energies another problem arises. The colliding particles do not behave like single objects, miniature cannonballs if you will. Instead they behave as if they were charges of buckshot. A proton is made not only of quarks but also gluons, additional particles that hold the quarks together, and at very high energies this composite structure becomes apparent. A proton and antiproton then might meet head on and simply pass through each other, with all their quarks and gluons mutually missing one another in their passage.

This means that, when such a particle collision indeed takes place, it generally involves no more than two quarks or gluons out of the entire ensemble. If one wishes to achieve an honest 1 or 2 TeV in a particle collision, the actual energy of the colliding proton and antiproton must be vastly higher, because each quark or gluon holds only a small fraction of the whole. Indeed, the total energy, within the colliding beams, would be as great as 40 TeV, 20 in each. No present-day accelerator can remotely approach this goal.

With this, one enters a realm of particle accelerators that exist only on paper. For the past decade, the American physics community has set its hopes on a true 40-TeV machine, the Superconducting Supercollider. As of October 1993, it was under construction near Waxahachie, Texas. Builders were digging an underground tunnel of 87-kilometer circumference that was to hold its magnet ring. But during that month, the roof fell in. This does not mean that the tunnel collapsed. Rather, Congress voted to cut off funding.

As a result, physicists today are not entirely clear what they will do next. CERN, ever helpful, has proposed to pick up the falling torch by building an entirely new magnet ring within the existing LEP tunnel of 27 kilometers. The Europeans then would take the lead with an entirely new accelerator, the Large Hadron Collider (LHC). It would collide beams of protons and antiprotons, which physicists refer to as hadrons. However, being smaller than the Supercollider, it also would achieve less energy: 15 TeV. It then could find the Higgs only if that particle's mass were under 600 GeV.

But with this, high-energy physics faces a different type of collision, as experimenters' plans collide with the restrictions of government budgets. The LHC is budgeted at $1.4 billion, compared with $11 billion for the late lamented Supercollider. This lower cost results from its use of the existing LEP tunnel, along with a need for fewer magnets. Still, even before the Americans dropped the Supercollider, Europeans had repeatedly put off approving the LHC as a formal project. With no one now spurring them to compete in seeking these new frontiers, they could well offer further delays.

Moreover, Rome was not built in a day, and neither are major accelerators. Optimists hope that LHC will win approval in 1994, but even if it does, it would start operating no sooner than 2002. Even then, it would be sheer folly for physicists to try to find the Higgs particle by searching blindly in the energy range of 60 to 1000 GeV; that simply is too large an extent. As was true when Carlo Rubbia was searching for the W and Z particles in the early 1980s, it will help greatly if physicists can begin with a reasonably clear idea of the specific energy that will create the Higgs.

Fortunately, such a prediction is achievable—if experimenters first succeed in discovering the top quark. The reason is that the theory offers a set of mathematical relations that link the masses of the W, Z, top, and Higgs. Two of these masses are presently in hand, for the W and Z, with the W mass being steadily sharpened in ongoing experiments. A well-defined mass for the top will then pin down the Higgs mass with good precision. In this fashion, recent history may repeat. Just as detailed studies of the Z settled the very different question of the number of quark families, so discovery of the top quark can open the way to finding the Higgs.

Indeed, searches for the top have been a specialty at Fermilab, and these searches are continuing. During 1989, researchers at that lab competed with colleagues at CERN in pursuing this discovery. Neither group succeeded, but they showed that the top would have a mass of at least 91 GeV. This prediction drew on measurements of the W mass, made at both laboratories. Then in 1991 new results from the LEP at CERN offered an estimate for the mass: 140 ± 30 GeV. This sparked a new search at Fermilab, beginning in 1992. So far, this effort has not succeeded.

Still, the work continues. John Huth, an experimental leader at Fermilab, states that in 1994 or 1995 his lab will be able to extend the search to energies of 170 to 180 GeV. Further improvements in the accelerator system, which could enter service later in this decade, would increase their limit to 240 to 250 GeV. Huth notes that an upper limit for the mass of the top, at the 95 percent confidence level, is 225 GeV, according to inferences from theory. Such an estimate does not guarantee success, of course, but it encourages physicists to continue to try.

To discover this quark demands highly sophisticated detectors, and these have kept pace with developments in the design of the accelerators themselves. As recently as 1974, when researchers at Brookhaven were working to discover the charm quark, their detector was the bubble chamber. It was quite advanced for its day and sufficiently important to win a Nobel prize for its inventor, Donald Glaser of the University of California at Berkeley. Still, it had some rather noteworthy limitations.

The Brookhaven chamber featured a large tank of liquid hydrogen that acted as a photographic emulsion that was both three-dimensional and reusable. When charged particles passed through this liquid—electrons, protons, particles formed from quarks—they produced lines of tiny bubbles to mark their paths. Cameras would take photos from several angles; then their film would advance for the next shot while the chamber reset itself and absorbed the bubbles.

The system could generate photos by the tens of thousands, and people had to go through all of them to look for interesting events. Lowly graduate students often handled this tedious work, but at times they were in short supply. When that happened, the job often fell to local housewives who needed the extra income. At Brookhaven the woman who supervised this work was a former YMCA cashier; she had gained her post because she had a talent for telling her co-workers what the physicists were looking for. When the lab recorded its first evidence for the charm quark in May 1974, the data came from a former switchboard operator. It was on frame 6967 of roll 27.

The bubble chamber was highly useful and featured strongly in the key discoveries of the era. However, it was quite unable to keep up with the floods of data than increasingly capable accelerators would generate. The answer lay in electronic detectors, massive arrays of circuitry that could cope with almost any number of particle collisions. The first of them went into operation at CERN and proved essential to Carlo Rubbia's discovery of the W and Z in 1983. They had the highly useful feature of generating their data in the direct form of electrical signals, rather than as photographs. Such signals lent themselves to computer processing. The computers then could ease the task of searching for new particles by rejecting many of the data sets that were entirely devoid of interest.

In turn, these computers have had no shortage of data. At Fermilab

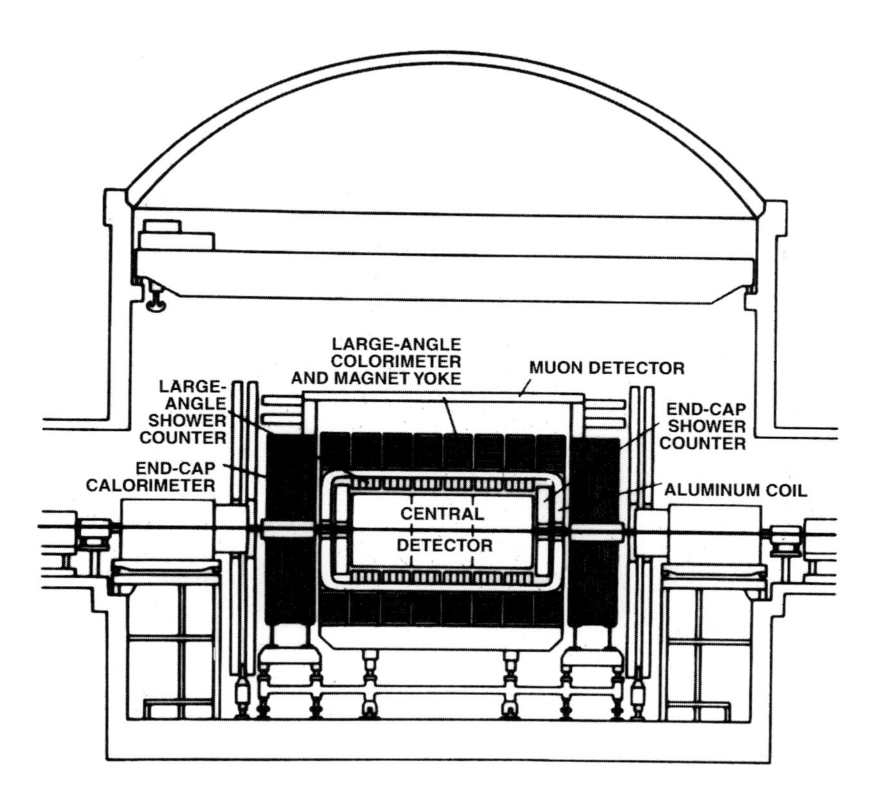

FIGURE 8.6 *Main detector at CERN, 1982. Side view of the UA1 detector shows it in place in the SPS beam line. Experimental components for various purposes, including the search for intermediate vector bosons, are labeled. (From "The Search for Intermediate Vector Bosons" by David B. Clive, Carlo Rubbia, and Simon van der Meer. Copyright © 1982 by Scientific American, Inc. All rights reserved.)*

the CDF detector can track the individual particles produced in each collision. CDF generates over 10,000 bits of information regarding each such particle. A typical collision produces 30 or more of them, and the Fermilab accelerator can generate over 100,000 collisions per second, with CDF faithfully keeping pace.

Further, with accelerators offering increasingly high energies, the sheer size and weight of detectors are also keeping pace. The CERN detector of 1983 came to 2000 tons (see Figure 8.6). Today's CDF weighs in at 5000 tons, while the main detector for the Supercollider was to reach 30,000. That equals the weight of a battleship, and the comparison is not accidental. Much of a battleship's weight comes from heavy armor plate. Similar armor plate, of iron, lead, and even uranium, guards the outer regions of a detector. It screens out less-penetrating particles, such as protons, making it easier to detect the highly penetrating ones that spray from collisions. But as these collisions rise in energy, in successively larger accelerators, the energy of collisional debris also rises. Hence, detectors need more and more shielding to accomplish this screening.

In addition, increasing collisional energy makes it desirable to increase the volume of a detector. The CERN accelerator of 1983, at 540

GeV, called for a detector measuring 5 by 10 meters long. For the Supercollider, operating at 40,000 GeV, one proposed detector was to have the dimensions of 18 by 30 meters (see Figure 8.7). Its need for shielding and its weight would have grown proportionately.

Still, if all goes well, there will be some very interesting experiments in the next decade or two. Fermilab, with the world's most powerful operating accelerator, could find the top quark and pin down its mass. That would permit a clear prediction of the mass of the Higgs. In turn, this mass might prove to lie within the energy range accessible to the LHC, spurring its construction. Discovery of the Higgs, along with detailed studies of this particle, then could shed light on the origin of mass.

None of this is guaranteed, to be sure. Diligent search might fail to find the top quark, and this would be a disaster for the Standard Model. The Higgs mass might turn out to be too high for the LHC. Even if it lies within its range, it nevertheless might evade detection. That would pose fundamental problems for the basis of electroweak theory.

And even if all goes well, if, in the year 2005 or thereabouts, we have

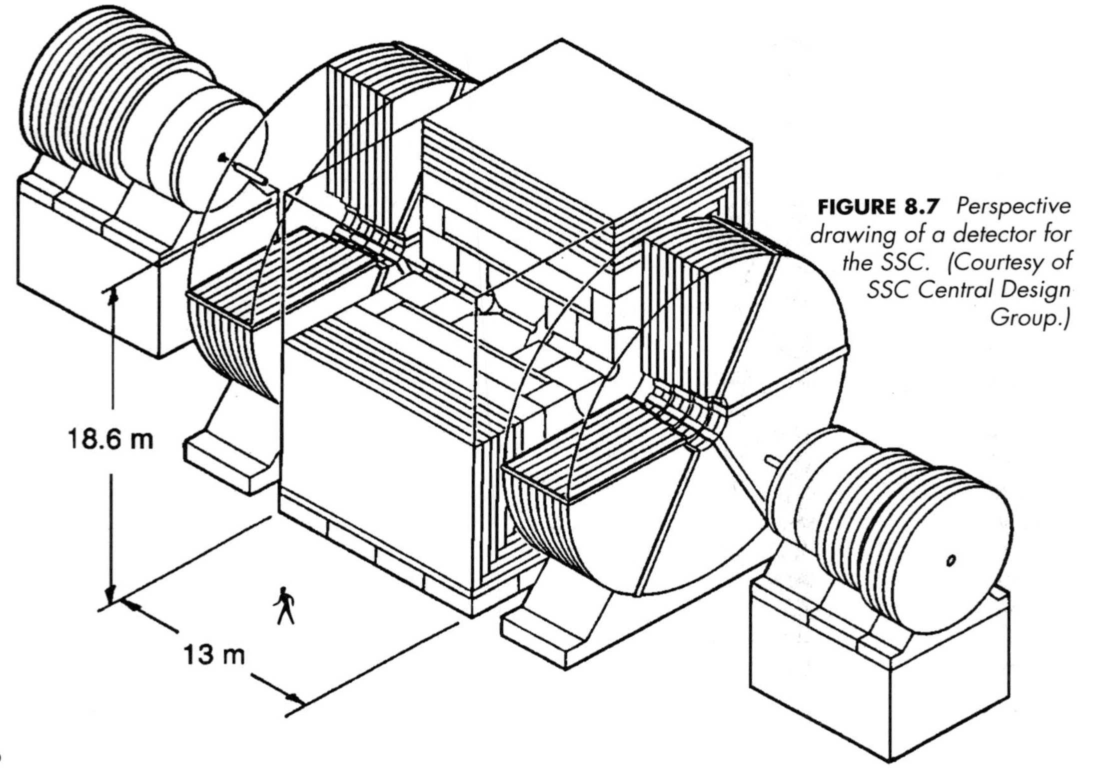

FIGURE 8.7 *Perspective drawing of a detector for the SSC. (Courtesy of SSC Central Design Group.)*

the top quark and the Higgs safely in hand, physicists will still face a host of issues. The question of the size of the universe, and why it is not curled up into the size of a football, is only one of them.

"I do not know what I may seem to others," wrote Isaac Newton. "But, as to myself, I seem to have been only as a boy playing on the seashore, and diverting myself in now and then finding a smoother pebble or a prettier shell than ordinary, whilst the great ocean of truth lay all undiscovered before me." In this spirit, rather than with the hope of final and ultimate insight, physicists may welcome new accelerators such as the LHC, as they usher in the agenda of the next century. Our researchers may divert themselves with the pretty shells of the top quark and even the Higgs, but Newton's ocean still conceals many of the truly deep issues: The origin of mass; the origin of the universe; the character of a truly ultimate theory of nature, if indeed a theory exists. For all our hope and all our work to date, we still must stand with Paul: "We know in part, and we prophesy in part. . . . Now we see through a glass, darkly."

BIBLIOGRAPHY

Breuker, H., H. Drevermann, C. Grab, A. A. Rademakers, and H. Stone. 1991. Tracking and imaging elementary particles. Scientific American (Aug.): 58–63.

Cline, D. B., C. Rubbia, and S. van de Meer. 1982. The search for intermediate vector bosons. Scientific American (March):48–59.

Crease, R. P., and C. C. Mann. 1986. The Second Creation. Macmillan, New York.

Feldman, G. J. and J. Steinberger. 1991. The number of families of matter. Scientific American (Feb.):70–75.

Fisher, A. 1979. Grand Unification: An elusive grail. Mosaic (Sept.):3–12.

Huth, J. 1992. The search for the top quark. American Scientist. (Sept./Oct.): 430–443.

Lederman, L. M. 1990. The Tevatron. Scientific American (March):48–55.

Myers, S., and E. Picasso. 1990. The LEP collider. Scientific American (July): 54–61.

National Research Council. 1986. Physics Through the 1990s: Elementary Particle Physics. National Academy Press, Washington, D.C.

Pagels, H. R. 1985. Perfect Symmetry. Simon & Schuster, New York.

Rees, J. R. 1989. The Stanford linear collider. Scientific American (Oct.):58–65.

Riordan, M. 1987. The Hunting of the Quark. Simon & Schuster, New York.

Schwarzschild, B. M. 1993. Decision approaching on CERN's proposed Large Hadron Collider. Physics Today (February):17–20.

Trefil, J. S. 1980. From Atoms to Quarks. Scribners, New York.

Veltman, M. J.G. 1986. The Higgs Boson. Scientific American (Nov.):76–84.

BOUNCING BALLS OF CARBON
The Discovery and Promise of Fullerenes

by Elizabeth J. Maggio

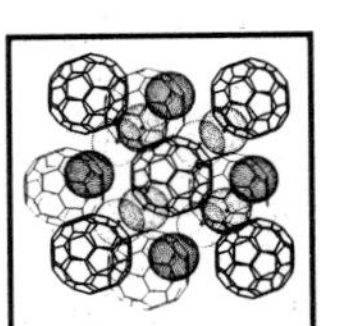

By early 1993, less than 3 years after the existence of a new form of carbon called buckminsterfullerene was confirmed, the scientific community's infatuation with these so-called buckyballs had turned into a love affair. Some might call it an obsession. Neatly packaged by Mother Nature into never-before-seen hollow molecules that resemble infinitesimally small soccer balls or geodesic domes, buckyballs now have celebrity status. *Science* magazine declared the buckyball molecule the 1991 "Molecule of the Year" because, said the editors, "rarely has one molecule so swiftly opened the door to a new field of science."

BUCKYBALLS AND THE OTHER FULLERENES KEEP ON ROLLING

The rotund carbon with the cutesy nickname has turned up on t-shirts and coffee mugs; it is the object of several patent proceedings; and it has spurred a commercial venture to produce buckyballs in bulk for the hundreds, if not thousands, of researchers around the world eager to get their hands on a precious gram or two. In 1992 the 10 most frequently cited papers in chemistry, according to *Science Watch*, were all on buckyball-related studies aimed at uncovering the molecule's secrets and potential commercial applications.

A buckyball molecule contains 60 carbon atoms, chemically designated as C_{60}, that bond into 12 pentagons and 20 hexagons with the same arrangement as the faces of a soccer ball (see Figures 9.1 and 9.2). Its shape also follows the same geometric principles that underlie the geodesic dome invented by American architect R. Buckminster Fuller, after whom it was named. Actually, buckminsterfullerene (a.k.a. buckyball) is only one—although the roundest, most abundant, and most popular—of a whole family of similar molecules generally referred to as "fullerenes." The second most common fullerene is C_{70}, whose 70 carbon atoms bond into what some say looks like a rugby ball (see Figure 9.1). "There are a lot of cousins in the carbon family that all share the same molecular form," said University of California, Los Angeles, chemist Robert L. Whetten. He and his colleagues at UCLA and UC-Santa Barbara have unlocked a number of fullerene secrets. They detected molecules with 76, 84, 90, and 94 carbon atoms, and other scientists have shown that giant versions with hundreds of carbon atoms exist.

Buckyballs and the other fullerenes constitute a third form of pure crystalline carbon, something entirely different from diamond and graphite, which were the only two crystalline forms of pure carbon thought to exist until scientists literally stumbled onto the fullerenes. The excitement they are generating is fueled not only by their aesthetic quality but also by their remarkable philosophical and commercial appeal.

For some researchers it has been a humbling experience to discover that modern science did not know everything about the very element that is critical for all earthly life. Apparently, nature has been making fullerenes all along wherever combustion is taking place. In fact, some scientists think they may be one of the most abundant molecules in the universe and that space may even be sprinkled with fullerenes, churned out by certain carbon-rich stars. There is even speculation that the incredibly stable buckyballs may have played a role in the development of organic life on earth.

On the practical end, the fullerenes are supremely tinkerable; being

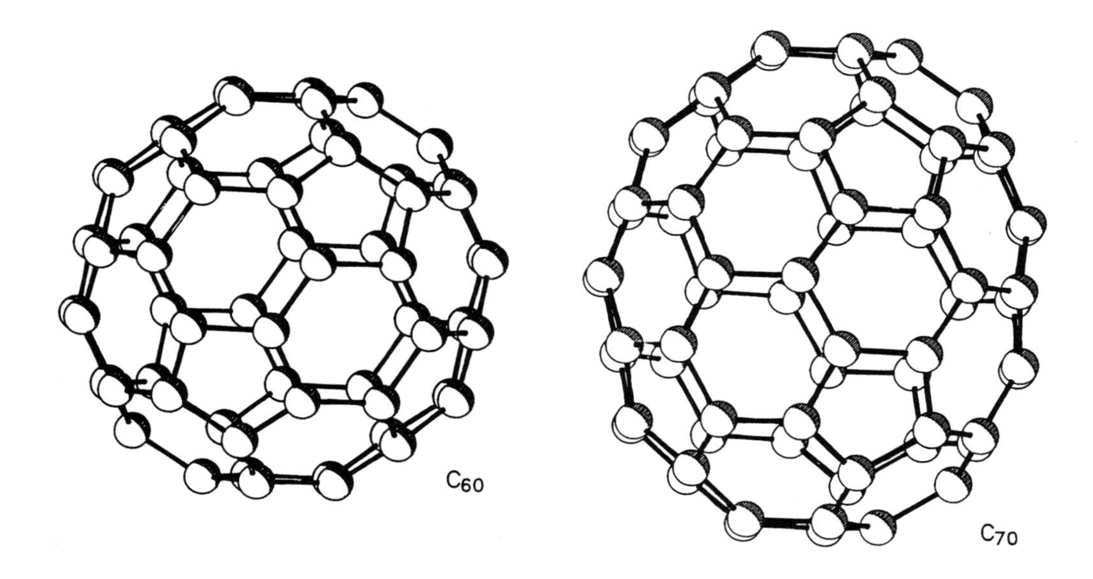

FIGURE 9.1 *Schematic drawing showing soccer-ball structure of a buckyball, C_{60} (left), and the rugby-ball structure of the next most common fullerene, C_{70}, (right). Clusters are drawn to scale. (Courtesy of IBM Almaden Research Center.)*

empty, round, microscopic cages, they offer the possibility of being stuffed, covered, and packed together with other atoms in experiments limited only by the scientist's imagination. "New branches of materials science and chemistry are opened up by these carbon molecules and their solids. Nobody has shown yet that there's a billion dollar industry in this, but some promising properties have already been discovered," said physicist Donald S. Bethune, of the IBM Almaden Research Center in San Jose. UCLA's Whetten has been on the trail of a particularly enticing fullerene property: superconductivity, perhaps the hottest subject in materials science today. And researchers such as Bethune and his colleagues at IBM have already succeeded in performing one of the neatest fullerene tricks so far—locking up one or more metal atoms inside individual carbon cages, creating what are called metallo-fullerenes whose properties are just beginning to be explored. Fullerenes may also have a role in the emerging field of nanotechnology in which electronic components, such as semiconductors, are miniaturized down to the atomic or molecular level.

"This is the best science story I've ever come across," commented Robert D. Johnson, a member of the IBM team. "If you had dreamed up fullerenes, it wouldn't be as fantastic as it is turning out to be in real life." Added Bethune, "Beyond the immediate challenge of making, characterizing, and finding applications for these new molecular structures, the most important thing that has happened in this field is the enormous expansion of people's sense of possibilities. Carbon atoms can be joined to each other and to other molecules, atoms, and clusters in ways that had not previously been imagined. Perhaps this shouldn't have come as such a surprise to beings whose own existence depends so heavily on the marvelous chemical versatility of carbon."

A CARBON PRIMER

Just how versatile is carbon? Of all the elements, not one forms more compounds than carbon; they number in the hundreds of thousands and are the basis for an entire subdivision of chemistry—organic chemistry.

Carbon is exceptional for a number of reasons but most of all for its versatility in bonding to other atoms (especially to other carbon atoms) by sharing its four available electrons in single, double, or triple bonds. Add to this the ease with which carbon can shift its bonding to make a more favorable arrangement. Other elements are not this flexible. For example, the formula for acetylene, a welding gas, is C_2H_2. The two carbon atoms are linked by a strong triple bond, and each carbon's remaining single bond attaches to a hydrogen atom. However, if tempted with two more hydrogen atoms, the carbons will gladly lessen their grip on each other and change their triple bond to a double one, so that each atom can take on another hydrogen, creating ethylene, C_2H_4, which is used in the manufacture of plastics. The carbons can accommodate yet another pair of hydrogen atoms by changing their double bond to a single one—the weakest of all—to form ethane, C_2H_6, a fuel and refrigerant.

Before the fullerene discovery, it was gospel among chemists that pure carbon had only two distinct crystalline forms: diamond and graphite. (Other carbon material, such as charcoal and coke, has an amorphous quality but appears to consist primarily of tiny graphite crystals.) Diamond is the hardest substance known, while graphite is so soft and slippery that it is the basis for pencil lead. Such different characteristics result from the very different ways in which the carbon atoms are arranged. Diamond is a lattice whose basic

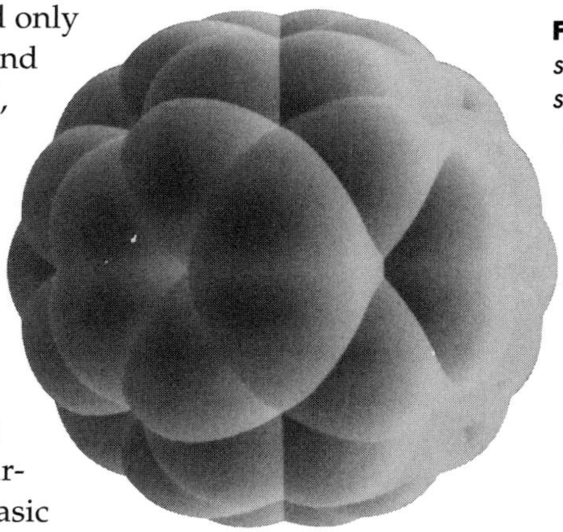

FIGURE 9.2 *Computer simulation of a soccer-ball-shaped buckyball (C_{60}). (Courtesy of University of Arizona/Dennis Lichtenberger.)*

Hollow Spherical Molecules "Predicted" in 1966

With tongue-in-cheek musing in the November 3, 1966, issue of the British magazine *New Scientist*, columnist D. E. H. Jones (writing under the pseudonym Daedalus) made some strikingly accurate predictions about hollow spherical molecules, predating the discovery of the fullerenes by nearly two decades. Some of that column is excerpted here:

> There is a curious discontinuity between the densities of gases around 0.001 relative to water, and liquid and solids from 0.5 to 25 or so; this week Daedalus has been contemplating ways of bridging this gap and has conceived the hollow molecule, a closed spherical shell of a sheet-polymer like graphite, whose molecules are flat sheets of benzene-hexagons. He proposes to modify the high-temperature graphite process by introducing suitable impurities into the sheets to warp them (rather like "doping" semiconductor crystals to cause discontinuities), reasoning that the curvature thus produced will be transmitted throughout the sheet to its growing edges so that it will ultimately close on itself.
>
> . . . Such fascinating new materials would have a host of uses, in novel shock-absorbers, thermometers, barometers and so on, and possibly in gas-bearings where the rolling contact of the spherical molecules would lower friction even further.
>
> Daedalus was worried that they might deform under pressure until he realized that if synthesized in a normal atmosphere they would be full of gas and resilient like little footballs: he is now seeking ways of incorporating "windows" into their structure so that they can absorb or exchange internal molecules, so as to produce a range of super molecular-sieves capable of entrapping hundreds of times their own weight of such small molecules as can enter the windows.

SOURCE: Reprinted, with permission, from *New Scientist* (Nov. 3, 1966).

framework is a carbon tetrahedron. This is a three-dimensional pyramid structure of five carbon atoms, one at each vertex of the tetrahedron and one in the center. The atomic arrangement of graphite is completely different. The carbon bonds into a two-dimensional hexagonal ring with an atom at each of its six points. Hexagons bond to more hexagons and eventually form flat sheets that stack one over the other. The sheets themselves are only weakly held together, giving rise to graphite's slipperiness and flakiness.

One hexagonal graphite fragment or one tetrahedral piece of a diamond lattice is not a stable chemical entity. Their carbon atoms still have unused "dangling" bonds that readily latch on to more carbon, leading to the growth of macroscopic crystals. In dramatic contrast, a fullerene is a closed carbon molecule. There are no edges to a fullerene sphere; no atoms left with "dangling" bonds. Once formed, it cannot add additional atoms and grow. It was this property that made buckyballs stick out like sore thumbs in mass spectra of carbon vapor and opened a new chapter on carbon chemistry.

AN UNEXPECTED DISCOVERY

The fullerene story is a relatively short one—less than a decade old—that, perhaps, could have unfolded years earlier if researchers doing the first theoretical studies on a new carbon structure had not written of their work in their native tongue—Japanese. Japanese scientists had made calculations showing that 60 carbon atoms *theoretically* could be bonded into a stable round molecule that looked like a soccer ball, a shape technically called a truncated icosahedron. (Strangely, the concept of hollow round molecules that very nearly match the description of a buckyball was put forth in a tongue-in-cheek column that appeared in the *New Scientist* in 1966. See Box on p. 262.) But the Japanese papers did not make a big hit in the international scientific community. "Language was one barrier," said Bethune, "plus nobody had any notion that this was really possible. It was a purely theoretical exercise."

Still, the synthesis of highly symmetrical molecules, especially of carbon, has been a classic quest in organic chemistry, and in the early 1980s researchers attempted, unsuccessfully, to synthesize the equivalent of buckyballs. They failed, said Bethune, because they relied on the classical approach used by chemists when trying to synthesize something new: start with an existing molecule and modify it through a logical sequence of steps, each accompanied by detailed analysis of intermediate products on which the next modification could be made. "It was time consuming," said Bethune, "akin to mountain climbing. You have to find a particular route up the mountain and along the way there are difficulties to overcome. If you're successful, however, you reach the top." But as those researchers would later discover, to their surprise, the path to fullerenes is amazingly easy: vaporize some graphite and the liberated carbon atoms spontaneously assemble themselves into hollow round balls of varying sizes. "The fact that the fullerenes were not synthesized by chemical intentional design," commented Whetten, "is a painful one for us chemists!"

First Hints of Something Strange

It was pure serendipity that fullerenes were stumbled upon at all. There were hints of their existence in the early 1980s from pioneering basic research being done at Exxon to study how vaporized atoms of different elements cluster back together. The Exxon scientists were running through the periodic table, said Bethune, and had noted strange behavior when they came to carbon. But the fullerenes remained theoretical oddities until 1985 when a landmark paper hit the British journal, *Nature*, written by British chemist Harold W. Kroto of the University of Sussex and Rice University chemists Robert F. Curl and Richard E. Smalley, along with Rice graduate students at the time, James R. Heath and Sean C. O'Brien. This U.S.-British team reported finding convincing evidence for the existence of a new form of pure carbon—a hollow round molecule, which they named "buckminsterfullerene." At the Houston campus of Rice University, Smalley had been doing experiments, similar to those at Exxon, using a supersonic cluster-beam device he had developed. Across the Atlantic in England, Kroto was heavily involved in his studies of how certain carbon-rich stars may be littering space with carbon molecules in the form of long chains. Eventually the two scientists got together to use Smalley's experimental setup to simulate carbon star chemistry and see what might be happening in outer space. The device vaporized the surface of a rotating disk of graphite using a pulsed laser. The liberated carbon atoms were carried away in a stream of helium gas that expanded as it exited a nozzle, allowing the floating carbon atoms to cool, react, and form clusters. The scientists then ionized the carbon clusters, accelerated them and measured their masses and relative abundances as they whizzed through a time-of-flight mass spectrometer.

The resulting mass spectra showed signals from a wide range of carbon clusters containing anywhere from 38 to 120 atoms, with a curious peak from a 60-atom cluster. The researchers theorized that the laser had torn off fragments of the graphite, liberating pieces of its flat network of hexagonal carbon rings, which then collided and linked up into variously sized clusters. (Bethune's work at IBM would later turn up evidence that the graphite is completely broken down into individual atoms before forming fullerenes.) Then things got exciting and confusing. By allowing more time for the carbon vapor to condense, the researchers discovered that the mass spectrum signals from most of the carbon clusters very nearly disappeared with the exception of that curious 60-atom cluster. Its peak shot straight up like the Washington

Monument, overpowering the readout. The only other noticeable peak, although much smaller, corresponded to a 70-atom cluster. "As the experiments progressed it became impossible to ignore the antics of the C_{60} peak, which varied from relative insignificance to total dominance depending on the clustering conditions," Kroto wrote in an article describing these experiments for the journal *Science*.

A Soccer-Ball Solution

There was no way to account for the extraordinary stability of the big C_{60} peak. Carbon is notoriously reactive. If this cluster were a flat sheet of carbon hexagons, it would have numerous free edges with which to bond with additional carbon atoms and continue to grow in size. What was stopping it at 60 atoms? Somehow the loose ends of the flat cluster would have to be tied up. The only plausible answer, concluded the researchers, would be if the cluster had closed into a sphere, and that was unheard of in chemistry.

Undaunted by the proposal, they consulted the work of R. Buckminster Fuller, the twentieth-century American architect/inventor who developed the geodesic dome, which is basically a closed network of five-sided (pentagonal) and six-sided (hexagonal) faces. The eighteenth-century Swiss mathematician Leonhard Euler elaborated the geometry behind such a sphere and showed that it had to have 12—and only 12—pentagons in order to close; the number of hexagons could vary, down to two. The soccer ball is a common example of such a sphere. It has 32 flat faces—20 hexagons and 12 pentagons—which was the exact structure proposed for C_{60}. The researchers theorized that as the carbon clusters were growing in the condensing vapor they rearranged their bonding to produce pentagonal as well as hexagonal rings of atoms that closed the clusters into soccer-ball-like spheres.

Suddenly, the C_{60} peak made sense. It was ". . . a most elegant and, at the time, overwhelming solution—the truncated icosahedron cage," recounted Kroto. It was too good not to be true. Not only would a geodesic structure for this cluster have 60 vertices or points marking the location of each carbon atom, but such an arrangement satisfied all the chemical requirements of the carbon atoms, such as bonding. With no loose ends to attract more atoms, this C_{60} molecule would be superstable. "This structure necessitated the throwing of all caution to the wind . . . and it was proposed immediately," wrote Kroto. With no independent confirmation that they were correct, the team went ahead anyway and named the new carbon molecule. After discarding such names as "ballene," "spherene," "soccerene," and "carbosoccer," they settled on "buckminsterfullerene," apologizing for the 20-letter, six-syllable moniker. Eventually, buckminsterfullerene became known as simply "buckyball."

All fullerenes have an even number of carbon atoms, which follows from their geodesic geometry, as well as the required 12 pentagons. The varying number of hexagons control their shape. Buckyballs, with 20 hexagons, are the roundest fullerene. The others become progressively out of round with a different number of hexagons in their structure. For example, the atoms in the next most common fullerene, C_{70}, are arranged into 12 pentagons and 25 hexagons, which make it slightly elongated, more like a rugby ball than a soccer ball (see Figure 9.1). In the report of their discovery, the researchers explain that C_{60} is uniquely stable and thus the most abundant form because its arrangement of pentagons and hexagons distributes strain perfectly. The second most stable arrangement is that of C_{70}, whose signal always appears along with C_{60} in mass spectra of vaporized carbon, although to a lesser degree. The researchers calculated that the other fullerenes are not as stable because the pentagons in their structures are located in strained positions that make them vulnerable to chemical attack. That's why they virtually disappeared from mass spectra taken after the carbon vapor was given more time to condense into clusters.

The 1985 *Nature* paper, titled simply "C_{60}: Buckminsterfullerene," sent a buzz through the scientific community. But even though other researchers were getting the same results as the Kroto-Smalley team, and theorists were coming up with ways in which such hollow spherical molecules should behave, doubts about the existence of buckyballs persisted. The problem was simply that scientists couldn't make enough of it to poke and prod to confirm the proposed structure. All they had were a relatively few transient molecules that could only be studied briefly as they were whisked along in the helium gas that carried them. They didn't hang around long enough or in great enough numbers to be analyzed with conventional methods.

Capturing Buckyballs

Bethune was one of many scientists intrigued by the reports coming out about fullerenes, and in late 1989 he began his own cluster research program at the IBM Almaden Research Center. "Here was this beautiful picture of fullerenes, an amazing new class of carbon compounds," recalled Bethune, "without any proof that it was right." At that point, only one spectroscopic result had been reported, and it gave very little useful information.

Bethune and his colleagues—Mattanjah de Vries and Heinrich Hunziker of IBM along with Gerard Meijer of the University of Nijmegen, The Netherlands—began working on a way to produce enough fullerenes to characterize. "We had this notion that if fullerenes were so stable and chemically inert, as scientists were claiming, you could collect them," said Bethune. "Why not?" The IBM group set up a system to vaporize graphite. However, instead of trying to analyze the resulting carbon vapor for the brief time it stayed around, they let it condense as soot on a substrate and then probed it for the presence of fullerenes. They placed a piece of copper about a half inch above the graphite, turned on the laser, and caught the escaping soot, which left a distinct dark gray haze on the copper substrate. A second laser was then used to gently knock the collected carbon clusters into an inert carrier gas so they could be analyzed with a sensitive mass spectrometer designed and built by Hunziker, deVries, and H. Russell Wendt.

When the results were read out by the laboratory's computer, both C_{60} and C_{70} made a remarkable showing and were accompanied by signals from other fullerenes up to around 200 atoms in size. The scientists had succeeded in trapping, storing, and accumulating fullerenes on a surface. "We felt like we had suddenly stumbled into a long lost treasure cave," said Bethune. "It was apparently unbelievably simple to produce and capture a whole family of fullerenes. Furthermore, they survived exposure to the atmosphere and room temperature." The IBM researchers estimated that they were making and—more importantly—collecting a nanogram (one billionth of a gram) of buckyballs with each laser pulse. Soon it would be possible to accumulate enough buckyballs for conventional chemical and physical analytical studies that would confirm their structure.

It was the summer of 1990, and the dam was about to crack wide open on fullerenes.

While the IBM group was working on collecting more fullerene-containing soot with its laser deposition method, a U.S.-German team of physicists reported evidence that they had produced soot containing a relatively large amount, roughly 1 percent, of C_{60} using an even simpler technique. Their fullerene generator, said Bethune, was little more than "a pencil lead and battery." A graphite rod was attached to the terminals of a battery or other power supply, and 100 amps of current was sent through the sample until it smoked and deposited soot on the inside of the helium-filled chamber. The researchers had taken an infrared spectrum of the collected soot, and it matched the theoretical calculations already made for a soccer-ball-shaped fullerene.

This work was a collaboration between Wolfgang Krätschmer and his graduate student Konstantinos Fostiropoulos at Germany's Max Planck Institute for Nuclear Physics in Heidelberg and Donald Huffman at the University of Arizona in Tucson. They hadn't set out to join the fullerene foray. In fact, one signature of the C_{60} and C_{70} fullerenes had appeared in optical analyses of a sample of soot they made back in 1983—2 years before the existence of fullerenes was proposed—only they didn't know it! Like the Kroto-Smalley team, the Krätschmer-Huffman group had been carrying out laboratory simulations of carbon chemistry in stars. They were vaporizing graphite, scraping off the resulting soot from inside the evaporation chamber, and analyzing it optically. They noticed that when the helium atmosphere in their device was at one particular pressure the resulting soot strongly absorbed light at deep ultraviolet wavelengths. A plot of the optical data showed a strange pair of "humps," something the scientists had never seen in all the years they had been analyzing the spectra of carbon dust. They dubbed this particular batch of soot their "camel" sample and put it away after not being able to figure out what the humps meant.

Then 2 years later, in 1985, they read about the proposed buckminsterfullerene carbon molecule in *Nature* and wondered if those curious humps in the absorption spectrum of the soot they made in 1983 had any connection. "The seeming unlikeliness of this hypothesis, together with some difficulty in reproducing the experiment, led the researchers to put the project on the back burner," recalled Curl and Smalley in a history of the fullerene discovery they wrote for *Scientific American*. By 1989, however, the Krätschmer-Huffman team became intrigued enough by the possible connection between the proposed fullerenes and their "camel" sample that they decided to give it a second look. They repeated the conditions of their 1983 experiment and this time successfully produced more soot with the double hump spectrum. Analysis of how the material absorbed infrared light not only agreed with what the theorists had already calculated for C_{60} but also indicated that the "camel" method produced soot with a relatively large quantity of the curious molecules—more than anyone had been able to generate before.

"When we learned of this result," said Bethune, "we immediately tried their recipe for making soot." The IBM team set up a similar "pencil-lead and battery" device and let the current run until the graphite glowed too bright to look at. The bar broke in about a minute. The researchers ran a soot sample through their mass spectrometer, and the unmistakable signature of the fullerene family appeared on their screen. They vaporized more graphite bars and collected about 20 milligrams of soot, which they put into a small cylinder and heated to nearly 850° F (500° C) to see if anything would sublime out. "We got this beautiful yellow thin film to form on a quartz substrate," Bethune recalled, "and when we looked at it with the mass spectrometer, we saw that the film was almost pure C_{60}." Heating the soot to an even higher temperature yielded another thin film; this time the mass spectrometer told them they had both C_{60} and C_{70}. "This seemed pretty convincing that the Krätschmer-Huffman team was correct. They had made C_{60}, and not just a little bit."

Now that the IBM group had large enough quantities of purified fullerenes captured as thin films on substrates, Bethune hurried to get samples to various team members who had been waiting to begin a series of sophisticated analyses that would prove—or disprove—the molecule's novel structure. They immediately tried obtaining a Raman spectrum, and began work on photographing the molecules using a scanning tunneling microscope (STM). The big push at IBM and elsewhere, however, was to do nuclear magnetic resonance (NMR) testing that would provide the final, critical piece of evidence for the proposed structure.

Bethune already had in hand the Raman spectrum, which agreed with a soccer-ball structure for C_{60}, when he left in early September 1990 to address the Fifth International Symposium on Small Particles and Inorganic Clusters in Konstanz, Germany. Excited and optimistic, he concluded his talk by saying: "While very soon many more measurements will be done to completely confirm these results, it already appears quite certain that the idea of molecular soccer balls is realized in nature, and that very soon you will be able to buy them from your favorite chemical company." Bethune's announcement was the first at any major scientific meeting of the production, partial purification and characterization of macroscopic quantities of both C_{60} and C_{70}. Upon returning home from the meeting, Bethune was met at the airport by an excited team member who took him directly to the IBM labs. New test results were in. Jet lag couldn't dampen the thrill when Bethune saw the first buckyball photographs and, more importantly, the NMR result everyone had been waiting for, the one that cleared up any lingering doubts that the C_{60} molecule looks like a miniature soccer ball.

FINGERPRINTING AND PHOTOGRAPHING BUCKYBALLS

Theorists had already calculated what the test results should be if the atoms in buckyballs really were arranged in soccer-ball fashion. The U.S.-German team had already obtained the first diagnostic result, an infrared absorption spectrum that matched the theoretical calculations and which the IBM team confirmed shortly afterwards. The remaining tests were increasingly more sophisticated, but they would inch researchers closer and closer to the final answer.

The Raman Spectrum

The IBM team was the first to get a key molecular fingerprint, called a Raman spectrum, from the thin films of C_{60} they had sublimed from soot prepared with the "camel" recipe. "Raman spectroscopy tells us some of the vibrational frequencies of a molecule," Bethune explained. "Just as a musical instrument can be identified by its characteristic

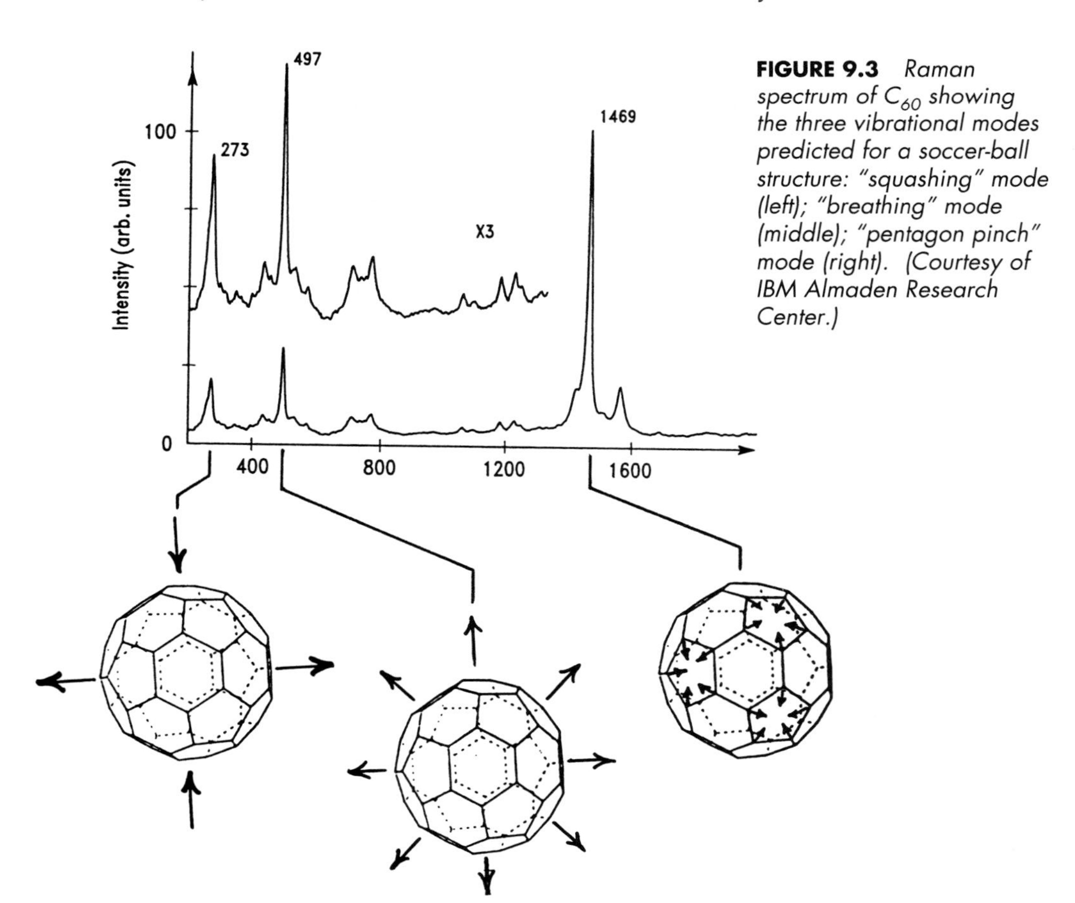

FIGURE 9.3 *Raman spectrum of C_{60} showing the three vibrational modes predicted for a soccer-ball structure: "squashing" mode (left); "breathing" mode (middle); "pentagon pinch" mode (right). (Courtesy of IBM Almaden Research Center.)*

[vibrational] sound, a molecule's vibrational frequencies tell us about its structure." The molecule's proposed high symmetry would drastically simplify this test. Team members Hal J. Rosen and Wade Tang made the measurements with their lab's Raman spectrometer. This device illuminated the thin film of C_{60} and then measured how much energy the light lost after scattering off the buckyballs. The energy loss is directly related to vibrations excited in the carbon molecules by the light scattering. The resulting Raman spectrum had three distinct peaks, each corresponding to molecular vibrational frequencies that the theorists had predicted for a soccer-ball structure. The first peak corresponded to a buckyball vibrating in a "squashing" mode, as if the molecule were being pressed between two hands and then released. The second peak corresponded to a "breathing mode," meaning that the C_{60} molecule was simply expanding and contracting. The last peak, corresponding to the "pentagon pinch" mode, told the researchers that the buckyballs' pentagons and hexagons were expanding and contracting out of phase with each other (see Figure 9.3).

The Raman spectrum, which also complemented the infrared absorption spectrum obtained earlier, was strong support—but not conclusive evidence—for a soccer-ball structure. That would come from the results of NMR studies of buckyballs.

NMR—The Chemist's Proof and a Surprise

Both the IBM group and Kroto's team at the University of Sussex succeeded in obtaining the decisive "single-line" readout from NMR analysis of C_{60}. To chemists, this was the ultimate proof of the molecule's structure. The test worked because a small amount of the carbon atoms in buckyballs are the isotope ^{13}C, which naturally occurs (at about 1.1 percent) along with ordinary carbon atoms, ^{12}C. The isotope has an extra neutron in its nucleus, making it detectable by NMR. Ordinary carbon, with an equal number of protons and neutrons, is not detectable by NMR.

IBM team member Bob Johnson was in charge of the NMR analysis. Sublimed films of C_{60} were first dissolved in solvent. The fullerene solution was then placed in the NMR device and exposed to a magnetic field. The C_{60} molecules with the ^{13}C isotope in their structure responded to the magnetic field by generating a tiny magnetic field of their own. This was measured to reveal the geometry and chemical environment surrounding the magnetically active isotopes and thus the entire molecule's structure. It didn't matter where, among buckyball's 60

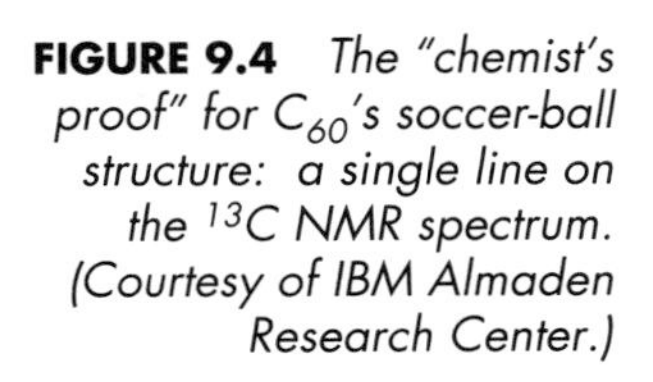

FIGURE 9.4 *The "chemist's proof" for C_{60}'s soccer-ball structure: a single line on the ^{13}C NMR spectrum. (Courtesy of IBM Almaden Research Center.)*

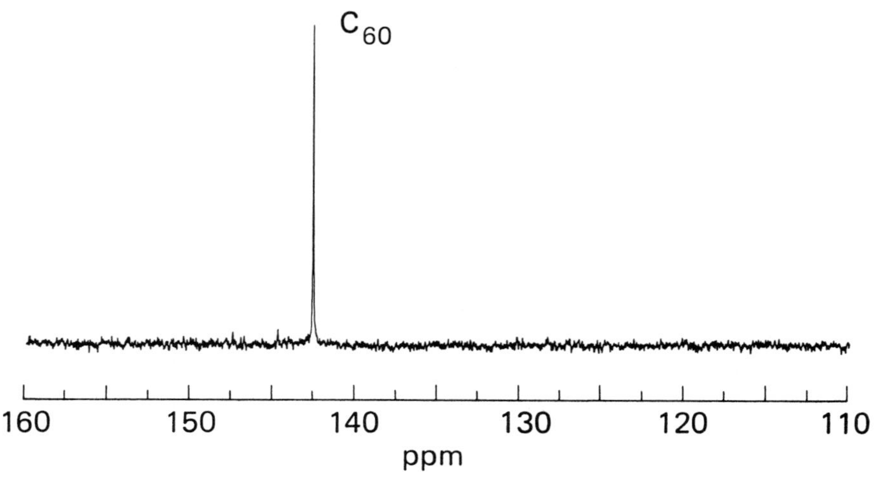

atoms, the magnetically active ^{13}C isotopes occurred, said Bethune. Their NMR reading would be the same for a ^{13}C in any location if buckyballs were truly as symmetrical as suspected. "Every carbon atom in the molecule is just like every other carbon as far as geometry and chemistry go." If C_{60} weren't symmetrical, the NMR spectrum would consist of multiple peaks. But the single beautiful line told the researchers they were looking at soccer-ball-shaped buckyballs (see Figure 9.4).

The British group also did NMR fingerprinting of the next most common fullerene, C_{70}. Because it has 10 extra atoms, giving it a slightly elongated, rugby-ball shape, this fullerene is not as symmetrical as a round buckyball. The molecule has five nonequivalent groups of atoms, and this produced an NMR spectrum with five distinct lines, a dead giveaway to the slightly out-of-round shape. The IBM group took an additional analytical step and subjected C_{70} to a more sophisticated examination called 2D "INADEQUATE" NMR. ("INADEQUATE" stands for "Incredible Natural Abundance Double Quantum Transfer Experiment.) This identified which carbon atoms are bonded to each other and, when combined with the NMR spectrum obtained by the British team, completely confirmed the rugby-ball structure for C_{70}. To do this more exacting test, however, Bethune and Meijer first had to enrich the natural ^{13}C isotope content of their fullerene sample. They did this by packing off-the-shelf ^{13}C into a hole bored into a graphite rod before burning it into soot and subliming out C_{70}.

Unexpectedly, this little chemistry trick performed for the 2D NMR analysis also turned up new insight into how fullerenes form in a condensing carbon vapor. When buckminsterfullerene was first proposed, the scientists theorized that carbon came off the vaporized graphite in clumps—essentially graphite fragments—that grew in size through collisions with other fragments and eventually closed into fullerene spheres. However, when the IBM researchers prepared their isotope-enriched samples of C_{70}, they found that the ^{13}C was randomly mixed in with the regular carbon atoms. "This implies that the graphite must have been completely atomized before forming fullerenes," said Bethune, "and argues against formation mechanisms that involve tearing multiatom fragments from the graphite."

Surprise: Buckyballs Can't Sit Still

While NMR analysis of C_{60} in solution allowed researchers to study the molecule's symmetry and bonding, the same kind of analysis done on solid fullerene (referred to as "fullerite") turned out to be a powerful tool for probing the detailed geometry of buckyballs and their properties. And, in fact, solid-state NMR analysis carried out independently by Costantino Yannoni of the IBM group and Robert Tycko at AT&T Bell Labs in Murray Hill, New Jersey, revealed a surprising fact: even when solidified into a crystal (when atoms and molecules normally are locked into place by their neighbors), buckyballs are about as sedate as a five year old in a dentist's chair; they're jittery molecules that don't like to sit still. "At room temperature, they nearly ignore each other," Bethune said, "and spin wildly with no correlation between the orientations of adjacent molecules." Such solids are referred to as "plastic crystals," and a few others exist in nature, such as a hydrocarbon called adamantane ($C_{10}H_{16}$). Its carbon atoms form a tetrahedron-like structure that is studded on the outside with the hydrogen. The sides of the molecule bulge out, giving it a somewhat roundish shape. Adamantane used to be the "classic solid rotator" that held the spin record for plastic crystals until buckyballs came along, said IBM team member Johnson, but the new carbon has clearly broken the record for slipperiness. Even though a buckyball is six times larger, it spins three times faster than adamantane, which is slowed by the hydrogen "bumps" on its surface, Johnson explained.

How fast can buckyballs spin in a solid? The IBM group clocked them at around 20 billion revolutions per second at room temperature and found that this is almost as fast as if the molecules were in gas phase, when there is no hindrance at all to rotation. "That's an incredible rate, pretty wild," said Johnson who added that perhaps buckyballs should be named the "new" classic rotator. Further research at IBM showed that the only way to slow buckyballs is to put them in deep freeze. At 8° F (–13° C), their rotation rate drops abruptly. "That's associated with suddenly the molecules beginning to care about the orientation of their neighbors, and the whole (crystal) lattice undergoes a phase transition where the relative orientation of all these things is specified," explained Bethune. Even then, the molecules aren't completely stationary. They continue to twitch, with a ratchet-like motion, between favorable orientations. Only when the temperature plummets to 77 degrees above absolute zero (around –200° C) will buckyballs stop fidgeting. Later, IBM's Yannoni used this deep-freeze method to keep buckyballs still while making the first NMR measurements of bond lengths between carbon atoms. Because of C_{60}'s symmetry, only two bond lengths had to be measured, said Bethune, to specify the position of every atom in the molecule and to calculate the diameter of the sphere: 7.1 angstroms.

Buckyball's propensity to spin about wildly even in solid form made it difficult for other researchers to do x-ray crystallography of their structure. Then scientists at the University of California, Berkeley, led by Joel Hawkins, hit on an ingenious idea to pin down the jumpy molecules. They used "molecular pliers," said Bethune. The researchers had added osmium tetroxide to a C_{60} solution and made crystals out of the chemical hybrid. Molecules of the metal oxide gripped individual buckyballs like clamps, breaking their slippery symmetry, and making it possible to crystallize them and do x-ray analysis of their atomic positions.

First Buckyball Photos

As 1990 came to a close, with test results undeniably supporting a soccer-ball structure, the world got to *see* buckyballs for the first time. Formal portraits were made by both the IBM team and a group led by the University of Arizona's Huffman using a scanning tunneling microscope (STM). The photos (see Figure 9.5) were released in back-to-back reports that appeared in the December 13 issue of *Nature*. Unlike a common light microscope, the STM visualizes individual molecules by recording changes in electric current as a pen-like instrument, whose tip is only a few atoms wide, gently moves over molecules that have been deposited on a specially prepared surface. The IBM researchers put their molecules on the surface of a single crystal of gold. Repeated scanning, done under ultra-high-vacuum conditions, revealed a picture of hexagonally arrayed, round molecules spaced about 11 angstroms apart (1 angstrom = 10^{-10}m) and which fit the predicted diameter for buckyballs, said Bethune. The STM images, made by team member Robert Wilson, also revealed some taller features that the scientists suspect are

FIGURE 9.5 *Smaller, dark-colored mounds are buckyballs (C_{60}) photographed with the scanning tunneling microscope. Larger, lighter-colored balls are believed to be C_{70}. (Courtesy of IBM Almaden Research Center.)*

the slightly larger C_{70} molecules mixed in with the C_{60}. While they could clearly identify buckyballs in the STM scans, they couldn't see the individual atoms that make up the hollow cages. They speculated that the molecules' propensity to spin may have blurred the view. Other researchers later found ways to bind the buckyballs more tightly to the surface, revealing in some cases internal structure, according to Bethune.

Buckyballs by the Bushel

Both the IBM team and Harry Kroto's Sussex group continued to concentrate their research efforts on analyzing the fullerene structure. During this time, the U.S.-German collaboration of Krätschmer and Huffman, now joined by University of Arizona physicist Lowell D.

Buckyballs: Over 84 Billion Sold

In less than a decade, buckyballs have been transformed from bizarre signals detected in mass spectra of vaporized graphite to a hot commercial product. In early 1993, pure buckyballs were selling for $1000 a gram, with a discount for orders of five grams or more.

Commercialization of buckyballs began on September 27, 1990, when a team of scientists led by Donald R. Huffman of the University of Arizona in Tucson, and Wolfgang Krätschmer of the Max Planck Institute for Nuclear Physics in Heidelberg, Germany, reported in the journal *Nature* a process they developed to make high-yield, fullerene-containing soot. That same day, Tucson-based Research Corporation Technologies (RCT) announced filing for worldwide patent protection for the process on behalf of Huffman and Krätschmer. It also filed patents for the various fullerene products produced from the soot.

RCT is a nonprofit technology transfer company that evaluates, patents, and commercializes inventions from research institutions. It subsequently licensed the Materials and Electrochemical Research Corp. (MER), also of Tucson, to produce useful research quantities of buckyballs and other fullerenes and to encourage the commercialization and development of new applications for this new form of carbon, according to MER Vice President for Business Development Chuck Hassen.

MER markets a range of fullerene products, in addition to pure buckyballs, at much less cost to scientists than if they had to set up their own fullerene generators, and then separate and purify the products they needed for their research, said Hassen. A popular seller is the solvent-extractable part of raw soot made with the Huffman-Krätschmer method. It contains 85 to 90 percent C_{60}; 10 to 13 percent C_{70}; and tiny amounts of C_{76}, C_{78}, and C_{84}. In early 1993, a gram of this extract was going for $200—and less than half that price for volume orders. That represents a 93 percent reduction from MER's first sale in January 1991, said Hassen, who expects even further reductions as improvements are made.

In other commercial developments, RCT has licensed Regis Chemical Company of Morton Grove, Illinois, to manufacture the Buckyclutcher I, the trademark for a chromatography column developed at the University of Illinois at Urbana-Champaign for fullerene separation. RCT has also filed international patents on a process developed at Northwestern University that uses C_{70} to greatly enhance the growth of diamond films used in a variety of commercial applications.

Lamb, reported another breakthrough in the fullerene saga in the September 27, 1990 issue of *Nature*: They had succeeded in mass producing large quantities of nearly pure C_{60} crystals. The mixture also contained a smaller amount of C_{70} crystals plus traces of larger fullerenes. They had mixed their "camel recipe" soot with the solvent benzene and then evaporated the liquid, leaving behind dark brown to black microscopic crystals, which the scientists subjected to x-ray diffraction studies and additional infrared and mass spectrum analyses. All tests turned up additional convincing evidence for the soccer-ball structure of the C_{60}.

When the soot was mixed with benzene, the clear liquid turned wine red due to the presence of both C_{60} and C_{70}, quite a startling color change for pure carbon soot! "When they saw the color red develop, they realized they were looking at the first concentrated solution of fullerenes ever seen," wrote Curl and Smalley. "They were the first to observe this roundest of all round molecules, and they knew that chemistry books and encyclopedias would never be quite the same. Now there were three known forms of pure carbon: the network solids, diamond and graphite, and a new class of discrete molecules—the fullerenes."

Currently prepared extracts of soot made with an optimized version of the Krätschmer-Huffman "camel" recipe are nearly 90 percent C_{60} (buckyballs) and around 10 percent C_{70}. The U.S.-British team that first proposed the existence of buckyballs later realized that they weren't able to produce more than a few fullerene molecules initially because their method didn't keep the vaporized carbon hot enough, long enough, for a larger quantity of the two ultrastable fullerenes to form. The accidentally discovered Krätschmer-Huffman method did. It is so successful at cranking out the carbon balls that it is being patented and has been licensed to a company that is now selling buckyballs in bulk to scientists around the world. (See Box on p. 276.)

Nonfullerene Fullerenes

Soon after scientists got used to the idea that nature, indeed, was packing carbon atoms into miniature soccer balls and geodesic domes, bizarre forms that Bethune calls "nonfullerene fullerenes" started popping up. In late 1991 Japanese researchers reported the discovery of what have been dubbed "buckytubes," hollow microscopic tubules that appear to be layers of graphite rolled into cylinders. The Japanese scientists, with the NEC Corporation, had been vaporizing carbon by attaching one graphite rod to the positive electrode of a power supply and a second rod to the negative electrode. The positive rod vaporized, liberating positively charged carbon ions that were pulled to the negative rod where they built up into a deposit. "If you break that off and put it in an electron microscope, what you see are tubular structures layered with different numbers of walls," said Bethune. "In some cases, you can see that the ends of some of them actually close to make closed-end tubes." The Japanese researchers speculate that the buckytubes are extremely strong and could be the basis for a new carbon fiber technology, for example, to reinforce composite materials.

Researchers in Switzerland made a surprise finding about buckytubes while studying them with an electron microscope running at a much higher power setting than normally used. The electron microscope works by bombarding a sample with electrons, and the high-intensity, high-voltage beam the researchers were using caused the buckytubes to transform into something else. "They would see these tubes begin to shrink," said Bethune, "and eventually anneal into something like 'buckyonions.'" These are spherical particles, structured like an onion, whose nested carbon atom layers are spaced apart like bulk graphite. Buckyonions have been seen with up to 10 million carbon atoms. The Swiss researchers suspect that buckyonions are more stable than buckytubes or even finite pieces of graphite. Their report appeared in *Nature*, and a picture of their "buckyonions" appeared on the cover with the title: "The Ultimate Fullerene?"

Perhaps the most unusual nonfullerene fullerene described by Bethune is a material called "schwarzite," computer modeled by a team from Cornell University. It consists of a network of graphite-like sheets that are both convex and concave, giving the material an undulating structure. "Locally, they look like two-dimensional sheets, but globally they're all linked together by various tubes and connections . . . leading to a porous 3D form of graphite," Bethune explained. The researchers calculated that each unit cell of crystalline schwarzite would consist of 80 six-carbon rings and 24 seven-carbon rings. Even more bizarre is random schwarzite, whose irregular form consists of hundreds of carbon rings containing not only five and six atoms each but seven, eight, and even nine atoms.

Schwarzite, speculate the researchers, could be a new form of strong yet lightweight materials, if it were real. Right now it exists in theory only, given form by computer modeling. No one has made schwarzite yet . . . or have they? Bethune wonders if Isaac Silvera, Nancy Chen, and Fred Moshary of Harvard, working in collaboration with the IBM team, may have made similar structures when they compressed tiny samples of C_{60} under the extraordinarily high pressures created by a diamond anvil cell. "When you squeeze it hard enough, the material undergoes a phase transition from a fairly dark material to a relatively transparent one," said Bethune. Before and after Raman spectra taken of the material also changed. "At first the lines in the spectrum looked like C_{60}, but the spectrum of the compressed material showed that it was *not* C_{60}, *not* diamond, *not* graphite, and apparently *not* ordinary amorphous carbon. So it's something different, but whether it's random schwarzite or not isn't yet known."

1001 USES FOR BUCKYBALLS—MAYBE

Even before researchers confirmed the existence of buckyballs and the other fullerenes in 1990, they were already speculating about what they could do with such neat carbon spheres. The last time chemists got this excited was over benzene, a hexagonally shaped ring of six carbon atoms, each attached to a hydrogen atom. Although benzene was isolated in 1825, it took 40 years to determine its structure. Once scientists learned how to manipulate the molecule, benzene became the chemical backbone for a phenomenal number of products, from drugs to plastics, that define modern industrialized society.

Fullerene scientists are hoping for at least as much, if not more, practical impact from their newly found carbon. "The physical stability and relative inertness of the fullerenes allow them to be assembled into very unusual solids," notes Bethune, "thus opening up a new branch of materials science." A particularly exciting aspect of fullerenes is that, since they are hollow spheres, chemistry can be done both on the molecule's exterior surface ("exohedral chemistry") and inside the sphere ("endohedral chemistry").

When talking about practical applications for the fullerenes, Bethune and other scientists enamored of the carbon balls are quick to caution that it is still a big unknown; there's lots of promise but nothing immediately practical. Bethune pulled out a piece of paper divided into two columns. The one on the left was marked *Forms of Carbon with Practical Uses or Significance*, and the one on the right was *Forms of Carbon with NO Practical Uses or Significance*. The left column was filled with such things as diamond, graphite, organic chemicals, and polymers. The right column was empty. "The question is," he asked, "where to put all the fullerenes. Obviously, statistics argue that they should go in the left column, based on the simple preponderance of evidence. But if they end up over in the right column with no practical use, it will still be interesting because they will be the only form of carbon there!"

From Insulator to Conductor to Superconductor

The electronics industry would be a big beneficiary of any practical applications to come out of fullerene studies. Scientists already know that, depending on its form, a buckyball solid can have different electrical properties. Unadulterated solid buckminsterfullerene is an insulator at low temperatures; an electric current can't flow through it like it does in a metal. Yet in 1991 researchers at Bell Labs used doping techniques to turn a buckyball solid into not only an electrical conductor but a superconductor as well. Superconductivity is one of the most sought after properties in the world of high technology. It means that a material offers no electrical resistance and lets a current flow through it unhindered. Superconducting materials have already been developed, but their big drawback to widespread commercial application, especially for such futuristic projects as magnetically levitated trains, is that they have to be cooled to temperatures not that much higher than absolute zero before they will work. The cooling requirements alone may be prohibitively expensive. Materials that become superconducting at higher and higher temperatures have long been sought.

What the Bell Labs scientists did was turn C_{60} into an artificial metal by slipping potassium atoms into the spaces between buckyballs that had been crystallized into a solid thin film (see Figure 9.6). In a report of their work in *Physics Today*, they described how the transformation from insulator to conductor took place right before their eyes: the pristine yellow color of the pure C_{60} thin film gradually turned a metallic gray with increased exposure to a potassium vapor. As the color changed, the thin film became more and more electrically conductive. The surprise came when the researchers chilled this artificial metal to about 18 degrees Kelvin (18 degrees above absolute zero) and discovered that it became superconducting. This transition temperature was surprisingly high by superconducting standards. An interesting aside is that in the mid-1960s scientists were able to make a superconductor by sandwiching potassium atoms between flat sheets of graphite, but its transition temperature was only a few tenths of a degree above absolute zero. Somehow, curling up graphite sheets into closed hollow spheres confers special superconducting properties on buckyballs when they are doped with an alkali metal.

A strange thing occurred, however, during the Bell Labs experiments. If the researchers exposed the carbon film too long to the potassium vapor, it lost its superconducting powers and reverted back to being an insulator. The researchers theorized that there is a limit to how

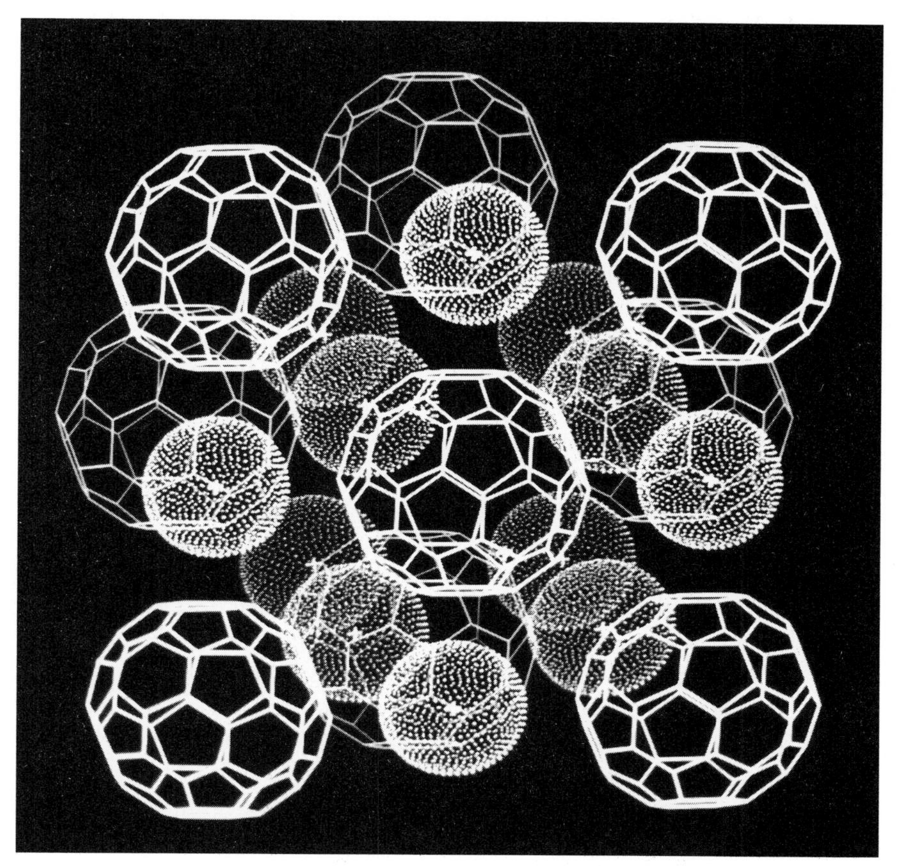

FIGURE 9.6 *Computer graphic showing how potassium atoms (smaller balls) fit in between larger buckyballs to make an artificial metal that becomes superconducting when chilled to 18 degrees Kelvin. (Courtesy of University of California, Los Angeles/State University of New York.)*

much potassium should be allowed to slip into the carbon network to make it superconducting, and they speculated that a three-to-one ratio of potassium to buckyballs was required. Whetten and his co-workers at UCLA began painstaking experiments to quantify the superconducting formula for this particular artificial metal and confirmed the Bell Labs group's suspicions that a three-to-one ratio is correct. Through this work the UCLA team also found a method to produce this artificial metal so that a large fraction of it ends up with the correct potassium-to-C_{60} ratio for superconductivity. The original material prepared by the Bell Labs team was only about 1 percent superconducting. Within months—attesting to the pace of fullerene advances—the UCLA team improved the material first to 60 percent superconducting and then all the way to 100 percent.

In the meantime, the Bell Labs and UCLA researchers were also trying to slip other atoms into the buckyball crystals to see the effect on the superconductivity. With rubidium the artificial metal became superconducting at 29 degrees Kelvin, described as "sizzling" by one of the scientists. Soon after, researchers got the transition temperature up to a stifling 33 degrees Kelvin by using a mixture of rubidium and cesium. There are other superconducting materials with much higher transition temperatures, for example, the so-called high T_c ceramic oxides, but they are planar materials that conduct electricity only in two dimensions. On the other hand, doped buckyballs (as of early 1993) are the high-temperature record holder for three-dimensional superconductors, according to the Bell Labs scientists, and are particularly good candidates for making such things as superconducting wires.

It is still too early to tell if fullerene superconductors will ever be practical. Already scientists have encountered a big obstacle: The material readily reacts with oxygen, and researchers have to be scrupulously careful to keep samples out of contact with air during experiments. "As the Bell Labs people are always careful to point out," said Bethune, "it's not clear whether a spontaneously combustible superconductor is going to be of much use!" Research is already under way at other laboratories to overcome this drawback as well as to improve the transition temperature. On a more positive note, Whetten's group at UCLA showed that once the doped fullerene is made into a superconductor it is stable; it won't loose its powers even after extended heating and mixing. This means it could be tough enough to withstand processing into useful components.

Prisoners in a Carbon Cage

To make superconducting fullerene materials, scientists packed "guest" atoms in between the C_{60} molecules. But what if they could somehow *stuff* "guest" atoms, especially metal atoms, into the carbon cages? What would such an exotic material do? Fascinated by the possibility, scientists had already put it on the top of their wishlist as soon as the hollow geodesic structure for the fullerenes was proposed. "Such species would be an interesting new class of molecules," said Bethune, "with tunable electronic, optical, vibrational, and chemical properties depending on what atoms you put inside."

What seems like a wild idea has turned out to be as easy to do as making plain vanilla fullerenes. Right after the U.S.-British team first detected the peak for C_{60} in the mass spectrum of carbon vapor and realized what they had, they tried an experiment to see if they could get a guest atom to take up residence inside some of the carbon cages by vaporizing a graphite disc impregnated with the metal lanthanum. Sure enough the mass spectrum of the carbon vapor had a strong peak where a substance containing both C_{60} and lanthanum would be. "They took that as evidence that they had produced a metal-containing fullerene," said Bethune. Such carbon-wrapped metals are called metallofullerenes, and since those initial experiments, researchers have succeeded in stuffing one or more different metal atoms into fullerene cages. To date, lanthanum, yttrium, scandium, calcium, uranium, hafnium, titanium, and zirconium have been successfully incorporated, and many other guest atoms are being tried. Researchers have proposed a kind of shorthand to describe how many atoms are trapped in which kind of fullerene molecule. For example, one lanthanum atom inside a buckyball would be designated "$La@C_{60}$," while "$Sc_3@C_{82}$" would mean three scandium atoms inside C_{82}, a metallofullerene recently made by the IBM group and by Shinohara and his colleagues in Japan.

Initial experiments on stuffing atoms into carbon cages produced a variety of metallofullerenes but in such minute quantities that the only indication of their presence were signals in mass spectra taken of the samples. That changed in early 1992, when Smalley's group at Rice reported a breakthrough that made it possible to produce macroscopic quantities of metallofullerenes. The IBM team then went on to make a milligram or so of C_{82} carbon cages containing single lanthanum atoms, enough to allow them to do spectroscopic studies and determine its electronic structure. They made this metallofullerene by pressing powdered graphite and lanthanum oxide into a rod, baking it at about 2500° F for 2 hours, and then vaporizing it. The researchers collected the soot and extracted the fullerenes and metallofullerenes from it with solvent. After evaporating the liquid, they were left with a black powder that contained about 2 percent $La@C_{82}$, the rest being empty C_{60} and C_{70} cages. They ran the sample through a series of tests that revealed this metallofullerene to be a truly exotic molecule: there was evidence that the lanthanum atom had transferred its three valence electrons to the fullerene cage. Electrically speaking, said Bethune, this metal-containing carbon cage looked like the C_{60} cages in the artificial metal that had been created by packing alkali atoms between crystallized buckyballs in a three-to-one ratio: both have three extra electrons. "So this might give a different route to making a fullerene superconductor." Calculations by researchers at IBM-Zurich suggest that the trapped atom isn't suspended in the center of the cage but is off to the side, attached fairly strongly to the wall of its carbon prison. But there have been other proposals, said Bethune, and the question of exact structure isn't settled yet.

Next, the IBM group tried making a metallofullerene with the metal scandium and got a surprise. Since this atom is smaller than lanthanum, more of them slipped into the carbon molecules—up to three. The case of $Sc_3@C_{82}$ may turn out to be particularly interesting. While some researchers think that the three scandium atoms are in the form of individual ions stuck to the inside of the carbon cage, Bethune speculates that they may have bonded into a molecule. If so, this fullerene confinement system may make it possible to spectroscopically examine molecular ions of this rare-earth metal at room temperature, instead of under the deep freeze conditions usually required because of scandium's high reactivity.

"Beyond characterizing these exotic molecules, we would really like to build materials with them," said Bethune. "This would require production in greater quantities and purification techniques. Purification has proved to be a difficult challenge. Nevertheless, the possibilities that these novel species offer, both for scientific study and for practical applications, are very exciting and still largely unexplored." For example, isolating clusters or molecules of different atoms inside fullerene cages or even tubes may make it possible to fabricate exotic new structures at the nanometer scale, said Bethune. A nanometer is one-billionth of a meter, a size so small that "nanotechnologists" work with building blocks that are only a few atoms in size to make the tiniest components physically possible. Shrinking components down to nano sizes, in theory, could lead to the development of speedier and more efficient electronic and optical devices, especially for computing and communications. The impact of nano-size transistors on computers, for example, could rival the impact that the first tiny transistor devices had when they replaced bulky vacuum tubes.

"The ability to manipulate matter on an atomic scale and create unique materials and devices with custom-designed properties has universal appeal. It marks a triumph of human ingenuity and imagination over the natural rules by which materials are formed," commented Yale solid-state physicist Mark A. Reed in an article about nanotechnology in *Scientific American*. The big problem, though, with such ultra-small structures is that any "defects" in them would be disastrous. Since fullerenes are unblemished spheres of carbon atoms, noted UCLA's Whetten, they could be perfect nanocrystal building blocks.

A Fullerene Ready for Market

One fullerene technology may soon make it out of the laboratory and into the marketplace, according to Bethune. It's a C_{70}-based treatment that has proved itself to be practical for enhancing the growth of diamond films. Because of their superior properties, including smoothness, hardness, high-temperature resistance, and chemical stability, diamond films are used to coat such things as cutting tools and to make electronic devices such as semiconductors as well as optical thin films for lenses and infrared windows. Worldwide demand for diamond films and coatings is projected to reach $260 million in 1995 and nearly $2 billion by the year 2000, according to Research Corporation Technologies. This Tucson-based technology transfer company has filed international patents on the C_{70} treatment method, developed at Northwestern University, and is leading efforts to commercialize it.

The fullerene is used to pretreat a surface so that the fine-grain polycrystalline diamond film can take hold and grow. Conventional pretreatments, such as diamond grit abrasion, are time consuming and expensive and not particularly suitable for anything other than relatively small, flat surfaces. The fullerene pretreatment overcomes this. An ultra-thin layer of C_{70} is first deposited on a surface and then bombarded with hydrogen and carbon ions. This opens up the fullerene cages, exposing free ends to which the diamond film can attach and grow.

Other Promising Uses and a Real Long Shot

Making nonlinear optical materials is another area in which fullerenes have a good potential for practical application, said Bethune. It has been shown that fullerenes, especially C_{60}, have the ability to react to light by changing their behavior, a very important property to the optics industry. A common example of a product made with nonlinear optical material are eyeglasses that automatically darken with increased exposure to sunlight. The metallofullerenes that the IBM group is working on may also have promising nonlinear optical properties.

Scientists have manipulated the fullerenes in a variety of ways and have come up with materials ranging from buckyball polymers to buckyball magnets. But there remains one buckyball property—incredible strength—that is aching for an application, and UCLA's Whetten has a long-shot suggestion: rocket propellant.

It turns out that the hexagon/pentagon geometry that makes geodesic domes such strong building structures also endows buckyballs with superior strength. The seams where the hexagonal and pentagonal faces of the molecule meet are highly resistant to compression, according to Whetten, who tested this by getting these carbon cages moving along at about 15,000 miles per hour and then crashing them into a stainless steel wall. The molecules simply rebounded, like bullets off Superman's chest, without so much as a dent. "They are resilient beyond any particle known," said Whetten. So what can you do with these superballs? It may seem like science fiction now, but Whetten fancies that they could be shot out of a spacecraft engine as some kind of exotic propellant!

BUCKYBALL POSTSCRIPT: HOW IT ALL STARTED

The entire incredible story of the discovery of fullerenes started as an effort to understand carbon chemistry in outer space. It is, says buckyball's British codiscoverer Harold W. Kroto of the University of Sussex, "a prime example of the way in which an interest in fundamental problems for their own intrinsic sake, irrespective of their predicted use, can yield results of applied significance."

Before getting caught up in fullerene fame, Kroto had been conducting a long-term research project to understand the origins of linear carbon chains detected in space. Signals from these molecules, which are mixed in with the gaseous matter between the stars, had been picked up by ground-based radiotelescopes and identified by comparison with similar signals obtained from laboratory-made molecules. This work was part of a long and fruitful collaboration between astronomy and chemistry that has turned up a rich menu of a variety of cosmic molecules, from ammonia to formaldehyde, by comparing laboratory data with astronomical observations. Kroto, in fact, had participated in the discovery of several different carbon chains, including the detection of one of the most complex molecules ever found in space, $HC_{11}N$.

At first, scientists suspected that the carbon chains were being formed by reactions between molecules and ions in interstellar gas. Kroto was interested in another possibility, that the carbon chains were forming in the outer atmosphere of certain carbon-rich stars, called red giants, that are constantly throwing out large amounts of dust. "In such stars the possibility that some symbiotic chain/dust chemistry [may be taking place], perhaps related to soot formation, seemed worth considering . . . ," he wrote in a summary of his work that appeared in *Science*.

To better understand the process, however, Kroto needed to carry out laboratory simulations of the chemistry going on in these carbon stars. Rice University chemist Richard Smalley had just the device: a machine that could vaporize graphite and then analyze the carbon clusters that formed. Kroto and Smalley got together, and the rest is history.

During their experiments, the U.S.-British team found that vaporized graphite not only formed fullerenes but also precursors of linear carbon chains. The two go hand in hand, according to Kroto, and the possibility that space is also full of fullerenes is being actively pursued. He speculated that the spontaneous formation of fullerenes out of a chaotic carbon vapor may be related to the mechanism that formed the first particles in space. "Indeed, that vital link in planet formation, the primordial solid particle, may well have been carbonaceous . . . ." In addition to linear carbon chains and related molecules, said Kroto, "a new character, C_{60}, has emerged, whose shadowy role, like that of the Third Man, has only now come to light."

In their now-famous *Nature* report announcing the discovery of buckminsterfullerene, Kroto and his colleagues wrote: "Because of its stability when formed under the most violent conditions, it [C_{60}] may be widely distributed in the Universe. For example, it may be a major constituent of circumstellar shells with high carbon content. It is a feasible constituent of interstellar dust and a possible major site for surface-catalysed chemical processes which lead to the formation of interstellar molecules. Even more speculatively, C_{60} or a derivative might be the carrier of the diffuse interstellar lines." Those lines have befuddled scientists for decades. They are signals from molecular material thinly spread out among the stars that scientists have been unable to identify. Kroto suggests that electrically charged buckyballs (ionized by starlight) may be a good candidate for at least some of the mystery molecules and adds that metallofullerenes—metal atoms trapped in the carbon cages—may also be worth looking for in space because of their ease of formation and resiliency.

In their report confirming the existence of buckyballs, the U.S.-German research team notes that both the visible and infrared absorption spectra they obtained from their C_{60} samples do not match up with any of the interstellar signals. They pointed out, however, that the data are from solid C_{60}, not the ionized buckyballs suggested by Kroto. "Nevertheless, these data should now provide guidance for possible infrared detection of the C_{60} molecule, if it is indeed as ubiquitous in the cosmos as some have supposed," said the researchers.

While no one has reported finding cosmic buckyballs or confirmed their possible role in planet formation, the rotund carbon has turned up in somewhat related places: 600-million-year-old earth rock and a tiny meteorite. A Russian geochemist reported finding both C_{60} and C_{70} in a rock called shungite, found only in a remote part of Russia near Finland. Shungite is a hard black rock made up of mostly coal-like carbon. How the fullerenes got inside is a mystery. Lastly, scientists found fullerenes while examining a dent in the formerly orbiting Long-Duration Exposure Facility (LDEF), which was returned to earth by the Space Shuttle. Apparently, a tiny carbon-rich meteorite had hit the satellite and left behind a fragment of itself in the resulting dent. The scientists were able to demonstrate that the fullerenes were not formed by the collision with LDEF but were truly cosmic—left behind by the meteorite.

BIBLIOGRAPHY

Baum, R. M. 1991. Systematic chemistry of C_{60} beginning to emerge. Chemical & Engineering News 69(50):17–20.

Bethune, D. S., G. Meijer, W. C. Tang, and H. J. Rosen. 1990. The vibrational Raman spectra of purified solid films of C_{60} and C_{70}. Chemistry and Physics Letters 174:219–222.

Dagani, R. 1992. Nanostructured materials promise to advance range of technologies. Chemical & Engineering News 70(47):18–24.

Hawkins, J. M., A. Meyer, T. A. Lewis, S. Loren, and F. J. Hollander. 1991. Crystal structure of osmylated C_{60}: Confirmation of the soccer ball framework. Science 252(April 12):312–313.

Hebard, A. F. 1992. Superconductivity in doped fullerenes. Physics Today 45(11):26–32.

Holczer, K., O. Klein, S.-M. Huang, R. B. Kaner, K.-J. Fu, R. L. Whetten, and F. Diederich. 1991. Alkali-fulleride superconductors: synthesis, composition, and diamagnetic shielding. Science 252(May 24):1154–1157.

Johnson, R. D., G. Meijer, and D. S. Bethune. 1990. C_{60} has icosahredral symmetry. Journal of the American Chemical Society 112:8983–8984.

Johnson, R. D., G. Meijer, J. R. Salem, and D. S. Bethune. 1991. 2D nuclear magnetic resonance study of the structure of fullerence C_{70}. Journal of the American Chemical Society 113:3619.

Johnson, R. D., M. S. de Vries, C. S. Yannoni, D. S. Bethune, and J. R. Salem. 1992. Electron paramagnetic resonance studies of lanthanum-containing C_{82}. Nature 355:239–240.

Johnson, R. D., C. S. Yannoni, H. C. Dorn, J. R. Salem, and D. S. Bethune. 1992. C_{60} rotation in the solid state: Dynamics of a faceted spherical top. Science 255:1235–1238.

Jones, D. E. H. 1966. Ariadne. New Scientist 245(Nov. 3):245.

Krätschmer, W., L. D. Lamb, K. Fostiropoulos, and D. R. Huffman. 1990. Solid C_{60}: A new form of carbon. Nature 347(Sept. 27):354–358.

Kroto, H. W., J. R. Heath, S. C. O'Brien, R. F. Curl, and R. E. Smalley. 1985. C_{60}: Buckminsterfullerene. Nature 318(6042):162–163.

Meijer, G., and D. S. Bethune. 1990. Laser deposition of carbon clusters on surfaces: a new approach to the study of fullerenes. Journal of Chemical Physics 93:7800–7802.

Meijer, G., and D. S. Bethune. 1990. Mass spectroscopic confirmation of the presence of C_{60} in laboratory-produced carbon dust. Chemistry and Physics Letters 175:1–2.

Moshary, F., N. H. Chen, I. F. Silvera, C. A. Brown, H. C. Dorn, M. S. de Vries, and D. S. Bethune. 1992. Gap reduction and the collapse of solid C_{60} to a new phase of carbon under pressure. Physical Review Letters 69:466–469.
Pennisi, E. 1992. Scaling chemistry's peaks. Science News 141(16):250–251.
Pennisi, E. 1992. Buckyballs combine to make giant fullerenes. Science News 142(10):149.
Reed, M. A. 1993. Quantum dots. Scientific American 268(1):118–123.
Stucky, G. D., and J. E. MacDougall. 1990. Quantum confinement and host/guest chemistry: Probing a new dimension. Science 247(Feb. 9):669–678.
Teresko, J. 1991. Buckminsterfullerenes. Industry Week 240(Nov. 4):38–44.
Wilson, M. A., L. S. K. Pang, G. D. Willett, K. J. Fisher, and I. G. Dance. 1992. Fullerenes—preparation, properties, and carbon chemistry. Carbon 30(4):675–689.
Wilson, R. J. , G. Meijer, D. S. Bethune, R. D. Johnson, D. D. Chambliss, M. S. de Vries, H. E. Hunziker, and H. R. Wendt. 1990. Imaging C_{60} clusters on a surface using a scanning tunneling microscope. Nature 348:621–622.
Yannoni, C. S. , P. P. Bernier, D. S. Bethune, G. Meijer, and J. R. Salem. 1991. An NMR determination of the bond lengths in C_{60}. Journal of the American Chemical Society 113:3190.
Yannoni, C. S., R. D. Johnson, G. Meijer, D. S. Bethune, and J. R. Salem. 1991. Carbon-13 NMR study of C_{60} in the solid state: Molecular motion and carbon chemical shift anisotropy. Journal of Physical Chemistry 95:9–10.
Yannoni, C. S., M. Hoinkis, M. S. de Vries, D. S. Bethune, J. R. Salem, M. M. Crowder, and R. D. Johnson. 1992. Scandium clusters in fullerene cages. Science 256:1191–1192.

RECOMMENDED READING

Amato, I. 1991. Doing chemistry in the round. Science 254(Oct. 4):30–31.
Crabb, C. 1993. More fun with buckyballs. Discover 14(Jan.):72–73.
Curl, R. F. and R. E. Smalley. 1991. Fullerenes. Scientific American 265(Oct.): 54–63.
Kroto, H. 1988. Space, stars, C_{60}, and soot. Science 242(Nov. 25):1139–1145.
Pennisi, E. 1991. Buckyballs still charm. Science News 140(Aug. 24): 120–123.
Taubes, G. 1990. Great balls of carbon. Discover 11(Sept.):52–59.

FOLD, SPINDLE, AND REGULATE

How Proteins Work

by David Holzman

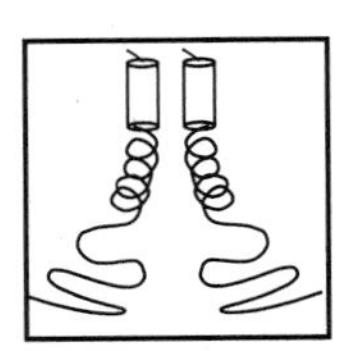

Protein molecules are the building blocks of living systems. Cartilage, skin, hair, nails, and eyeballs are formed from the same basic substance. So are the shell of a lobster, the wings of a beetle, feathers, the horn of a rhinoceros, and most everything else that makes up the bodies of creatures. In addition, the molecular machines that perform, or catalyze the building, tearing down, and maintenance work that keeps all organisms functioning are virtually all made of protein. There are literally hundreds of thousands of molecular tools and machines in a human body.

Not surprisingly, then, an understanding of how proteins work could lead to a host of useful applications. The symptoms of genetic disease are caused by defects in proteins. For example, sickle cell anemia results from a defect in the blood-transporting hemoglobin molecule. "If we could manipulate proteins, we could fix genetically defective proteins," says Tom Alber, an enthusiastic young molecular biologist who was then at the University of Utah and who presented his research at the 1992 "Frontiers of Science" symposium.

But this, said Alber, now at the University of California, Berkeley, is only the beginning of the benefits that could accrue. "We could design proteins that are useful to people but don't necessarily occur in nature." For example, pharmaceuticals that require no refrigeration could bring the benefits of modern medicine to the world's isolated and impoverished populations. Novel catalysts could be designed to mediate chemical reactions that are not found in biological systems, reactions that might be useful in the food or pharmaceutical industries.

THE IMPORTANCE OF THREE-DIMENSIONAL STRUCTURE

Although evolution has been building with proteins for 600 million years, scientists have only begun to understand the rudiments of how these substances work, and years or even decades may elapse before engineers will be able to systematically design proteins for specific tasks.

Three-dimensional structure determines function, Alber explained. Enzymes—the catalysts of living substances—fit their substrates as precisely as keys fit locks. For example, it is shape that allows the different enzymes that reside in the gastrointestinal tract to catalyze the disassembly of food molecules into the building blocks that the body uses for nourishment. But there are different enzymes to break down different types of protein and still others to dismantle starch into its constituent sugar molecules.

And it is shape that prevents the starch-digesting enzymes from allowing us to eat grass or bark. Cellulose, the woody material that prevents humans from grazing with the cattle, and from dining with termites on old, dead trees, is composed of chains of sugar molecules, just like starch. The difference between a two-by-four and potatoes or pasta is largely the result of slight differences in the way the sugar molecules are strung together. These slight differences render cellulose indigestible by starch-cleaving enzymes.

Like enzymes, cells of the immune system that adhere to invading microbes also fit their targets with the specificity of lock and key. Recognition is so precise that there are different antibodies for each disease.

"The problem [is] where does the structure come from?" says Alber. "Our work is aimed at understanding the architectural principles that determine the shapes of proteins. We're starting to understand some of the principles that govern shape."

ANATOMY OF PROTEINS

To illustrate the magnitude of the challenge, Alber projected a diagram of the cellular catalyst ATCase, a large protein consisting of roughly 30,000 atoms. Like many proteins, its convoluted topology is roughly as complex as a tangle of telephone wire, yet it has the structural grace of a suspension bridge or a bicycle wheel.

The structure of proteins is maintained by chemical bonds that form within proteins. Sometimes these are covalent bonds, the powerful forces that bind atoms together to form molecules, but more often they are the much weaker forces such as electrostatic charges that cause some atoms or molecules to associate with one another in solution. Then there are the water-hating side chains that also attract one another and—surprise—repel water. These are also referred to by scientists as "hydrophobic" or just plain "greasy." The difference between covalent bonds and the other forces is analogous to the difference between the strength of the ties of family and friendship.

Both the covalent bonds and the various weaker forces are properties of the building blocks of proteins, small molecules called amino acids. Proteins are chains of amino acids, which are strung together sort of like poppet beads. But instead of uniform inert beads, there are 20 different kinds of amino acids that occur naturally as well as some synthetic ones. The various forces within proteins are properties of the side chains of amino acids.

It is possible to melt the structure of ATCase by adding urea to the solution containing the protein. Melting overcomes the forces between the chemical groups, destroying ATCase's structure, so that the protein chain now flops around like boiling spaghetti. But remove the urea and the chemical groups pull the protein back into its proper shape.

THE SEARCH FOR STRUCTURE: X-RAY CRYSTALLOGRAPHY AND NUCLEAR MAGNETIC RESONANCE

Christian B. Anfinsen was the first to try this experiment, on the enzyme ribonuclease, in the late 1950s. That the protein regained its shape and its natural enzymatic activity in minutes suggested that all the information that directs folding is contained in the amino acid sequence and not in templates, such as the DNA template that directs the replication of DNA and the transcription of RNA from DNA. For this work, Anfinsen received the Nobel prize.

Theoretically, then, it should be possible to predict a protein's structure from its amino acid sequence. In fact, to researchers who study protein folding, this is the holy grail, and hundreds are currently attempting to discover the rules that would make this possible. But so far, such efforts have proven frustrating, to put it mildly. This is not surprising.

For a protein composed of just 100 amino acids, a rather small protein, there may be as many as a google (10^{100}) alternative structures that the protein could form, more structures than the number of atoms in the universe, Alber explained. It is amazing, he said, that the protein itself finds the right confirmation, "spontaneously, in a few milliseconds," even though a random search through all possible structures "would take longer than the age of the universe."

To understand how a protein folds, it is necessary to know that protein's three-dimensional structure. Until recently, piecing structure together was such a difficult process—impossible in some cases, taking years in others—that this goal was a dream. Now new developments in protein imaging technologies, x-ray crystallography, and nuclear magnetic resonance (NMR) have enabled researchers to piece structures together within a year, where formerly it could take longer than a decade. Recombinant DNA techniques have made it possible to perform experiments that are beginning to yield some rules of protein folding. And with this information coming in, the dream of predicting three-dimensional structure from amino acid sequence is becoming a tantalizing possibility on the distant horizon of the field.

X-ray crystallography is an ancient technology by scientific standards, dating back to the first half of the century. Rosalind Franklin's crystallographic work was vital to cracking the structure of DNA, and many observers believe that for this contribution she should have shared the Nobel prize with James Watson and Francis Crick. This alleged slight may have had to do with prevailing attitudes about gender.

Crystallography is often the only way to figure out the molecular structure of a protein. The reason is simple. At best, crystallography provides a picture of the entire protein. It accomplishes this by revealing the location of electron densities throughout the protein. These correspond to chemical groups on amino acids.

Nonetheless, the technique frequently does not succeed so marvelously. There are built-in errors of about 5 percent. Furthermore, usually parts of the protein crystallize poorly or not at all, so that the corresponding electron densities leave no signature in the diffraction pattern.

Because of these limitations, structures derived from x-ray diffraction patterns are prone to error, and results are considered uncertain unless researchers know the sequence, which they can use to verify a structure.

But recent advances have made the technique much easier to use and more successful. New technologies have reduced the time it takes to collect a set of x-ray diffraction data from as much as a month to less than a day; they have greatly increased accuracy as well. No longer must machines distinguish approximately among shades of diffraction spots on photographic paper. Electronic film can literally count photons. Whereas photographic film provided a dynamic range of accuracy of 3 to 5, the modern devices are accurate over a dynamic range of 100,000.

Compared to x-ray crystallography, NMR spectroscopy has a few advantages and a variety of disadvantages. The big advantage is that to use NMR it is unnecessary to crystallize protein, which is an arduous and often impossible undertaking. Moreover, portions of proteins that do not crystallize are often important for activity, and, not only can NMR detect these, but it can follow their movements. On the down side, the largest proteins that NMR can decipher are an order of magnitude smaller than for crystallography.

Like x-ray crystallography, NMR has undergone dramatic advances in the last 5 years or so. These advances have made the technique easier to use and more powerful. The size of the proteins that NMR can study has increased from 80 amino acids to 150 to 200 residues.

NMR detects atoms by magnetizing them. Of atoms that occur naturally in amino acids, only hydrogens can be magnetized. It is possible to set up NMR so that it detects atoms in pairs, because one atom can magnetize another provided the two are within several chemical bond lengths of each other. When one hydrogen is a member of more than one pair, researchers can begin to get spatial information.

Though hydrogens are the most abundant atoms in protein, they are not structural elements in its backbone. If a spinal column represents this backbone, and each vertebra represents an amino acid, hydrogens are analogous to appendages to the vertebrae that do not contribute to the connections between them. For reasons related to this, it is difficult for researchers using hydrogen-based NMR to "walk"—in the parlance of that technique—down the backbone from one amino acid to the next. Furthermore, the differences in the NMR signals of hydrogen are so small that the number of individual hydrogens that can be distinguished in a single protein is small.

However, in the mid to late 1980s, researchers began inserting isotopes of carbon and nitrogen into proteins. These isotopes made NMR more powerful because they are structural atoms in protein, which made it easy to walk the length of the chain. In addition, since the isotopes transferred their magnetic spin to other atoms more efficiently, it became possible to magnetize triplets, even quadruplets of atoms, resulting in obtaining bigger chunks of information at once. A final advantage is that the spectra of the isotopes are about 20 times wider than the spectrum for hydrogen.

THE STRUCTURE OF PROTEIN IS RESILIENT

Today, researchers are using both techniques for structural analysis of several dozen proteins. T4 lysozyme, a 164 amino acid protein that viruses use to break open bacterial cell walls, is one of the most thoroughly studied. It was the subject of Alber's postdoctoral research under Brian Matthews of the University of Oregon. (Lysozyme is a powerful antibacterial and is present in tears to protect the eyes.)

Matthews first turned his attention to lysozyme in the early 1970s. His colleague, the late George Streisinger, had created about 100 versions of the protein, using a traditional genetic technique that introduces random mutations. "We thought it would be a wonderful opportunity to take advantage of that genetic information," says Matthews, to find out how the mutations affected three-dimensional structure and function.

There was an intrinsic advantage to the use of lysozyme for the study of structure. An important characteristic of proteins is their stability. Heat a protein and eventually the structure falls apart as the weaker noncovalent bonds are breached. The measure of stability is the temperature at which the structure melts. Some proteins are hard to work with because they coagulate in the melted state, like an egg white, losing the ability to reform their proper structure—but not T4 lysozyme.

From 1974 to 1979, Matthews spent most of his time identifying interesting mutants, purifying the proteins, and analyzing their three-dimensional structure with the help of his colleague Rick Dahlquist, an expert on NMR. At that time, "There was a perception that protein structures were very delicately poised between folding and unfolding . . . and that a single mutation might tip the balance," says Matthews. But he doubted this. He had noted that, while the hemoglobins of horses and humans are structurally similar, they differ in amino acid sequence by nearly 50 percent. Still, he says, "what you don't know in looking at these naturally occurring variants is whether the 50 percent that are different are different for functional reasons or just because they don't matter."

But his late colleague's mutant lysozymes reinforced his doubts. Although a few of the mutations destroyed the protein's structure, most did virtually nothing to change it. Nonetheless, Matthews knew that it would have been dangerous to generalize from his results since Streisinger's techniques had created mutations in but a small fraction of lysozyme's 164 amino acid positions. "For large parts of the protein there was no information at all."

The advent of recombinant DNA made it possible to extend this research to the rest of the protein. One could now substitute any amino acid at any position and crank out copious copies of the mutant. By the mid-1980s, lysozyme had been cloned, and Matthews' studies would soon take off on a wave of technical progress. When Alber came to Matthews's laboratory in 1982, Matthews had studied five mutants. By the time he left in 1987, 50 had been characterized. The total has now reached 500. And as the experimental record grew, it continued to support the hypothesis that most single mutations do not change the structure of proteins.

In one of his first experiments to take advantage of recombinant DNA, in 1984, Alber took one of Streisinger's mutants, which had an amino acid substitution at position 157, and created a series of additional mutants by inserting virtually every other amino acid into that position, in turn. He performed the first x-ray crystallography that had ever been done on any of these variants, and the surprising result was that all were very similar to one another. The stability of each mutant was then measured by heating it until it melted. The mutants were all quite stable.

To further test these ideas, Matthews's postdoctoral student Xue-Jun Zhang systematically substituted the amino acid alanine at each position in the protein, one by one. Alanine is one of the simplest of all the amino acids and lacks the reactive chemical groups that interact with those on other amino acids. In subsequent experiments, Zhang substituted 10 alanines at 10 consecutive positions at once, repeating the process all over the protein. The structure of lysozyme was impervious to these changes at all but a few sites.

The experiments with alanine, and extensive substitutions with other amino acids, such as those at position 157, led to two general conclusions. First, "Probably less than 50 percent of the amino acids in the protein are critical to structure," says Matthews. And different proteins can have the same basic three-dimensional structures despite large differences in amino acid sequence.

To Alber, this makes perfect sense. "Just as a house is held together by nails, glue, screws, beams, and other binders, many forces give a protein its shape: van der Waals forces, hydrogen bonds, disulfide bonds, and electrostatic attractions and repulsions." But as few single fasteners or frame elements are critical to the integrity of a house, so most amino acids can be changed without destroying a protein.

The second result was that the most destructive substitutions almost always occurred in the core of the protein, while those on the surface were predominantly benign. The logic of this is that amino acids in the core, packed together as tightly as they are, are well placed to interact with each other, while those on the surface are not. (Surface amino acids might well interact with other substances in the surrounding solution or the solution itself.) "We now believe that for folding it is the core that is critical," says Matthews.

DESIGNING SUPERSTABLE PROTEINS

If that were true, Matthews speculated at the time of these earlier experiments, it should be possible to increase a protein's stability by creating additional bonds or stronger bonds in the core. Since the strongest structural bridges in proteins are the covalent disulfide bridges, Matthews and his postdoctoral researcher, Masazumi Matsumura, created disulfide bonds in the core of lysozyme. This meant substituting cysteine, the sulfur-containing amino acid, in place of other amino acids on either side of a fold in the protein.

Matsumura, now at Scripps Research Institute, built the bridges. Not all of them worked. "In many cases, the introduction of these bridges actually makes the protein less stable," says Matthews. The reason for this is that the geometry of these bridges needs to be very precise, more so than generally is possible. Otherwise, "you distort the protein, and that makes it less stable."

Nonetheless, the scientists discovered one approach to covalent bridge building that worked well. Certain parts of the protein are naturally flexible. Here, precision was unnecessary. The flexing prevented the bridges from distorting the rest of the protein; yet a disulfide bridge could increase stability typically by 5 to 10°C. Three bridges in one protein raised its melting temperature by 25 degrees.

Matthews' group also discovered that noncovalent bonds could more easily be used to increase stability since the force of their attraction was not great enough to cause the destabilizing distortion.

Tying pieces of protein together with opposite charges could also increase stability, the researchers discovered, performing the experiment in an alpha helix, which is a common structural element of many proteins, including lysozyme. (Screws, slinkies, and stairways that wind around to save space are all examples of helices, as is DNA.)

One end of an alpha helix is positively charged, while the other end is negatively charged. By placing an opposite charge one turn of the helix from one of the end charges, the researchers found that they could raise the melting temperature of lysozyme by about 2° C. Several similar substitutions added further stability.

THE LEUCINE ZIPPER AND GENE REGULATION

From Oregon, Alber went to the University of Utah, where he was soon caught up in studies of the leucine zipper, collaborating with Peter Kim of the Whitehead Institute and several other researchers.

Leucine zippers are alpha helices that direct gene regulation. Gene regulation is the process that determines when the body will produce each protein. For example, eat a donut and the gastrointestinal tract must crank out the enzymes that digest starch. Then, as sugar enters the bloodstream, the pancreas must begin to manufacture insulin. Leucine zippers manage the regulatory process.

The work on leucine zippers ultimately would support Matthews's findings that, although most amino acids exert little influence on shape, that of core amino acids can be profound. Alber got caught up in the leucine zipper work as Kim et al. were piecing together the structure.

But it was Steven McKnight, of the Carnegie Institution of Washing-

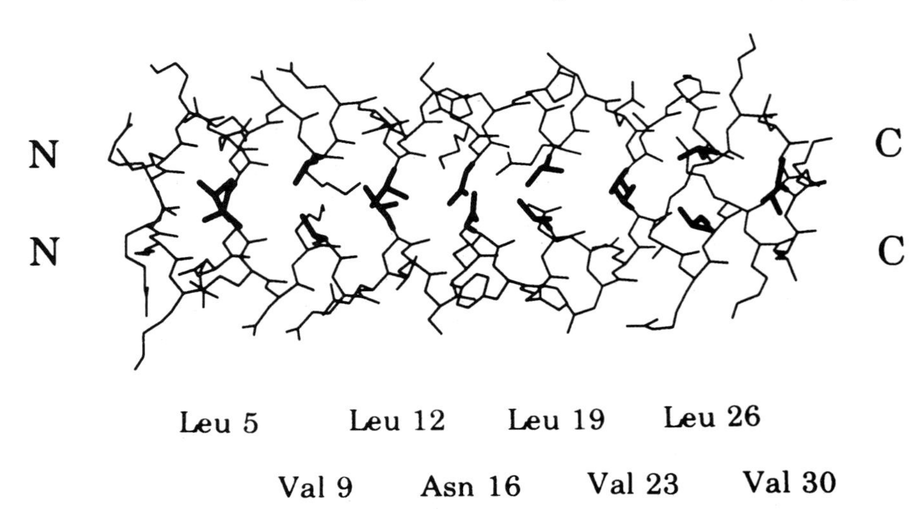

FIGURE 10.1 *Side chains (thick lines) make contacts between the helices in a schematic drawing of a leucine zipper.*

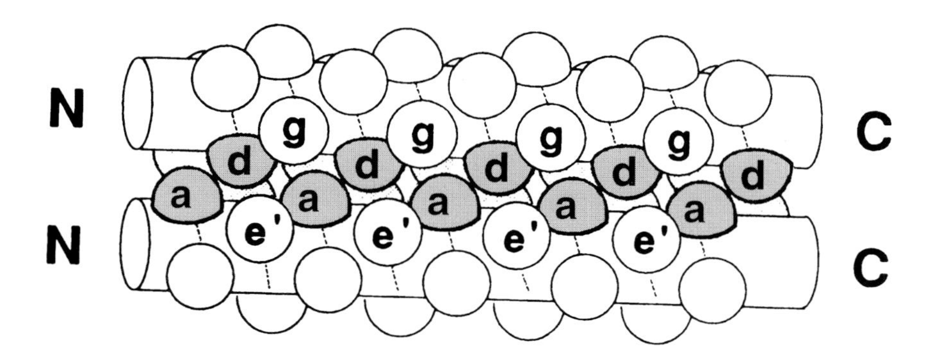

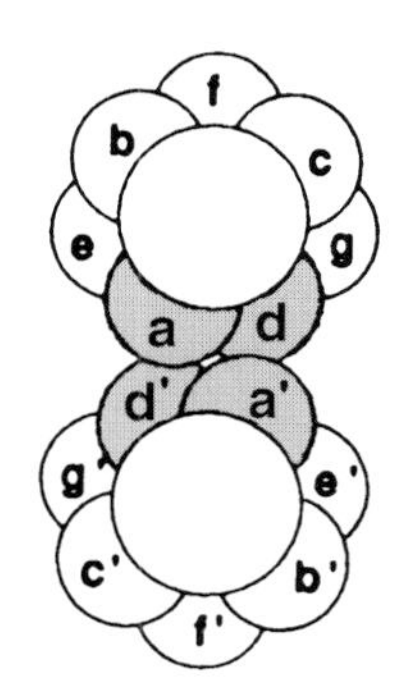

FIGURE 10.2 *An amino acid sequence in helices of a leucine zipper contains a 7-residue repeating pattern. Scientists label the positions a through g. The amino acids in the a and d positions influence whether helices come together in pairs, triplets, or quadruplets. (See Figures 10.6 and 10.7.)*

ton in Baltimore, who actually discovered the leucine zipper, in 1987, around the time that Alber departed for Utah. McKnight recounted his discovery in the April 1991 issue of *Scientific American*.

Leucine zippers are alpha helices, which are named for the fact that the amino acid leucine appears in the helix at intervals of seven amino acids, or slightly less than two turns of the screw. Moreover, the leucines, and another, similarly hydrophobic amino acid, which also appears at intervals of seven, form a zipper-like structure that binds two alpha helices together (see Figures 10.1 and 10.2). The pairs are called dimers. (The nomenclature here is counterintuitive. A leucine zipper is only one alpha helix, comprising only half of the zipper-like structure that binds two helices together into the dimer. To make things more confusing, scientists usually refer to both single and paired leucine helices as leucine zippers. However, in this account, leucine zipper refers only to the unpaired alpha helix; dimer or leucine zipper dimer refers to the pairs.)

More than 80 leucine zippers have been discovered during the past 4 years, and about 100 are known. Most are very similar to one another. Each leucine zipper is attached to a regulatory protein. Pairing brings the regulatory proteins together, which is necessary in order for them to turn genes on or off. Free ends of each helix then wrap part way around the DNA, fitting nicely along the groove of the DNA helix, thereby placing the regulatory proteins in the proper position to control expression of the gene(s) (see Figure 10.3).

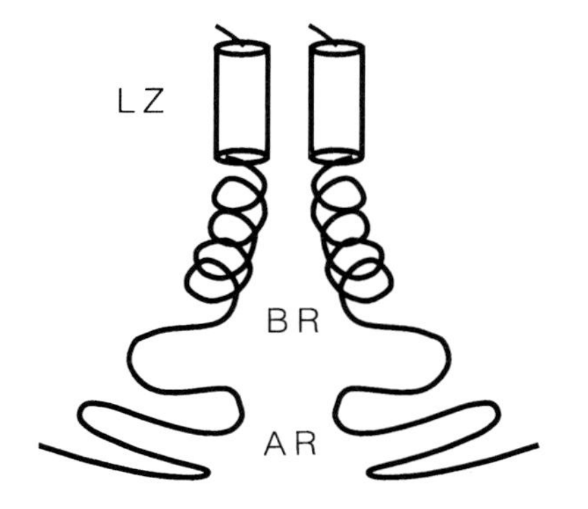

FIGURE 10.3 *Schematic drawing of a leucine zipper protein. The leucine zipper (LZ) is the pair of cylinders. The amino acid tails include the basic region (BR), which binds DNA, and the activation region (AR), which turns genes on.*

It is the structure of the helices that determines which regulatory proteins come together. Slight differences in the amino acid sequence of the helices result in different pairings.

Different combinations of regulatory proteins regulate different genes in different ways. Some leucine zipper dimers turn genes on; others turn them off. For example, when the leucine zippers bind the regulatory protein jun to itself, the resulting complexes turn on many different genes. But the regulatory protein fos sunders the jun/jun pairs, and then binds with jun, turning off those same genes.

Some leucine zipper dimers control just a few genes, while others control huge batteries of genes. The GCN4 leucine dimer, which Kim et al. studied extensively, "is like a switch hooked up to 30 lights," says Alber. "You flick one switch, and all the lights go on."

"Gene regulatory proteins," McKnight recounted, "were discovered as an outgrowth of research into the structure of DNA and, later, into the organization of genes." His discovery of the leucine zipper grew out of his efforts to learn more about how DNA binding proteins—of which the leucine zipper dimer is one class—activate genes. The first such proteins had only recently been identified.

The big clue that the leucine zipper was important came when McKnight and his colleagues sequenced a regulatory protein that goes by the evocative acronym C/EBP and discovered via computer search that a segment of 60 amino acids was very similar to segments of the regulatory proteins fos and myc. (Fos and myc are proto-oncogenes, genes that normally play important roles in the body but that contribute to the development of cancer when they undergo certain mutations. More on that later.)

Then, McKnight set out to model the structure of the leucine zipper. "We knew that parts of many proteins fold into an alpha helix, a kind of coil," McKnight wrote. "We therefore began by considering whether the segments relating C/EBP, fos and myc might adopt that structure."

A frequent characteristic of alpha helices is that the two sides of a helix have different chemical properties. (Imagine the slinky with lines running the length of it, on opposite sides.) One side might be soluble in water, or hydrophilic, while the other side might be hydrophobic—and never the twain shall mix. The amino acids on each side of the helix determine these properties. And so it was with the C/EBP leucine zipper. The leucines, which are extremely oil soluble, were "aligned in a plane along the length of the helix, forming a prominent ridge," McKnight wrote. He had originally proposed that the hypothetical leucine ridges bound the helices together with each facing in opposite directions, the leucines forming interlocking teeth, exactly like the teeth of a zipper.

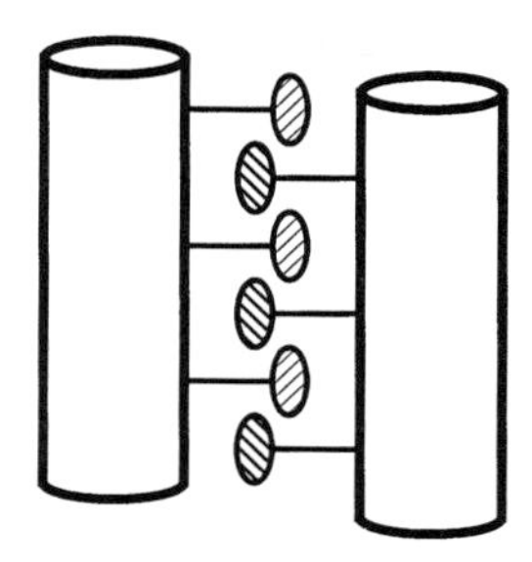

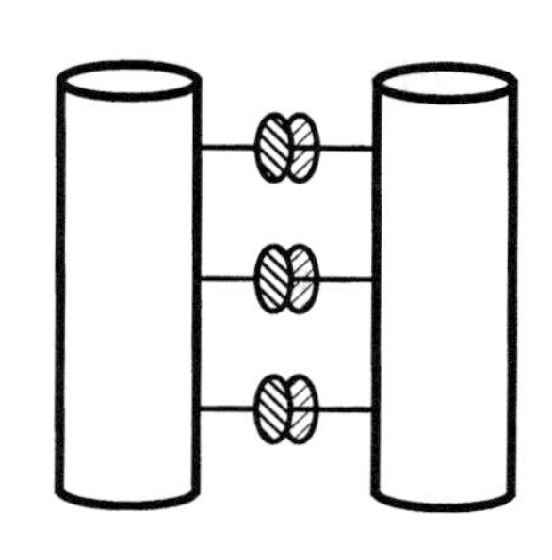

FIGURE 10.4 *The old and new models of the leucine zipper. The side chains do not interdigitate (left), but come together side by side (right).*

Compelling as it was, the model of two cylindrical helices side by side, fastened together with a molecular zipper proved somewhat inaccurate. Instead, the leucine zipper is a structure well known to experts in protein folding, called a coiled coil, which Linus Pauling and Francis Crick had first proposed to be the structure of the protein keratin, independently of each other, nearly 40 years earlier. Rather than lying side by side, the two helices of a coiled coil wind around each other in a super helix, like a pair of slinkies wrapped around each other or like the twining of rope.

Instead of one zipper running the length of the two helices, the structure of the binding mechanism is more like a series of very short zippers forming individual connections between the two helices (see Figure 10.4).

THE LEUCINE ZIPPER IS A COILED COIL

It was a freshly minted graduate of Smith college, who was working as a technician in Peter Kim's lab at the Massachusetts Institute of Technology, who spotted McKnight's mistakes. Her insight ultimately led to her Ph.D.

"McKnight's paper came out in June, 1988," says Alber. "The week it came out, [Erin] O'Shea read it. Peter and I had come back from a meeting, and Erin ran up to us in the hall, gave him the papers, and said, 'Read these by tomorrow, the peptide is on the synthesizer'." O'Shea, a sure-footed young woman of 26, is more modest about her achievement, explaining that she was quite familiar with coiled coils. "I had majored in biochemistry, and my advisor worked on coiled coils." So it was not surprising that she could spot a distorted one.

"It was evident from what I read that McKnight's [model] couldn't be correct," says O'Shea. "It takes 3.6 amino acids to make one full turn of a protein helix," she explains. "If the helical repeat were 3.5, the leucines, spaced every seventh amino acid, would fall on top of each other in the coil," as McKnight had described it. "One-tenth of a helix doesn't sound like much, but it is enough to make the ridge of leucines spiral around the helix, instead of lying flat."

To hear Alber tell it, Kim and O'Shea already had a great circumstantial case. But if circumstantial evidence is considered to be a weak basis for conviction in a court of law, in science it almost never passes muster. So the scientists would approach the problem from many perspectives over the next 4 years in order to nail down the structure.

One of the Kim lab researchers' first steps was to synthesize the leucine zipper—minus the DNA binding part—on a protein-synthesizing machine and see if it would fold. At that time, the dogma had been that small fractions of proteins, such as the leucine zipper, could not fold by themselves. But Kim had been challenging this dogma and had already shown that a small piece of a protein called trypsin inhibitor could fold. Now the leucine zipper folded into its normal helical form, and Kim had another example.

Next, Kim and O'Shea synthesized the leucine zippers of fos and jun, and demonstrated that these disembodied sequences could also fold into helices that could then form dimers. This, says Alber, "set the stage for all the subsequent discoveries, because it gave people confidence that what was being studied was relevant for [the] regulation of genes."

Using a relatively crude but simple imaging technique called circular dichromism, O'Shea was able to detect the helices in the leucine zipper. But she was unable to determine that they coiled around each other, let alone view any of the molecular details of the structures. That proof should have had to come from the x-ray crystal structure that O'Shea would finish later, but serendipity provided a quicker path to proof.

A CANINE PINCUSHION POINTS THE WAY

One day in Utah, one of Alber's dogs, a chocolate labrador retriever, returned from the adjacent foothills to the lab with his nose looking like a pincushion. Marley had tangled with a porcupine. Quills must be removed with extreme care, because they are designed to tear the tissues during extraction. This gave Alber plenty of time to think. Quills are made from keratin, as are fingernails, the scutes of turtle shell, hair, and other common proteins. Of all of them, quills make the cleanest x-ray diffractions because the fibers are especially well lined up. A clear image of the diffraction pattern of a quill had been published in the British journal *Nature* in 1943, but no one had derived the structure from the pattern.

In 1949 Pauling had proposed that some proteins would form alpha helices. He knew how amino acids fit together, and he could take a protein full of them and manipulate the topology in his mind, and now he and Crick were in a race to find an actual protein that had this structure. Each figured out what kind of diffraction pattern a helix would produce, and at roughly the same time both men deduced that the helix could explain the diffraction pattern of keratin if two helices were coiled around each other, like the fibers in a rope.

(The political story, says Alber, is that although Pauling, an American, submitted his paper to *Nature* first, toward the end of 1952, the British-born Crick's paper was published first, in 1952, while Pauling's was held until 1953.)

Anyway, as he was extracting quills from his dog's nose, Alber realized that he now had a means to quickly compare the diffraction patterns of the leucine zipper and keratin. "We had known from looking at the leucine zipper," says Alber, "that it was qualitatively similar to the published pattern of keratin." However, the nearly half-century-old x-ray provided no detail on the molecular level. But once they had a clear diffraction pattern for the quill, they could see that the diffraction patterns of that and the leucine zipper, which O'Shea had already obtained, were very similar to one another.

Besides proving that the leucine zipper is a coiled coil, the comparison of diffraction patterns showed that the structure of the zipper is similar in detail to that of keratin. But at the time, no one had ever deciphered the fine structure of keratin from the diffraction pattern. NMR and O'Shea's work on crystallography would fill in those details.

At that time, says Alber, "We only knew that some part of the peptide was helical, from the circular dichromism and from the original x-ray pattern." "The NMR showed exactly which residues were in the helix." NMR also proved O'Shea's hypothesis that the two helices did not zip together quite as McKnight had described. Lawrence McIntosh and Terry Oas performed the NMR in the winter of 1989, "practically overnight," says Alber. Ironically, it was a property of NMR that is usually a limitation that made the work go so quickly.

When two or more hydrogens lie in identical chemical environments, only one shows up on the NMR spectrum. This is because hydrogens in identical chemical environments produce the same peak in the same place on the NMR spectrum. (The forces that are relevant in NMR to an atom's chemical environment are only those that are located within a couple of chemical bonds from the atom. See Box beginning on p. 304.)

McKnight's zipper hypothesis required the leucine zippers to face in opposite directions; otherwise, the chemistry of the binding would not work. Although the leucine zippers that McIntosh and Oas were studying were identical to one another, had they faced in opposite directions,

text continues on page 308

Deciphering Structure: The Methodology of Molecular Modeling

Of all imaging techniques, x-ray crystallography can usually provide the most complete picture of a protein's structure. In a sense, crystallography takes a photograph of the entire protein (though this can be clear or blurry depending on how much information is generated in the form of x-ray photographs). It does so by revealing the density and location in space of a protein's electron clouds. This provides information on where the different atoms of a protein might be located, because different atoms have different electron densities.

The technique's biggest limitation is that the protein must be crystallized. "You might get a crystal in a day, or a week, and sometimes you never get a crystal at all," says Alber. In other words, you might spend months walking down a long road only to find yourself at a dead end in the middle of a desert.

"It's a process of trial and error to find the right conditions to make a crystal," says Alber. Whether you get a crystal or not is very sensitive to conditions of temperature, pH, salt concentration, the buffer salt you use, and how much protein you have. So the basic strategy is to get close, and then you vary each of the conditions in turn, and all the sudden you have a crystal.

"By chance, with the GCN4 leucine zipper we tried conditions that were close enough that we knew we were on the right track," says Alber. "We changed each little variable in the conditions in turn, and we discovered that the crystallization was incredibly sensitive to all the conditions. In that sense it was very unusual. Many proteins can crystallize over a range of 2 pH units or more, and some over 7. In this one you have to be within a couple of tenths of a pH unit. We were very lucky to hit close enough to begin with." They had their crystal within 3 months.

Sometimes parts of the protein crystallize poorly or not at all, so that electron clouds of those atoms do not contribute to the diffraction pattern. Another limitation is the need for sequence data. It is not an absolute necessity, but it helps, in order to be able to verify the structure once it has been derived.

The problem is that the electron density data are prone to errors of up to 5 percent. These errors inevitably crop up in the counting of literally billions of photons, as well as from background radiation, which is hard to accurately subtract from the picture, and from some of the mathematical equations used to generate the positions of the electron clouds, which are approximations, says Alber. For example, when Alber and O'Shea deciphered the structure of the leucine zipper, "We could only see one-half of it clearly," says Alber. "But the half that we could see was enough to bootstrap the rest."

The first step, once a protein has been crystallized, is to take x-ray pictures of the crystal. Often thousands of x-rays are taken, from every possible angle encircling the crystal. Until recently, the x-ray patterns were recorded on ordinary film, and the machine judging of how strong the x-ray diffraction had been by the shade of spots on the film was imprecise. Now, photographic film has been replaced by electronic film that can literally count photons, greatly increasing accuracy.

In fact, it is almost ironic how sophisticated the equipment for deciphering crystals has become, compared to the intellectual process of piecing together the structure. Using the diffraction data, a powerful computer maps the electron densities in space and displays them on a TV screen. The computer allows the crystallographer to turn the map around in order to peer at the pattern from every angle.

If the crystallographer has a sequence, it can be generated on the screen as well, and, by twirling a set of dials, the crystallographer can manipulate each atom in three dimensions, among the electron clouds. Without a sequence, the crystallographer simply tries fitting different atoms or chemical groups among the electron clouds. Poor Watson and Crick had to manipulate their models of DNA's structure in their heads and on paper, and then they had to build them.

Sequence or none, there is method to the madness of deciphering structure. Due to the various built-in errors, the electron clouds are ambiguous indicators. This can leave the crystallographer in the position of a rat trying to run a three-dimensional maze. The most important work is to track the backbone of the protein. But researchers can get sidetracked onto side chains, particularly when two connect. In sum, crystallography is a time-consuming and exacting process of trial and error, so much so that there is a whole field dedicated to making model fitting methodical.

Where x-ray crystallography fails, nuclear magnetic resonance (NMR) can sometimes succeed (and vice versa). Proteins that fail to crystallize may yield their structural secrets to NMR. The mobile fractions of proteins that cannot be crystallized can be viewed through the lens of NMR. Furthermore, NMR can reveal the different configurations of such proteins. Researchers are beginning to use NMR to track the pathways of folding proteins. And, finally, NMR can reveal interactions between charged sidechains that stabilize a protein.

continued

Deciphering Structure—*continued*

But where x-ray crystallography is like taking a photograph and getting the big picture, NMR is more like assembling a jigsaw puzzle—putting the structure together from tiny pieces of disembodied information. Not surprisingly, then, for NMR, having the sequence is like having a picture of the entire jigsaw.

But recent advances have made NMR much easier, says Thomas James, a researcher at the University of California, San Francisco. These advances have also allowed researchers to use the technique to examine much bigger proteins: with 150 to 200 amino acids, up from about 80 amino residues.

Prior to the advances, NMR was limited. It is sort of like a game of connect the dots. NMR detects atoms by magnetizing them. Magnetized atoms can magnetize other atoms if they are close enough together—within 5 angstroms, or the equivalent of a couple of chemical bonds. NMR can be set up to reveal the pairs and the distance between them. Stick a third atom into the mix, one that is within 5 angstroms of both of the others, and now one can deduce the angles between the hydrogens as well. Instead of one dimension, one is now working in two. In the same way, a fourth hydrogen within 5 angstroms of the other three can add the third dimension. But if the third hydrogen is more than 5 angstroms from, say, the second hydrogen, NMR cannot reveal the distance between them, or the shape of the triangle outlined by the three, unless some clue from another source is available.

Another limitation of the old hydrogen-based NMR is that hydrogen atoms are deadends in the covalent structure of protein. Moreover, although hydrogens are attached to most of the atoms in this chain, certain carbons within the backbone of each amino acid lack hydrogens. The result is that the smallest hydrogen-hydrogen distance across the carbon gap via the backbone exceeds the 5-angstrom limit, forcing researchers to trace their way from one amino acid to the next through space, a more difficult technique.

Yet another problem was that the NMR spectrum for hydrogen is narrow, which had much to do with placing the 80 amino acid limit on the size of proteins that could be studied. Imagine shrinking the FM dial to one-tenth its size. It might still be able to accommodate all the stations available in a small city like Bismarck, North Dakota, but forget about trying to squeeze in the airwaves of New York.

This was the state of the art until the mid-1980s. Researchers had long realized that NMR could be much easier and more versatile if certain isotopes of carbon and nitrogen that can be magnetized were incorporated into protein. But no one had tried it until then.

One advantage of the isotopes was that their NMR spectra are about 20 times wider than that of hydrogen. This contributed to increasing the size of proteins that the technique could study. Another was that the use of carbon-13 provided a new, more direct way to connect atoms along the length of the protein chain—through the bonds rather than through space. Still another advantage of isotopes is that it became possible to connect triplets and even quadruplets of atoms directly, instead of just pairs. Researchers could trace larger pieces of structure at once, speeding up solving the puzzle.

In theory, it would have been possible to connect triplets under the old hydrogen-based system, because when hydrogen magnetizes a neighboring hydrogen the neighboring hydrogen can magnetize a third atom. In practice, hydrogen magnetizes its neighbor so inefficiently that only about 1 percent of the magnetic "spin" would get transferred to hydrogen #3, too little to show up on NMR. But the carbon and nitrogen isotopes transfer roughly 90 percent of the spin to their neighbors.

NMR picks up additional information on structure from so-called torsional angles. All atoms have preferred bonding angles, and scientists could have used these to deduce the spacial confirmation of a protein. However, the forces that give the protein its shape bend the bonding angles in somewhat the way a load of cargo bends the springs of a car. Torsion angles between molecular bonds can be deduced from NMR spectra.

Ultimately, all this information is processed by using computers, using several different methods to deduce a protein's structure. Since experimental error is inevitable, the structure is analyzed by a program called the distance geometric algorithm, which determines what parts of the protein's structure have been accurately defined, and what parts have been poorly defined.

The data are delivered in the form of a "family of structures," which the researcher can compare to one another. Accuracy is determined by the degree of congruence among the structures in the family. But an area where the structures are different may not mean poor definition. Instead, it may mean that that part of the molecule is mobile and that the different structures correspond to different real confirmations of the molecule.

the chemical environments for the hydrogens from each helix would have been slightly different, and so both sets of hydrogens would have appeared on the NMR spectrum. Instead, McIntosh and Oas saw spectral peaks corresponding to only one helix. The zipper hypothesis was becoming increasingly untenable.

X-RAY CRYSTALLOGRAPHY: FINDING THE MOLECULAR DETAILS

Only the high-resolution structure would divulge the location of all the atoms in the protein. This was a job for x-ray crystallography. The researchers began their attempts to crystalize the protein in November 1988. This and the rest of the preparatory work, through collection of x-ray data, took nearly a year. Then, at the end of 1989, O'Shea went to Alber's laboratory to take advantage of his skills and facilities.

It was amazing, says O'Shea, how quickly she generated the structure. Within 9 months she had confirmed her original model. Technological developments in data collection and software made this speed possible, in particular, programs written by Wayne Hendrickson. O'Shea had gotten stuck, but once she began using his program, which exploited the various symmetry relationships that exist in many proteins, she was able to finish the job in 2 months.

The solution of the crystal structure not only proved beyond a doubt that the leucine zipper is a coiled coil but was also a milestone in the study of protein structure. The list of suspected coiled coils had included many important and ubiquitous proteins, including muscle proteins; dynein, which forms molecular motors; intermediate filaments, which make up some of the girders that give cells their structural integrity; and oncogene products. But O'Shea had uncovered the first complete high-resolution structure of a coiled coil. This "made the models of myosin, keratin, and all these other coiled coils much more believable," says Alber.

THE COILED COIL: RULES OF FOLDING

Meanwhile, in 1991, Pehr Harbury, one of Peter Kim's graduate students, set out to determine how structure influences function in the leucine zipper.

Leucine and the other hydrophobic amino acids in the zipper are spaced as evenly as possible within an odd-numbered repeating sequence: Four positions from leucine to the other hydrophobe, and back to leucine again is three, and so on. Since there are 3.6 amino acids in one turn of the helix, this places the other hydrophobic amino acids nearly in line with the leucines.

Harbury made certain changes in the sequence, substituting different hydrophobic amino acids at the positions of leucine and the other hydrophobe that forms the zipper mechanism. Then he asked how the amino acid substitutions changed the structure of the leucine zipper, the way it bonded, and its stability.

One set of substitutions barely altered the structure. But other changes caused the leucine zipper to form tetramers (quadruplets) or trimers (triplets) instead of dimers (see Figure 10.5). (To visualize the structure of these, imagine winding three and then four fibers around each other to form a rope). In the latter cases the main difference between the normally occurring hydrophobes and their replacements was shape. The substituting amino acids had side chains that branched at the first carbon atom; leucine does not.

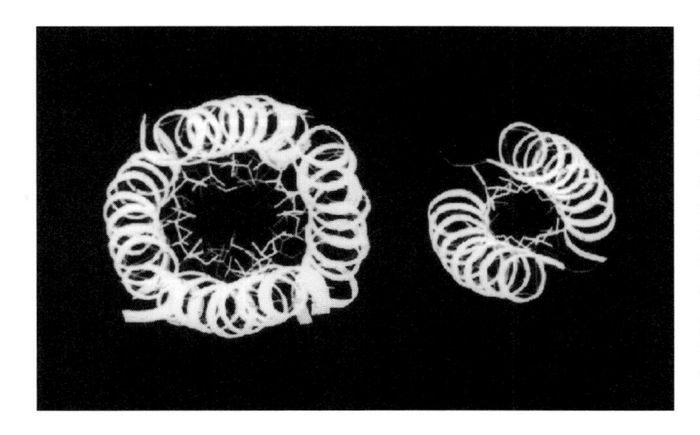

FIGURE 10.5 *Dimer and tetramer coiled coils viewed end-on. The amino acid side chains of the coils pack very tightly between the coils, and their shape, which can be changed by changing amino acids in certain positions within the helix, determines the number of coils in a structure.*

It was surprising that a couple of substitutions could cause such drastic changes in bonding patterns, Alber said at the "Frontiers of Science" symposium, "because intuitively, you know that sequences that are similar generally form the same structure. That's a very big idea in protein structure. If you have two similar sequences, they have the same fold. This is an exception to that rule."

"These sequences are in fact 75 to 87 percent identical, and yet they form different arrangements," Alber continued. "Why is that? The only thing that we've changed in these sequences is the geometry of these [amino acids that connect the helices]. So it must be the packing of the interface that determines the number of strands. We've figured out how that can happen by determining by x-ray crystallography the three-dimensional structure of this tetrameric coil." (see Figure 10.6.)

The other reason the result seemed surprising is that researchers had

FIGURE 10.6 *The different shapes of these three amino acids, in the a and d positions of the helices, influence whether coils group together as dimer, trimer or tetramer.*

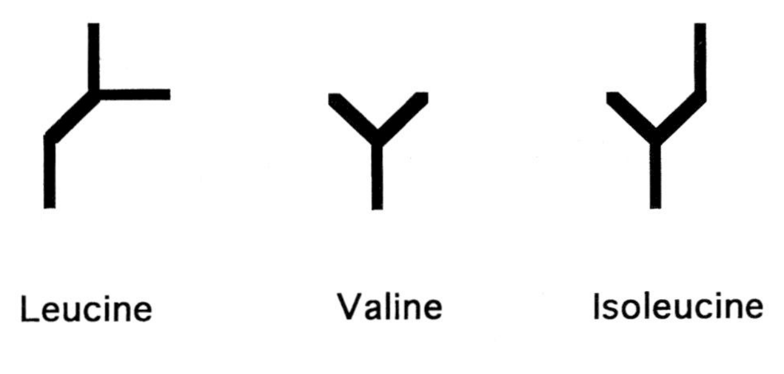

based their predictions on chemical forces and the size of amino acid side chains, ignoring their shapes, and for the most part this had worked. (But note that the result should not seem quite so surprising as all that because the amino acids causing the changes in structure are the interacting ones, which are comparable to those in the core of a protein.) But now Harbury had demonstrated that protein folding was still more complex than this. And in particular, says Alber, how the amino acids fit together in the core of the coiled coil determines the number of strands.

The rule of folding that the results suggest, says Alber, is that amino acids with side chains that branch at the first carbon do not fit the holes in the so-called knob-in-hole packing that leucines occupy in the leucine zipper dimer.

This, at any rate, is the current hypothesis. Harbury has since studied the sequences of many helices that come together as twins or triplets and has found that these rules generally hold (see Figure 10.7). The next step will be to design and synthesize idealized trimers and tetramers and see if they come together as predicted.

FIGURE 10.7 *The specific amino acid changes in the a and d (circled and boxed) positions that determine the number of coils that come together.*

a d a d a d a d a

RMKQ[L]EDK(V)EE[L]LSK(N)YH[L]ENE(V)AR[L]KKL(V)GER

Residues at (a)	Residues at [d]	Number of strands in the coiled coil
Valine (N16)	Leucine	2
Isoleucine	Leucine	2
Isoleucine	Isoleucine	3
Leucine	Isoleucine	4

THE ASPARAGINE RULE

Oddly enough, it might seem, among all the hydrophobic amino acids that bind leucine zippers in pairs, there is one water-loving amino acid that is found in virtually every leucine zipper helix: the amino acid asparagine. "So it's a really important thing for function," says Alber. Since such an entity could only weaken the zipper, the researchers were mystified as to what its purpose might be. To answer the question, they substituted in turn four different amino acids for asparagine and then determined the structure and bonding strength variant. Chavela Carr, a graduate student in Kim's lab, studied the stabilities, and Russ Brown, a postdoctoral student with Alber in Utah, deciphered the structures. Two things happened. First, the new helices bonded so strongly that even boiling didn't sunder the zipper. Second, these asparagine-free leucine zippers formed both pairs and triplets, and Alber speculates that they may have formed skewed dimers as well (see Figure 10.8).

In hindsight, it was easy to deduce what the purpose of the destabilizing asparagine might be. Any biological switching device must be capable both of holding on to and letting go of its substrate. For example, hemoglobin must bind oxygen strongly enough to pull it from the lungs as blood cells pass through the alveolar tissue but weakly enough to release it throughout the rest of the body. Carbon monoxide molecules are poison because they adhere so tightly to blood cells that they rarely let go, thereby preempting oxygen. Similarly, a leucine zipper would be useless if it bound so tightly to one partner that it could never let go. Furthermore, too strong a propensity to bond would allow the helices to pair in an improper alignment. An improperly aligned leucine dimer would be unable to bind to the DNA because the DNA binding ends of the leucine zippers would be askew.

For biological switches to function, then, the strength of the binding must be finely tuned: Strong enough to come together but weak enough to do so without misaligning the molecules and weak enough to break apart when the time comes. In the leucine zipper, asparagine accomplishes this fine tuning.

Most coiled coils are not as regular as leucine zippers, and frequently another hydrophobic amino acid occupies the "leucine position." Harbury hopes to develop a set of rules for predicting the structures and bonding arrangements of this variety of coiled coils, and he is asking such questions as how many substitutions of leucine by isoleucine, one of the so-called beta-branching hydrophobes that fail to fit in the knob-

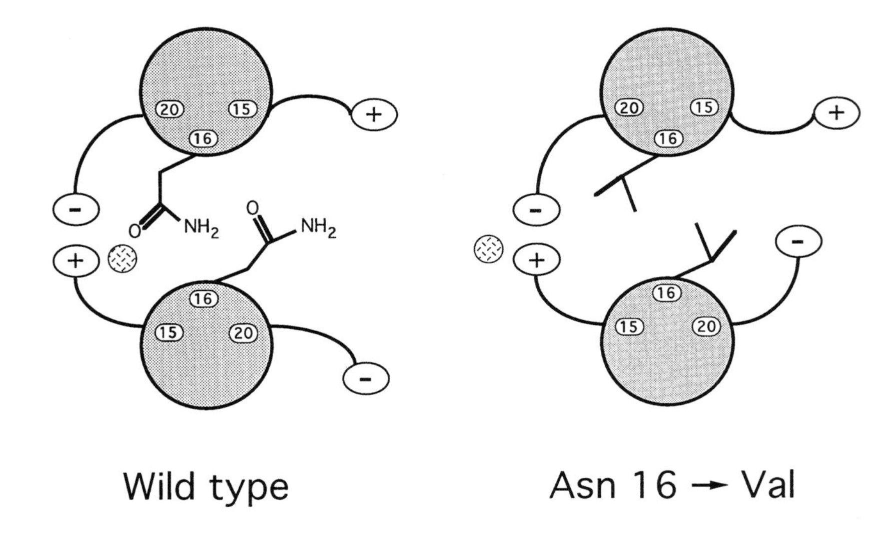

FIGURE 10.8 *When experimenters replaced asparagine with valine, in the 16th amino acid position of the GCN4 leucine zipper, the result was a much stronger, more stable dimer than occurs naturally.*

in-hole fastener, it takes for alpha helices to come together in trimers, instead of dimers. The results of experiments now under way will tell.

ANOTHER APPROACH IN THE QUEST FOR THE HOLY GRAIL

Despite the growing knowledge of the mechanics of lysozyme and leucine zippers, the Holy Grail of protein research, predicting structure from sequence, remains elusive. This was driven home to Harbury and his colleagues when, prior to synthesizing his experimental leucine zippers, he tried to predict what their structures would be. "Because the leucine zipper is such a tiny repetitive structure, Pehr could use effectively the most sophisticated methods for doing this calculation," says Alber. Nonetheless, "The calculation failed dramatically."

"We know the laws of chemical physics, so why can't we compute structure, given sequence?" David Eisenberg, professor of molecular biology at the University of California, Los Angeles, demands rhetorically. One cannot describe the folding of protein from first principles, he explains, because protein is too big and complex to be fitted into Schroedinger's equations, which are the mathematical expression of the chemical physics laws.

Another hypothetical approach would be to find the conformation with the greatest stability. Stabilities of different conformations can be calculated for small molecules, and, with certain exceptions, the conformation with the greatest stability is the correct one. The problem with this approach is that for proteins there are far too many potential conformations to begin to do this, as Alber explained when he put the number at a google for ATCase. "It's not just that we don't have the computers to do that now," says Eisenberg. "We won't have them in 10 years."

Eisenberg has been taking the opposite approach to the problem of the relationship between sequence and structure. In his experiments he asks, as Stanford University engineering professor Eric Drexler did first in 1981: Given a specific structure, what amino acid sequences are compatible? But Eisenberg and his postdoctoral student Jim Bowie have simplified this approach in two ways.

First, they ask what known amino acid sequences are compatible with the structure. "That limits us to 50,000 sequences, whereas Ponder and Richards [researchers who also had previously tried the same approach] had looked at every conceivable sequence. Why should we care if there are sequences we don't know? Our job is to take information from the human genome and find the structure for each protein sequence. That limits the dimensionality of the search."

"The second major simplification is that instead of working with a three-dimensional structure, we have simplified that into a one-dimensional string which we can compare to amino acid sequences." ("String" is a computer term meaning things that have been strung out in one dimension.)

Eisenberg replaces three-dimensionality with the details of the chemical environment of each amino acid position. For example, "What are the amino acid side chains around that position, and do they prefer an apolar or a charged environment?" He has divided the chemical environment into 18 different environmental classes, and each amino acid in a protein is assigned to one of them. "We call the string of classes the three-dimensional profile. By looking at the environment, we are looking at the footprint, rather than the foot."

Eisenberg predicts the job will be manageable because he believes the number of types of protein folds is limited. "When we learn the structure of a new protein, it often has the structure of a known protein. But we can't necessarily predict that from the sequence, because the sequence may have diverged too much. This suggests that there may be a finite number of folds."

In early 1992 Cyrus Chotia, of the MRC Laboratory in Cambridge, England, estimated that there may be only 1000 to 1500 distinctive folds. "If that is true, then the job of the person studying protein structure will be to assign sequences to one of these folds," says Eisenberg. "That is what our method is aimed at."

Three-dimensional protein structures are being compiled in a data base at Brookhaven National Laboratory, and sequences are stored at Georgetown University and the National Institutes of Health. There are about 1200 three-dimensional structures in the Brookhaven data base, and Georgetown and NIH have a combined total of about 60,000 different sequences. "We used those data bases in our research," says Eisenberg. "We started with the structure of lysozyme, and computed a three-dimensional profile, and we were able to find all the sequences in the sequence data base which are compatible with the lysozyme structure."

But so far this method has had its own limitations and failures. A coiled coil that Eisenberg had predicted should be a dimer turned out to be a trimer. And Alber says, "Our finding that the shape of an amino acid makes a difference to the structure suggests that something is missing from his calculations, because he doesn't consider shape at all."

"It's not that his method is wrong," Alber hastens to add. "It's just not fully developed." Will the two methods eventually merge somehow in a way that will generate general principles, much as theoretical physicists hope to combine the four forces into a grand unifying theory?

The best answer that either researcher could give was Alber's response that on occasion his predictions agree with Eisenberg's. For example, in the case of one coiled coil sequence, both methods correctly predicted a trimer. This weak response suggests that both methodologies remain embryonic.

APPLICATIONS: A CANCER DIAGNOSTIC

While a grand unified theory of protein folding is probably years in the future, the basic research is already yielding useful developments. So far, most of them involve research methods, but Alber and Ray White, of the University of Utah, are working on a diagnostic system for colon cancer.

White and Bert Vogelstein, of Johns Hopkins University, had discovered mutations in the so-called APC gene. That gene is present in many people who have colon cancer. The mutations truncate the APC protein but leave a coiled coil intact. One of White's first-year graduate student advisees at Utah who had been working with him brought Alber and White together. Now Alber is designing a coiled coil to be used in the diagnostic system.

The diagnostic would work as follows. Doctors would take cells from patients' colons, use laboratory techniques to separate the proteins by size, and label APC proteins using Alber's coiled coil—radioactively tagged—as a hook. These would then be analyzed to see if truncated APC proteins were present.

CONCLUSION

No doubt the quest for the Holy Grail will continue for years, if not decades. Researchers are only beginning to design novel proteins while sitting at the keyboard. The amino acid chain remains a Rosetta stone with the wealth of information on how proteins fold largely undeciphered. Nonetheless, the empirical data that researchers have collected are beginning to yield patterns that are providing valuable clues as well as insights into the mechanisms of gene regulation, which is one of the most fundamental processes in biology. The coiled coil, says Alber, "although simple, has been especially rich. On the one hand, we are finding out how protein-protein interactions control gene expression. On the other hand, we've described a simple structural motif in enough detail that we now think we know how to design this motif. This is a small step in the direction of being able to design proteins."

BIBLIOGRAPHY

Alber, T. 1992. Structure of the leucine zipper. Current Opinion in Genetics & Development 2:205–210.

Alber, T. 1993. How GCN4 binds DNA. Current Biology 3:182–184.

Brändén, C.-I., and T. A. Jones. 1990. Between objectivity and subjectivity. Nature 343:687–689.

Crick, F. H. C. 1952. Is keratin a coiled coil? Nature 170:882–884.

Crick, F. H. C. 1953. The packing of a-helices: simple coiled coils. Acta Crystallographica 6:689–697.

Harbury, P. B., T. Zhang, P. S. Kim, and T. Alber. 1993. A switch between two-, three- and four-stranded coiled coils revealed by mutants of the GCN4 leucine zipper. Science, in press.

Mattherw, B. W. 1993. Structural and genetic analysis of protein stability. Annual Review of Biochemistry 62:139–160.

McKnight, S. L. 1991. Molecular zippers in gene regulation. Scientific American 264(4):54–64.

Olson, A. J., and D. S. Goodsell. 1992. Visualizing biological molecules. Scientific American 267(5):76–81.

O'Shea, E. K., J. D. Klemm, P. S. Kim, and T. Alber. 1991. X-ray structure of the GCN4 leucine zipper, a two-stranded, parallel coiled coil. Science 254:539–544.

Pauling, L., and R. B. Corey. 1953. Compound helical configurations of polypeptide chains: structure of proteins of the a-keratin type. Nature 171:59–61.

Richards, F. M. 1991. The protein folding problem. Scientific American 264(1):54–63.

Wüthrich, K. 1989. Protein structure determination by nuclear magnetic resonance spectroscopy. Science 243:45–50.

APPENDIX A

Abstracts of Additional Sessions of the Frontiers of Science Symposia

MATERIALS SCIENCE

Many applications of lasers and fiber optics depend on molecules and crystals that show nonlinear optical activity. This can involve doubling or tripling the frequency of laser light; a laser of standard type then can produce light with wavelength better suited to its intended use. Another type of activity rotates the plane of polarization of light in response to an applied electric field (the Pockels effect), to modulate or switch an optical signal.

In preparing materials that show nonlinear optical activity, a standard approach has featured the use of polarizable molecules. These have an electron donor at one end and an electron acceptor at the other, separated by a bridge. A bridge is an organic structure containing both single and double carbon-to-carbon bonds.

Seth Marder, of the Jet Propulsion Laboratory, has been seeking better molecules by optimizing properties of the bridge. He defines a quantity, bond length alternation (BLA), as the difference in length between adjacent bonds within the bridge.

In conventional materials, BLA = 0.10 angstrom, approximately, and Marder finds that this value is too large to optimize the proper performance of the donor-acceptor dye. Using computational chemistry, he finds that an index of nonlinear activity has maximum value when BLA = 0.04 angstrom. Standard techniques of organic chemistry then permit synthesis of molecules whose bridges show this value.

Commonly used polymers include stilbenes such as dimethylamino nitro stilbene (DANS), with an index value of 466. Marder has prepared counterparts, using thiobarbituric acid acceptors, with an index as high as 19,000. In implementing the Pockels effect, this increase would reduce the necessary electric field from 200 volts to 4 or 5.

For engineering improved bulk crystalline materials, following a suggestion in the literature from Gerald Meredith (now at Dupont), Marder has studied organic salts. He finds particular promise in dimethylamino methylstilbazolium tosylate (DAST); its frequency-doubling efficiency is 20 times that of lithium niobate, a commonly used crystal. Moreover, its electrooptical properties are superb and may lead to entirely new types of devices. This enhancement results from the almost ideal orientation of the molecules in the crystal lattice.

A separate topic in materials science features the search for principles of chemistry that can permit construction of an organic ferromagnet. The rationale lies in observing that physicists today identify some 14 distinct magnetic states of matter, representing modes of ordering of electron spin. Five of these magnetisms are known classically, having the prefixes ferro-, antiferro-, ferri-, para-, and dia-. Most of the rest have been discovered only since about 1975 and merit further research. Because organic chemistry offers great freedom in modifying molecular structures, it suggests the prospect of creating families of related materials that display novel types of magnetism in varying degrees. An organic ferromagnet then represents a first step, an initial problem in this area.

Difficulties exist in the chemical syntheses, and characterization often demands liquid helium temperatures. Nevertheless, standard rules permit selection of candidate molecular radicals that display the fundamental ferromagnetic property of having two unpaired electrons with spin parallel. Dennis Dougherty of the California Institute of Technology, working with colleagues, has tested these candidates' suitability by combining them with trimethylenemethane, which also has spin-parallel electrons. This work shows that m-phenylene and cyclobutane offer particular promise.

As a next step, these chemists have inserted units of m-phenylene, a 1,3-substituted benzene, within long-chain molecules of polyacetylene. The resulting short runs of polyacetylene can be chemically treated—doped—to remove one electron. The treated segments are called polarons, and they also have spins. In addition, they are stable at room temperature. A variant of this material shows properties indicating that within each such polymer molecule, and in the presence of an external magnetic field, an average of nine polaronic spins align ferromagnetically. This stands as a step toward a true ferromagnet, for which all spins within a bulk sample would align, even without the external field.

Dougherty's polymers are paramagnets rather than ferromagnets. Researchers in Japan have created a true organic ferromagnet, but it is a molecular, rather than a polymeric, system that is ferromagnetic only below 0.65 Kelvin. By contrast, iron shows ferromagnetism up to 1044 Kelvin. Thus, while progress is being made, many challenges remain for the field of organic magnetism.

OZONE HOLES

A protective ozone layer exists within the stratosphere. If brought to sea level, the earth's ozone would form a band only 3 millimeters thick. Nevertheless, this diffuse gas suffices to screen out solar ultraviolet, which is hazardous to life. Hence there has been a great deal of concern over the ozone hole, a seasonal reduction in atmospheric content of this gas, over the Antarctic, by as much as 95 percent at some altitudes. This reduction was first seen in ground-based measurements, around 1985, and has since been confirmed repeatedly through satellite observations.

Several theories have sought to account for this hole, which reappears annually, during the south polar spring. It might result from atmospheric dynamics, with ozone-poor air rising from below to dilute the air of the stratosphere. Alternately, the ozone hole might result from increased solar activity during the 11-year sunspot cycle, for energetic solar particles might stimulate formation of nitrogen oxides, which destroy ozone. A third theory holds that the ozone hole results from human activities. Chlorofluorocarbons (CFCs) rise slowly into the stratosphere, where molecules break apart from solar radiation, releasing reactive chlorine. This reactive chlorine then goes on to attack ozone in catalytic cycles.

Mario Molina of the Massachusetts Institute of Technology, a coauthor of the original 1974 paper on the CFC theory, notes that these three alternate explanations have been subject to observational test. The atmospheric dynamics theory fails because upwelling air would contain significant amounts of trace constituents such as methane and nitrous oxide, which are not found. The solar cycle theory also fails. It would imply that the antarctic stratosphere should contain excessive amounts of nitrogen oxide. In fact, the measured amount is less than expected, which rules out that theory as well.

The third theory, that of CFCs, holds particularly that chlorine oxide, ClO, should serve as a reactive intermediary in catalyzing the breakdown of ozone. Each ClO molecule should destroy a substantial number of ozone molecules, while being regenerated following each such molecular reaction. Aircraft observations indeed show that in late August, during the antarctic winter, elevated concentrations of ClO exist at high southern latitudes. Then, 3 weeks later, the zone of high ClO also shows drastically lowered concentrations of ozone. This gives strong support to the CFC theory.

In making these observations, ground- and satellite-based work finds a complement in flights of the ER-2, an instrumented research airplane that can cruise at 20 kilometers altitude. David Fahey, of the National Oceanic and Atmospheric Administration's Aeronomy Laboratory, notes that such flights have also been of value in studying ozone in both the northern and southern hemispheres. Various changes have been noted. In the northern hemisphere, for example, the reduction has approached 5 percent in the temperate latitudes during winter. At Oslo, Norway, it has at times topped 30 percent.

Fahey emphasizes that rapid ozone destruction in polar regions involves the presence of polar stratospheric clouds, which form below temperatures of 195 Kelvin and provide surfaces on which chlorine-liberating reactions take place. Where such temperatures exist, as he puts it, "you see ClO turn on like a lightbulb, reaching one and a half parts per billion"—a very high concentration.

These findings have contributed to major efforts aimed at doing away with CFCs. They are valuable commodities, which have seen wide use as air-conditioning and refrigeration coolants, in producing styrofoam and similar materials, and in cleaning electronic circuit boards. However, international agreements now mandate that they must be phased out. Replacements could include chemically related substances that break down more readily when released to the atmosphere, so that far fewer of their molecules would reach the stratosphere. Some of these replacements are to be chlorine-free, further reducing the danger. There also is interest in avoiding chemicals of this type altogether. Thus, in pressurized spray cans, hydrocarbons have replaced the CFCs that formerly served as propellants.

This does not mean the ozone problem is at an end or will be soon. CFC molecules are extremely long lived and stable, while the processes that transport them to the stratosphere, and that destroy them once they are there, take decades to operate. Indeed, Fahey notes that atmospheric concentrations of CFCs are projected to stay above "pre ozone" values until mid to late next century.

TOPOLOGY

A topologist has been described as a mathematician who can't tell the difference between a coffee cup and a doughnut. Both are three-dimensional shapes that have a single hole, and which hence can be deformed or mapped into each other, point for point. This relationship is called homotopy equivalence. In topology a basic problem is to list all spaces that are homotopy equivalent.

A point of departure lies in work by J. H. C. Whitehead, who introduced the concept of a CW complex: a space that can be constructed out of n-dimensional spheres. A torus is such a space; it then can be mapped to either the doughnut or cup. Whitehead proved a fundamental theorem: every "interesting" topological space is homotopy equivalent to a CW complex. It follows that a solution to the basic problem lies in writing down all maps from spheres to other spheres, in n dimensions.

We require all spaces to have a basepoint, which remains invariant under mapping. This permits defining an addition of two maps; this additive property then leads to definition of groups $\pi_k(S_n)$. Here S_n is the n-sphere and π_k is the set of maps from S_k to S_n, with two maps regarded as identical if they are homotopic. For fixed k and $n \geq k$, the groups $\pi_{n+k}S_n$ are all the same and are known together as the kth stable homotopy group of spheres.

A focus of attention then is the chromatic tower, which breaks the homotopy groups of spheres into "monochromatic" layers, one for each integer $n \geq 0$. The first such layer draws on work by Hurewicz, who introduced a method for constructing maps between spheres. The part of the homotopy groups of spheres constructed with this method is called the image of J, and this appears as the $n = 1$ monochromatic layer.

The second monochromatic layer has been calculated by the topologist Katsumi Shimomura in 1986. Some of its properties are encoded within a diagram, which has a bottom half that is symmetric with the top half, and with numerical information in both halves that is related in a simple fashion, through a shifting factor. Analogous diagrams exist for the nth monochromatic layer. They are also symmetric and have known shifting factors.

Nevertheless, this offers no more than a partial solution to the basic problem, which again is to list all spaces that are homotopy equivalent. Michael Hopkins of the Massachusetts Institute of Technology notes that even within the Shimomura diagram "there are many suggestive patterns, but there is no really good theory. One of the most important problems in homotopy theory today is to find a theory that predicts the qualitative features of this diagram, and its analogs for the higher chromatic layers."

Another topic in topology involves knots and unknots. An unknot is a simple closed curve in three-space that can be deformed into a round circle. Alan Hatcher has shown that the deforming or untangling is continuous and involves no choices. A knot lacks this property; it cannot untangle into the simple loop of a circle.

Michael Freedman and Zheng-Xu He define the energy of a closed curve, E, by introducing a $1/r^2$ potential. Two similar curves have equal energy, and if two strands come close, as if to cross, then E blows up. A curve then can untangle by following the gradient of E. Freedman and He find a theorem: If $E < C$, where C is a constant, the curve is an unknot. $C \geq 22$, but C is not well known; Freedman says it could be about 70. For a round circle, $E = 4$; hence, only unknots exist for $4 \leq E \leq C$.

A related issue is the average number of crossings that a curve makes, when one projects its shape onto a randomly oriented plane. If the curve forms a knot, the number is at least 3; you can see this with a loop of string. Both for knots and unknots, Freedman and He find that this number has an upper bound in terms of energy: $11E/12\pi + 1/\pi$. This theorem draws on the work of Gauss.

The energy criterion for an unknot recalls a similar result due to John Milnor. Milnor defines a total curvature of a loop, T. For a round circle, $T = 2\pi$. Milnor has shown that for $T \leq 4\pi$, the loop or curve is an unknot. This then raises a question: In addition to the energy and total curvature, do other integral quantities exist whose values distinguish a knot from an unknot?

QUANTUM CONFINED SEMICONDUCTORS

In semiconductor physics a significant topic involves preparation of materials in unusually small volumes and thin layers and using them to fabricate new devices. At AT&T Bell Labs, Louis Brus has directed studies of crystallites, nanometer-size particles containing 10^3 to 10^4 atoms. These have high percentages of surface atoms and offer properties intermediate between those of single molecules and bulk crystals.

Michael Steigerwald has produced crystallites of CdSe, a II–VI compound, by conducting the synthesis within a solution of surfactant or soap. The crystallites form within soap bubbles; their sizes are controllable and range from 15 to 60 angstroms, ± 15 percent or better for any batch. When these crystallites are coated with small organic molecules, they can be removed and exhibited.

Other investigators have prepared similar crystallites of other II–VI compounds, as well as of III–V compounds such as GaAs and I–VII compounds such as AgCl. A general rule holds: smaller particles give larger "band gaps." In Steigerwald's CdSe, for instance, this leads to changes in the color of bulk crystallite powder, from yellow to orange to red with increasing particle size. Current experiments feature particles that are all of the same size, and Steigerwald has synthesized macromolecules, for example, containing lattice arrays of 20 nickel atoms and 18 telluriums, that have this property.

At Caltech a team of researchers have discovered several different routes to the synthesis of GaAs and silicon nanocrystals based on gas-phase nucleation. In certain cases, whole new effects occur as size is reduced. For example, some silicon nanocrystals luminesce. This is remarkable because bulk silicon does not exhibit photoluminescence. According to researchers involved with this project, the ability to make individual optically active monocrystals may help to resolve another mystery: the origin of optical luminescence from porous silicon that is generated by chemical etching.

In practical device fabrication it often suffices to grow polycrystalline layers on a substrate. Metal conducting paths are examples. But one achieves specialized electronic properties using epitaxial thin films, which take their crystal structure from that of the underlying substrate. Epitaxial growth demands a close crystallographic match between layer and substrate and tight control of substrate temperature, vacuum, and deposition rate. The substrate also must be atomically clean. Meeting these conditions, however, permits growth of multilayer structures featuring different materials, which have the crystallography of single crystals.

Two basic classes of deposition techniques are in use: physical vapor deposition (PVD) and chemical vapor deposition (CVD). Liquid-phase epitaxy is a third method for depositing thin films or layers of semiconducting crystals on substrates or other crystals. PVD is well suited for processing individual wafers. It demands high vacuum and delivers atoms or molecules directly to the substrate. A laser may evaporate deposited materials (laser ablation); an energetic beam of inert atoms may sputter this material from a block, or it may boil away to form a molecular beam, which the user turns on and off by opening a shutter.

CVD techniques take place at near-atmospheric pressures and are well suited for batch processing of numerous wafers. CVD delivers chemical precursors in gaseous form. These are organic compounds that break down at the surface to yield the desired species. Both CVD and PVD can grow epitaxial layers at rates of some one atomic layer per second; layers one atom thick are achievable. Resulting devices, which rely on epitaxial films, include very fast transistors, tiny solid-state lasers, and advanced solar cells.

Such fabrication techniques also permit creation of novel semiconductor structures. Quantum wells, known since the 1970s, attract electrons and confine them in two-dimensional sheets. Quantum wires are the electronic analog of a single-mode optical fiber. There also are quantum dots, 10 to 20 nanometers in diameter, which confine electrons in zero dimensions. These clusters of atoms have quantized electron energy states.

A topic of current research is the pursuit of bulk semiconductor materials possessing large densities of quantum wires or dots, which are to be approximately uniform in size.

At Caltech another approach to fabrication of quantum structures takes the precise thickness control for one-dimensional layers that has been developed for formation of quantum wells and extends it into two and three dimensions to fabricate quantum wires and quantum dots. Kerry Vahala says that nanocrystals can be coaxed into locally nucleating on a specially prepared substrate. These nuclei become quantum dots or wires that can be easily embedded into a higher-bandgap host material during the growth process. Vahala's group is currently working toward incorporation of arrays of dots and wires in an optical wave guide that would be suitable for use as a laser or optical amplifier. Such devices should exhibit greatly improved efficiency and have a broad range of applications.

MASS EXTINCTIONS

Mass extinctions have occurred repeatedly over the past 600 million years. Major events occurred at the Permian-Triassic boundary, some 250 million years before the present (myr), and at the Cretaceous-

Tertiary (KT), 65 myr. Lesser extinctions include the late Cambrian, 510 myr to 520 myr; the Frasnian-Famennian at 365 myr; the Triassic-Jurassic, 210 myr; the late Cenomanian at 91 myr; and the late Eocene, 34 myr.

It is not clear how many have been due to asteroid or comet impacts. The Permian-Triassic event, for example, was the most severe. It wiped out 90 percent or more of all species, including half the families of marine invertebrates. However, very little is known about this mass extinction.

At the Cretaceous-Tertiary boundary (KTB), by contrast, extinctions were somewhat less severe. Yet there is abundant evidence for a bolide impact: iridium, shocked quartz, spherules, and tektites. In addition, the extinctions evidently were quite sudden. Planktonic organisms, notably coccoliths and Foraminifera, flourished right up to the boundary before being cut off. The paleontologist Peter Sheehan also finds no falloff in the abundance and diversity of dinosaurs during the last 3 million years of the Cretaceous. At the KTB, though, the dinosaurs also go extinct.

Jan Smit of the Free University of Amsterdam, along with Walter Alvarez and his colleagues Alessandro Montanari and Nicola Swinburne of the University of California at Berkeley, have explored the candidate impact site: a 300-kilometer-diameter crater centered near the town of Chicxulub in northern Yucatan. Alvarez and co-workers propose that the impact took place on land but produced a kilometer-high tsunami in the adjacent Gulf of Mexico, as ejecta fell into the water. They find evidence for a backwash from this immense wave, in a thick deposit at the KTB boundary that contains abundant plant remains. Smit describes these as driftwood from coastal swamps, swept into the sea. They also find ripple marks indicative of surface waves—at depths below 400 meters. This suggests that the great wave sloshed back and forth within the Gulf. Alvarez describes these findings as consistent with the impact of an object with a diameter of 10 kilometers, traveling at tens of kilometers per second. The bolide would have struck with an energy of 10^8 megatons, some 10,000 times greater than that in the world's nuclear arsenal.

The astronomer Piet Hut, of the Institute for Advanced Study, proposes that the impactor was probably a comet, not an asteroid. He notes that there indeed are two craters that appear to have the right age, 65 myr: Chicxulub in Yucatan and Manson in Iowa. He describes such an impact sequence as resulting from the breakup of the comet as it rounded the sun, producing closely spaced fragments resembling a swarm of buckshot.

Hut also notes that there is danger even today from similar impacts. Once a year, on average, a rock strikes with an energy of 10 kilotons. Once a century the earth experiences an event such as Tunguska in Siberia (1908), in the megaton range. About every million years a kilometer-size body strikes, with energy exceeding all nuclear weapons together. That would produce worldwide crop failures and kill some 10^9 people, yet it would not qualify as a mass-extinction event. These would be considerably rarer.

Problems remain, nevertheless, in showing just how the KTB mass extinction occurred. Certainly, a 10-kilometer bolide would have produced vast environmental stress. Blast and tsunami would have affected only part of the earth, but the impact could have produced great quantities of nitrogen oxides, yielding a rainfall resembling nitric acid. That could have killed forests and dissolved shells of plankton. Dead forests then would have burned in continent-wide wildfires, pouring massive amounts of soot into the air. This soot, along with atmospheric dust, could have blanketed the planet, cutting off sunlight, shutting down photosynthesis, and setting the stage for further death of forests. Next would come additional wildfires and still more soot. There even could have been a long-term greenhouse effect, for thick limestone formations underlie the Yucatan. Their vaporization, due to impact, would have dumped a great deal of carbon dioxide into the atmosphere.

Evidence exists in isotopic anomalies for large-scale shutdown of photosynthesis, while a sooty layer at the KTB indeed points to massive fires. But such a sweeping catastrophe raises the question of, not why so many species went extinct, but why vertebrate life persisted at all. Plants have seeds that resist damage. But why did turtles, crocodiles, amphibians, and fishes—but not marsupial mammals—come through the KTB little the worse for wear? And if the KTB extinctions were less severe than those of the Permian-Triassic, why was this so? These are among the critical questions that paleontologists and paleoecologists will have to answer through further high-resolution studies of the geological record. Such studies will certainly lead to a better understanding of the behavior of the world's ecosystem under high-stress conditions.

THEORETICAL COMPUTER SCIENCE

In computer science a significant set of themes involves complexity and its applications. One topic includes the checking of very long proofs in mathematics, wherein a single mistake can yield an erroneous result. Transparent proofs offer a solution. A transparent proof is a lengthened counterpart of the original one, structured in such a manner as to cause any error to propagate throughout it. Then, by checking only part of this new proof, one has a very high probability of finding an error. The lack of such an error then validates the original proof. Furthermore, the longer one makes the transparent proof, the more broadly an error will propagate and the shorter is the computation needed to detect it. Indeed, the checking of a transparent proof can take much less time than to merely read the original one. At modest cost in additional checking time, one can reduce the probability of a mistake—of failing to catch an error and hence of accepting an incorrect proof as valid—to vanishingly low values such as 10^{-30}.

An initial formulation, according to Laszlo Babai of the University of Chicago, has the user define two quantities: ε, a measure of increased length of the transparent proof, and δ, the probability of accepting a false theorem. The original proof has a length of N characters. The transparent proof then has a length of $N^{1+\varepsilon}$. The length of its verification is $(\ell n\ N)^{2/\varepsilon} \times \ell n(1/\delta)$, where ℓn is the natural logarithm. More recent work shows that the length for verification can be a constant, independent of N.

The complexity of graphic detail is at the center of issues in computer graphics. This field features such applications as real-time interactive graphics used in flight simulation for pilot training; presentation of data in multiple variables or dimensions, including solutions for partial differential equations; and generation of computational grids.

In all these areas a central problem is the rapid division of many-sided polygons into constituent triangles. Such polygons often model shapes or scenes to be portrayed graphically. The computer then renders such objects using texture and a description of the available lighting, displaying them with assistance from a z-buffer in hardware, which is used to remove hidden lines and surfaces. Triangulated polygons also serve in data display and in representing computational boundaries, such as the wing of an aircraft.

Given a polygon with N vertices, then, a key problem is the development of rapid algorithms for triangulation. Since 1978 a number of techniques have come into common use that carry out the triangulation in time proportional to $N\ \ell n\ N$. In 1988 Robert Tarjan of Princeton University and Chris Van Wyck of AT&T Bell Laboratories introduced an algorithm offering $N\ \ell n\ \ell n\ N$ time. Then in 1990 Bernard Chazelle, also of Princeton, achieved an algorithm linear in N. Even so, Maria Klawe of the University of British Columbia notes that $N\ \ell n\ N$ algorithms remain the ones in standard use. That is because they are simple, whereas the newer ones are complex and offer insufficient advantage to users.

Complexity in computer science also offers problems in the design of hardware as well as software. Such a problem is the optimal interconnection of parallel processors having large numbers of computational modes. Networks derived from hypercubes form the basis for architectures of such systems as the BBN Butterfly, the IBM RP3 and GF11, the Intel iPSC, the Connection Machine, and the NCUBE. However, such architectures introduce worst-case communication problems for which the run time scales as the square root of N, where N is the number of processors.

Tom Leighton, of the Massachusetts Institute of Technology, notes that randomly wired interconnected networks represent a useful alternative. Such networks are not wholly random; the randomness is subject to constraints. Even so, they can outperform traditional well-structured networks in several important respects. The worst-case problems disappear; all problems offer run times that scale as $\ell n\ N$. In addition, randomly wired networks have exceptional fault tolerance because they offer multiple redundant paths. They are well suited for both packet-routing and circuit-switching applications.

APPENDIX B

Symposia Programs

THIRD ANNUAL SYMPOSIUM ON FRONTIERS OF SCIENCE
NOVEMBER 7–9, 1991

Astrophysics:
Looking Inside The Sun

Organizer:
William Press, Harvard-Smithsonian Center for Astrophysics

Helioseismic Observations of the Solar Interior
Ken Libbrecht, California Institute of Technology

Physical Inferences from Solar Oscillations
Doug Gough, Cambridge University, UK

Solar Neutrinos–Probing the Sun or Neutrinos?
John Wilkerson, Los Alamos National Laboratory

Biology:
Basic Science Related to AIDS

Organizers:
Eric Lander, Whitehead Institute
Kevin Struhl, Harvard Medical School

Cell/Virus Interactions Involved in Retrovirus Replication
Eric Hunter, University of Alabama-Birmingham

Games that HIV and Molecular Geneticists Play
Richard Young, Whitehead Institute

Computer Science:
Aspects of Theoretical Computer Science

Organizer:
Robert Tarjan, Princeton University

Transparent Proofs
Laszlo Babai, University of Chicago

Geometry and Graphics: Challenges in Complexity
Maria Klawe, University of British Columbia

The Role of Randomness in Computer Architecture
F. Thomson Leighton, Massachusetts Institute of Technology

Geosciences:
Earthquake Prediction

Organizer:
Marcia McNutt, Massachusetts Institute of Technology

The Physics of Earthquake Recurrence
Thomas Heaton, US Geological Survey, Pasadena

Clocks in the Earth
William Ellsworth, US Geological Survey, Menlo Park

Discussion panel:
Thorne Lay, University of California-Santa Cruz
Joann Stock, Geological Museum, Harvard University
Kerry Sieh, California Institute of Technology
Duncan Agnew, Scripps Institution of Oceanography

Geosciences:
Mass Extinctions

Organizers:
Marcia McNutt, Massachusetts Institute of Technology
Richard Muller, University of California-Berkeley

The Link Between Large-body Impact on Earth and Biological Mass Extinction
Walter Alvarez, University of California-Berkeley

The Geological Record of Mass Extinctions
Jan Smit, Free University of Amsterdam, The Netherlands

Astronomical Mechanisms for Multiple Impacts
Piet Hut, Institute for Advanced Studies, Princeton

Discussion leader:
Alessandro Montanari, University of California-Berkeley

Material Science:
Quantum Confined Semiconductors

Organizers:
Robert Cava, AT&T Bell Laboratories
Mark Davis, California Institute of Technology

Quantum Wires and Quantum Dots
Kerry J. Vahala, California Institute of Technology

Molecular Particles of Semiconductor Solids
Michael L. Steigerwald, AT&T Bell Laboratories

Atomic Layer Growth of Thin Films
David Rudman, National Institute of Standards and Technology, Boulder

Mathematics:
Topology

Organizer:
William Thurston, Princeton University

A Lecture on the Energy of a Knot (or an Unknot)
Michael H. Freedman, University of California-San Diego

Recent Progress in Algebraic Theory
Michael J. Hopkins, Massachusetts Institute of Technology

Physics:
Manipulating Atoms by Laser Atom Trapping and Scanning Tunneling Tips

Organizers:
Sylvia Ceyer, Massachusetts Institute of Technology;
Robert Cava, AT&T Bell Laboratories

Neutral Particle Manipulation with Light
Steven Chu, Stanford University

Atomic and Molecular Manipulation with the Scanning Tunneling Microscope
Donald M. Eigler, IBM Almaden Research Center

FOURTH ANNUAL SYMPOSIUM ON FRONTIERS OF SCIENCE
NOVEMBER 5–7, 1992

Atmospheric Science:
Ozone Holes—Causes and Effects of Polar Ozone Depletion

Organizer:
Mark Davis, California Institute of Technology

The Science of Ozone Depletion: Results from Aircraft Measurements in the Polar Stratospheres
David Fahey, National Oceanic and Atmospheric Administration, Boulder

An Overview of Polar Ozone Depletion
Mario Molina, Massachusetts Institute of Technology

Discussion leader:
Ralph Cicerone, University of California-Irvine

Biology:
The Cell Cycle

Organizers:
Kevin Struhl, Harvard Medical School;
Robert Tjian, University of California-Berkeley

The Replication of DNA and the Cell Cycle
Bruce Stillman, Cold Spring Harbor Laboratory

How Cell Division is Controlled During Growth of an Embryo
Patrick O'Farrell, University of California-San Francisco

Biology:
Structure and Function of Proteins

Organizers:
Robert Tjian, University of California-Berkeley;
Kevin Struhl, Harvard Medical School

The Leucine Zipper: New Twists in Gene Regulation and Protein Structure
Tom Alber, University of California-Berkeley

Protein Structure and Dynamics in Solution
Rick Dahlquist, University of Oregon

Chemistry and Physics:
Buckminsterfullerenes—A New Form of Carbon

Organizers:
Shirley Chiang, IBM Almaden Research Center;
Sylvia Ceyer, Massachusetts Institute of Technology

Experimental Studies of Fullerenes and Metallofullerenes
Donald Bethune, IBM Almaden Research Center

The Fullerenes and the Idea of Perfect Nanocrystals
Robert Whetten, University of California-Los Angeles

Geosciences:
Venus—Results from Magellan

Organizer:
Marcia McNutt, Massachusetts Institute of Technology

Venus—Dead or Alive? The View from Magellan
Robert Grimm, Arizona State University
Suzanne Smrekar, Massachusetts Institute of Technology

Materials Science:
New Materials—Conducting Polymers, Organic Magnets, Composites

Organizers:
Robert Cava, AT&T Bell Laboratories;
Nathan Lewis, California Institute of Technology

A Chemist's View of Nonlinear Optical Materials
Seth Marder, California Institute of Technology

Magnetic Organic Materials
Dennis Dougherty, California Institute of Technology

Mathematics:
The Wavelet Transform

Organizers:
David Donoho, Stanford University/University of California-Berkeley;
Vaughan Jones, University of California-Berkeley

Wavelets and Signal Analysis
Ingrid Daubechies, Rutgers University

Wavelets and their Applications
Yves Meyer, Université de Paris

Discussion Panel:
Gregory Beylkin, University of Colorado-Boulder
Michael Frazier, Michigan State University
Stéphane Mallat, Courant Institute-New York University
Victor Wickerhauser, Washington University

Physics:
Current Topics in Hadron Collider Physics—Prelude to the Superconducting Supercollider Era

Organizer:
Jim Siegrist, SSC Laboratory

The Search for the Top Quark
John Huth, Fermi National Accelerator Laboratory

Testing Electroweak Theory: Present and Future
Kevin Einsweiler, Lawrence Berkeley National Laboratory

APPENDIX C
About the Authors

Marcia Bartusiak

After obtaining an advanced degree in physics, Marcia Bartusiak has been covering the fields of astronomy and physics. Her work has appeared in a variety of national publications. A contributing editor of *Discover* magazine, she is also the author of "Thursday's Universe," a layman's guide to the frontiers of astrophysics and cosmology, and "Through A Universe Darkly," a history of astronomers' centuries-long quest to discover the universe's composition. In 1982, Bartusiak was the first woman to receive the prestigious Science Writing Award from the American Institute of Physics. She now lives in Sudbury, Massachusetts.

Barbara Burke

Barbara Burke entered Harvard intending to study no more science than required, and certainly no math. Her opinion that science was dull changed as a result of a biology course and by two years as a science writer at MIT. Later, as a reporter for the *Ithaca Journal*, she won an American Association for the Advancement of Science-Westinghouse science writing award for a series on acid rain. Marriage to mathematician John Hubbard of Cornell convinced her that there was more to mathematics than what she had glimpsed from high school courses. She became intrigued by the challenge of communicating the ideas behind the formulas to people who think that math and related topics—physics, engineering, information processing—are boring, scary, or simply inaccessible without years of advanced training. She is now at work on a book in French on Fourier analysis and wavelets and has recently completed an article for the French Scientific American on the KAM theorem and its connection to longstanding problems in celestial mechanics. She currently resides in Ithaca, New York.

Andrew Chaikin

Andrew Chaikin is a Boston-based science writer who specializes in astronomy, planetary science, and space exploration. His articles have appeared in *Popular Science*, *Air & Space/ Smithsonian*, *Discover*, *OMNI*, and other national publications. His book on the Apollo lunar astronauts and their experiences will be published by Viking/Penguin in Spring 1994. A former editor of *Sky & Telescope* magazine, he is the co-editor of "The New Solar System," a compendium of planetary science issues now in its 3rd edition. He also authored a children's book, "Where Dinosaurs Walked," for World Book/Childcraft. Before entering the field of science writing in 1980, he served as a research geologist at the Smithsonian Institution's Center for Earth and Planetary Studies in Washington, D.C. He holds a bachelor's degree in geology from Brown University.

Addison Greenwood

Addison Greenwood is the author of six books, most recently "Science at the Frontier" (1992), published by National Academy Press. His magazine credits range from *BrainWork* to *Rolling Stone*. He writes about science for radio and television, and also lectures about the brain and mind. Greenwood lives in Washington, D.C., and is currently at work on a novel.

T. A. Heppenheimer

T.A. Heppenheimer holds a Ph.D in aerospace engineering, and has held research fellowships at the California Institute of Technology and the Max Planck Institute in Heidelberg, Germany. He is an associate fellow of the American Institute of Aeronautics and Astronautics. Three of his hardcover books—"Colonies in Space" (1977), "Toward Distant Suns" (1979), and "The Man-Made Sun" (1984)—have been selections of the Book-of-the-Month Club and its affiliates. As a free-lance writer, he has contributed to most of the nation's popular magazines that deal with science and technology and is a frequent contributor to *American Heritage*. He also has written extensively on Air Force and naval technology. He lives in Fountain Valley, California.

Michelle Hoffman

Michelle Hoffman writes and edits articles about the life sciences with a particular emphasis on molecular and cell biology, biotechnology, immunology and genetics. She started her publishing career as a staff editor and the cover editor at *Cell*, and moved on to write for *Focus*, a weekly newsletter produced at Harvard Medical School. Later, she joined the news staff of *Science* as the magazine's northeast correspondent. In addition to her magazine work, she edited the molecular biology textbook *Genes IV*. She is now an associate editor for *American Scientist* and lives in North Carolina.

David Holzman

David Holzman is currently Washington editor of *BioWorld*, a daily fax newsletter on biotechnology. As a staff writer on science and medicine for *Insight* magazine, a position he held for five years, Holzman won first prize in feature writing in 1989 from the American College of Radiology for his cover story, "Imaging Unveils Medical Mysteries." Holzman has also freelanced articles for *Science*, *Forbes*, *Mosaic*, *Business and Health*, and numerous newspapers, including the *Washington Post*, the *Los Angeles Times*, and the *Cleveland Plain Dealer*. A cycling enthusiast, Holzman bicycled across America following his graduation from the University of California, Berkeley, in 1975. He resides in Washington, D.C.

Elizabeth J. Maggio

Elizabeth Maggio is a Los Angeles-based freelance writer. Her career in science journalism began in 1969 after graduation from New York University with a degree in geology. Maggio polished her journalism skills as a medical writer with a Manhattan public relations firm before joining the *Arizona Daily Star* (Tucson) where she served as science editor for nine years. She then left newspaper work for the University of Arizona to write about campus science news and founded what was to become a nationally recognized research magazine. In her spare time she earned a degree in Italian which led to a three-year consulting contract in Turin as the bilingual science editor for a major Italian aerospace company. Maggio's work has earned numerous state and national honors including the American Association for the Advancement of Science-Westinghouse science writing prize.

Anne Simon Moffat

Anne Simon Moffat is a Chicago-based writer specializing in the life sciences, with occasional forays into physics, chemistry and horticulture. She is the midwest correspondent for *Science* magazine, a regular contributor to *The Sunday Times of London* and *Genetic Engineering News* and for many years wrote for *Mosaic*, the magazine of the National Science Foundation. She started her writing career while doing public relations work for the Rockefeller University, Cornell University, and the University of California, San Francisco, and became a full time freelancer nine years ago. Moffat is a member of the National Association of Science Writers and the American Society of Journalists and Authors. She is married to a scientist, has a 12-year-old son, and, in her free time, practices applied biology in the back garden.

INDEX

J

K

L

M

N

Q

R

S

T

U

V

W

X

Y

Z